D1121472

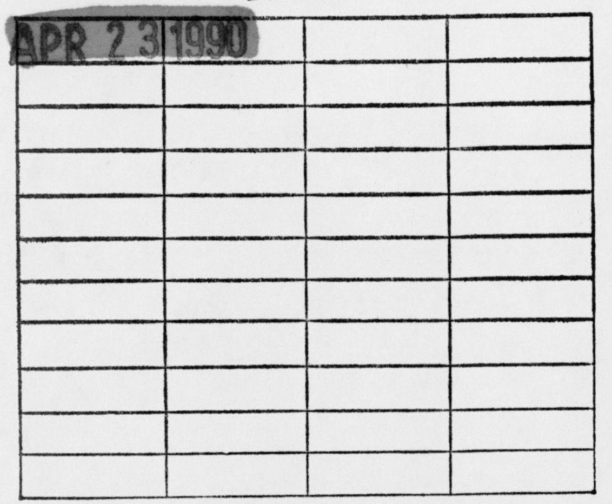
Date Due
APR 23 1990

NEW MEXICO
EASTERN
UNIVERSITY
E
N
M
U
LIBRARY

UNDERWATER ACOUSTICS

CONTRIBUTING AUTHORS

RICHARD STERN, Assistant Professor

School of Engineering and applied Science
University of California, Los Angeles

B. M. BROWN, Head of Department and Professor of Physics

Southwestern University
Georgetown, Texas

UNDERWATER ACOUSTICS

LEON CAMP

Director of Acoustic Research
Bendix Corporation
Electrodynamics Division
North Hollywood, California

Research Associate in Engineering
University of California at Los Angeles

WILEY-INTERSCIENCE
a Division of John Wiley & Sons
New York · London · Sydney · Toronto

Library of Congress Catalogue Card Number: 78-109428

ISBN 0-471-13150-4

Printed in the United States of America

10 9 8 7 6 5 4 3 2 1

PREFACE

The motivation for and the character of the contents of this book are best understood from a description of its origin. For a number of years a "short course" in underwater acoustics has been offered by Engineering and Physical Sciences Extension, University of California, Los Angeles. This short course consists of an intensive presentation in the form of two three-hour lecture sessions daily for 10 days. The students are a mature group of engineers, most of whom have been confronted in their jobs with the expanding applications of sonic techniques. Although the time spent in class and the material discussed are equivalent to those of a conventional semester course, the participant lacks the necessary opportunity to fix these experiences as a part of his professional background. It is the task of the instructor therefore to provide a panorama of the subject and tools in the way of notes and references to serve as a guide for the continued use of the student. He also faces a special problem that results from the diverse educational and professional backgrounds of his audience.

The subject of applied acoustics, no matter what the medium, involves several disciplines to a degree that considerable controversy exists regarding the department in the engineering college to which it belongs. So it is readily understood that a related problem exists in the selection of material for a text. Any other than an encyclopedic effort is certain to result in a scanty treatment of sections, depending on the point of view, most relevant to a readers interest. The alternative to a multivolume edition by several authors is either a general introduction to fundamentals or a more complete discussion of limited topics.

A general review of fundamentals is best suited to the needs of the average engineer who has completed his formal training but is faced with new problems. There are now available several excellent books on fundamentals, but the fact that they are rather far removed from present-day practices and particularly from the hardware in use requires a supplement.

Sound sources and receivers are the prime movers and sensors of acoustic systems and therefore occupy the center stage of this presentation. They are representable on paper by mathematical models and equivalent circuits that serve as important aids to performance analysis. The tendency, however, is to emphasize them at the expense of mechanical models. For the optimum utilization of materials and the achievement of reliability required of high power sources a complete mechanical model showing displacements, stresses, and areas subject to fatigue is essential.

It is intended that this book give the student an understanding of transduction techniques and the ability to design and to evaluate the most commonly used transducer elements. There is also an introduction to array design, transmission problems, reception, and signal processing. Obviously there can be no advanced discussion of these subjects, since a serious treatment of any one of them would require a separate volume.

I am indebted, first, to Engineering and Physical Sciences Extension, University of California, Los Angeles and to the Electrodynamics Division of the Bendix Corporation for the time and opportunity to explore and practice in this field of continued education. Also I appreciate the contribution made by Dr. Stern, who supplied the classical background of acoustics, and we are all obligated to the Navy Department and to Tracor for supporting Dr. Brown in the study that resulted in the chapter on sonar signal processing and for permitting its publication.

LEON CAMP

Santa Monica, California
April, 1970

CONTENTS

CHAPTER 1

The Mechanics of Vibration

Nowhere is the relation of sound to vibration more clearly demonstrated than in underwater acoustics where its production or reception involves developing a vibratory motion with a mechanical device. An interesting and somewhat surprising item is the magnitude of the vibrating displacements required to produce the sort of sound levels in which we are interested. A peak displacement amplitude three times the wavelength of visible light develops a fairly high level, whereas a weak sound level represents distances smaller than atomic spaces. Periodic and nonperiodic sound are employed in water, but the concept of frequency is useful in both. Because of the properties of the medium and the ability of instruments to respond to them, the possible frequency range is much greater than in air.

This first chapter introduces the basic mathematics of simple vibrating systems. The treatment is limited to electromechanical devices because these are by far the most widely used and lend themselves to many purposes. The discussion of methods of excitation is postponed to a later chapter. According to function, we may group an electromechanical source into three sections; an electrical section with two or more terminals; a mechanical section with two or more terminals; and a section between, which has the capability of converting events at one set of terminals into related events at the other set. The performance of this midsection must be judged by data from the two sets of terminals.

The greater ease of measurements at the electrical terminals makes this the popular approach to the system, and once a system has been designed and the equivalent conversion factors determined an equivalent electrical circuit is a powerful aid to its analysis. But, first, the design of the middle section is a problem in mechanics, and therefore it is appropriate to begin the discussion with the mechanics of a simple harmonic vibrator.

A person accustomed to the small size and flimsy structure of sound receiving and transmitting instruments used in air may be surprised at the relatively massive devices performing like functions in water. Two sources of this contrast in design have to do with extreme differences in physical properties of the two media, and a third results from a practical utility of greater power in water. The transmission of equal amounts of power by plane waves through equal areas at a given frequency requires an acoustic pressure in water that is about 60 times that required in air, but with an amplitude of motion in the reverse order. Hence the loudspeaker with large compliance and displacement amplitude is suitable as an air sound source, and the stiff piezoelectric bar with small displacement, as a source in water. The difference in sound velocity in the two media also calls for a difference in size of radiating surfaces because a meaningful unit for defining the size of a radiating surface is the wavelength, in the medium, of the sound being radiated. Since the velocity of sound in water is about four and one-half times that in air, an equivalent radiator in terms of directivity in water would have an area that is a multiple of that in air by the square of four and one-half.

1.1 SYMBOLS AND UNITS

As the discussion proceeds an attempt is made to stay with the symbols and to define acoustical terminology used and understood by the majority of workers in the field. The expression *an attempt* is used because it is often difficult to determine where the majority of workers agree. It is assumed that the publications of the American Standards Association are the best sources of information of this sort. It should not be necessary to belabor the argument for a uniformly accepted and consistent set of symbols and of units. The author who attempts a new set, however logical, will produce something quite troublesome to read. The fact that we are involved in the three disciplines of mechanics, electricity, and acoustics creates a more difficult problem than exists in any one of them alone; for example, I is a symbol that long use has fixed with a definite but quite different meaning for each of these three sciences; hence a certain area of confusion will develop should two or more of these show up in one discussion.

Adoption of the MKS system of units by acousticians has been slow. A probable explanation for this is that a major branch of the science has long been working with a system related to reference levels based on somewhat arbitrary definitions of physiological effects. The more recently developed technology of underwater sound began with this system. Caution must be observed in reviewing data from the past which may be difficult to interpret because they are often expressed in a mixed set of units not clearly defined.

The advantages of the MKS system are most convincingly demonstrated by the ease of transformation from mechanical, within the system, to analogous electrical quantities, or vice versa. Conversion factors are not needed for transduction coefficients, or electromechanical transformers. Unfortunately agreement has not been reached on the choice of a reference for acoustic pressure. In common use have been dynes/centimeter^2 or microbars, 0.0002 dynes/centimeter^2, newtons/meter^2, and micronewtons/meter^2. Frequently we may find pressure levels one yard from a source expressed in one of these references from the metric system. The consensus appears to favor the microbar and this unit is used in this book as the pressure level in expressing, transmitting, and receiving responses. In general, however, the newton/meter^2 is used to express acoustic pressure.

1.1.1 The decibel

In acoustics the *level* of a quantity has a specific meaning. It is the logarithm of the ratio of that quantity to a reference quantity of the same kind. The decibel is a unit that is one-tenth of the level as defined above with ten as the logarithm base. The reference quantity must be stated or clearly understood. One of several advantages of the decibel notation may be demonstrated by the problem of plotting sound pressure levels as a function of some other variable where the ratio may vary from one to one to one ten-thousandth. Expressing the pressure level in decibels makes the plot simple and revealing, whereas on a linear scale it would be quite impractical.

In practice the convenience of the decibel scale is a temptation to extend its use; for example, in a transmission problem in which a plane wave passes through a boundary from medium 1 to medium 2 we may properly express the ratio of the intensities I_1 and I_2 as a difference in decibels as

$$10 \log \frac{I_1}{I_2} = 10 \log I_1 - 10 \log I_2$$

and as

$$I_1 = \frac{p_1^2}{\rho_1 c_1} \qquad I_2 = \frac{p_2^2}{\rho_2 c_2}$$

where p is effective (rms) pressure, ρ is the density, and c the velocity of propagation of sound in the respective media; we may also write

$$10 \log \frac{I_1}{I_2} = 20 \log \frac{p_1}{p_2} + 10 \log \frac{\rho_2 c_2}{\rho_1 c_1}$$

unless $\rho_2 c_2 = \rho_1 c_1$, the pressure squared is not proportional to the intensity and expressing the pressure ratio in decibels is meaningless.

1.2 SIMPLE HARMONIC OSCILLATOR

This fundamental vibrating system has been chosen as an instrument for the introduction of the basic concepts utilized in the description of sound. It is doubly useful because most of these concepts are common not only to the vibrating mechanism producing the sound, but also to the sound wave produced by this mechanism. The simple oscillator is an abstraction conventionally represented by a mass m concentrated at a point and attached to a spring which has the other end fixed. In addition to being weightless, the spring obeys Hooke's law; that is, an elongation per unit length bears a linear relation to the force producing it. Therefore, when the mass is displaced from its equilibrium position by an applied force, the attached spring exerts a force in a direction opposite to the displacement and proportional to its magnitude. We may establish a coordinate x as a measure of the displacement from equilibrium, x being positive when the spring is extended, and negative when the spring is shortened. A stiffness factor s is the force required to extend or compress the spring unit distance. If the displacing force f is just sufficient to hold the mass stationary at the point x, the sum of forces on m is zero:

$$sx + f = 0 \qquad f = -sx \tag{1.1}$$

or the force exerted by the spring is opposing the displacing force, and therefore is in a direction opposite to the displacement. At a point x let the force f be reduced to zero leaving $-sx$ as the only force acting on the mass m. This condition is described by the linear second-order differential equation:

$$m\frac{d^2x}{dt^2} = -sx \tag{1.2}$$

which by setting $\omega_0^2 = s/m$, may be written as

$$\frac{d^2x}{dt^2} + \omega_0^2 x = 0 \tag{1.3}$$

One solution to this equation is the sinusoidal function

$$x = A\cos\omega_0 t \tag{1.4}$$

Equation 1.4 may be represented pictorially by a plot on an (x, t) coordinate system, or, as shown in Figure 1.1, by the projection of a radius of length A on the horizontal diameter of a circle; that radius rotating at an angular velocity ωt. The projection of A represents a vibratory motion of the mass about the equilibrium position O. The motion is periodic; that is, for each revolution, a cycle, requiring a fixed time interval, the period T, the motion is repeated. The number of cycles completed per second for frequency f has

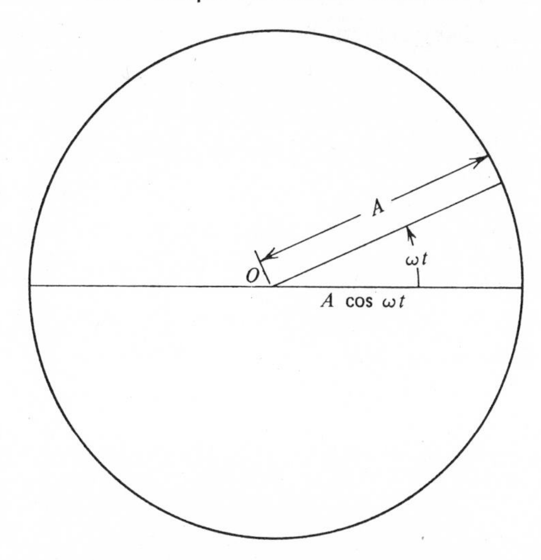

FIG. 1.1 Simple oscillatory motion.

recently, by popular usage, come to be termed Hz after the German physicist. Thus we have Hz cycles per second, and kHz kilocycles per second, which is about as far as we need to go in underwater acoustics. Angular velocity ω, the number of radians covered per second, is also $2\pi f$ and, since $\omega_0 = 2\pi f_0 = \sqrt{s/m}$, $f_0 = (1/2\pi)\sqrt{s/m}$, where f_0 is called the natural frequency of vibration. Also the inverse of f_0 is the period $T = 2\pi\sqrt{m/s}$. From (1.4) the maximum displacement of the mass from its equilibrium position is A, the *displacement amplitude*. The natural frequency of vibration f_0 is independent of A.

1.2.1 Phase

In underwater acoustics *phase* plays a more vital part than in other branches of acoustics. The angle $\omega_0 t$ is a measure of the *phase* of the motion with reference to its starting position. We could have chosen some other starting time when the motion had an initial *phase angle* φ. The displacement would then be given by

$$x = A \cos (\omega_0 t + \varphi) \tag{1.5}$$

This expression contains the two constants necessary to a complete solution of a second-order differential equation. They also describe how the motion began. By taking derivatives of (1.5) we may see the phase relations between the displacement, the velocity, and the acceleration of the mass m. The velocity v is

$$\frac{dx}{dt} = -\omega_0 A \sin (\omega_0 t + \varphi) \tag{1.6}$$

which is ahead of the displacement by a phase angle of $\pi/2$, as can be shown by expanding $A\cos(\omega t + \varphi + \pi/2)$. The acceleration of m is

$$\frac{d^2x}{dt^2} = -\omega^2 A \cos(\omega t + \varphi) \tag{1.7}$$

which is out of phase with the displacement by π.

The constants of (1.5) are determined from the initial conditions as follows: Let

$$x = x_0 \quad \text{and} \quad \frac{dx}{dt} = v_0 \quad \text{at} \quad t = 0$$

then

$$A\cos\varphi = x_0 \qquad A\sin\varphi = -\frac{v_0}{\omega_0}$$

and

$$x = A\cos(\omega_0 t + \varphi) = A\cos\varphi\cos\omega_0 t - A\sin\varphi\sin\omega_0 t$$

or

$$x = x_0\cos\omega t + \frac{v_0}{\omega_0}\sin\omega_0 t \tag{1.8}$$

As we shall see later in dealing with sound sources, the total quantity of energy that can be put into a system per unit displacement, and the rate of its dissipation are important properties of the system. The total energy is certainly never greater than that represented by the work done by the initial displacing force, which extends the spring a distance x_0, thereby storing in it potential energy E_0 as the integral of

$$E_0 = \int_0^{x_0} f\,dx = \int_0^{x_0} sx\,dx = \tfrac{1}{2}s{x_0}^2 \tag{1.9}$$

So far we are discussing a conservative, or nondissipative oscillator having a constant energy content given by the sum of energy of motion

$$E_k = \tfrac{1}{2}mv^2$$

and energy of position

$$E_p = \tfrac{1}{2}sx^2$$

From (1.5) and (1.6)

$$E_p + E_k = \tfrac{1}{2}sA^2\cos^2(\omega_0 t + \varphi) + \tfrac{1}{2}m{\omega_0}^2A^2\sin^2(\omega_0 t + \varphi) = \tfrac{1}{2}sA^2 \tag{1.10}$$

where x and v are instantaneous values at any time t.

1.3 COMPLEX NUMBERS

The advantage of using complex numbers in the solution of differential and integral equations is so important that we shall review their basic properties in conjunction with further discussions of the harmonic oscillator. Beginning with Figure 1.2 we have the complex plane with real and imaginary axes. A complex number **A** is represented by a point in the plane. The line of length A joining the point to the origin is called the modulus and the angle ωt which this line makes with the real axis is called the argument of the complex number. The point **A** has coordinates (a, jb) in the plane and is defined as the vector sum of two directed lines, a and jb. From Figure 1.2

$$a = A \cos \omega t \quad \text{and} \quad b = A \sin \omega t$$

so the complex number **A** is defined by the figure in the following ways:

$$\mathbf{A} = a + jb \tag{1.11a}$$

$$\mathbf{A} = A \cos \omega t + jA \sin \omega t \tag{1.11b}$$

$$\mathbf{A} = Ae^{j\omega t} \tag{1.11c}$$

If we think of the argument ωt as a linear function of time, **A** is the terminus of a line rotating with the constant angular velocity ω, and the projection of this line on the real axis is $A \cos \omega t$. Again, as in our discussion in Section 1.3.1, an *initial phase angle* φ may be chosen by the operation

$$e^{j\varphi}Ae^{j\omega t} = Ae^{j(\omega t+\varphi)} \tag{1.12}$$

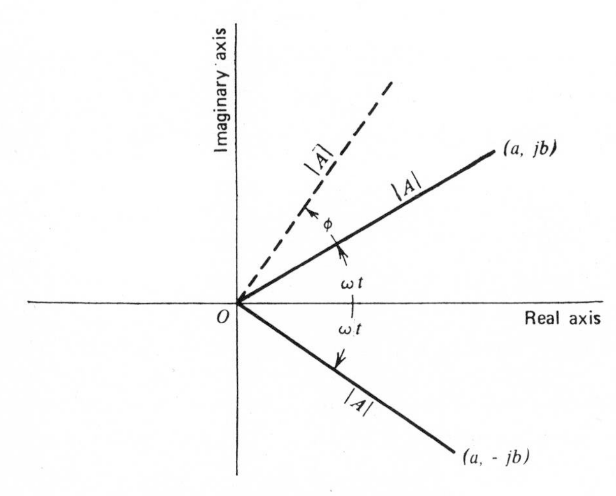

FIG. 1.2 The complex plane.

giving a number advanced in phase over **A** by the angle φ. From

$$Ae^{j(\omega t+\varphi)} = A\cos(\omega t + \varphi) + jA\sin(\omega t + \varphi)$$

the real part of the rotating complex number containing an initial phase angle is a complete solution to the differential equation of harmonic motion. A more concise expression for the complex number is

$$Ae^{j(\omega t+\varphi)} = \mathbf{A}e^{j\omega t} \tag{1.13}$$

where

$$\mathbf{A} = Ae^{j\varphi}$$

Inspection will show that the right-hand member of (1.13) will solve an equation similar to that defining harmonic motion, if x and its time derivatives are complex. However, displacement and velocity are real numbers, and for the sake of brevity, when the text clearly deals with real physical quantities, and a complex number is introduced, its real component only is meant, although that fact is not explicitly stated.

Although it is assumed that the reader is familiar with the algebra of complex numbers, some frequently used identities are worthy of mention. The factor $e^{j\varphi}$ is an operator which by acting on the complex number **A** rotates it through the angle φ. Therefore

$$e^{j\pi/2} = j \qquad e^{j\pi} = -1 \qquad e^{-j\pi/2} = e^{j3\pi/2} = -j$$

In addition to these identities, Figure 1.2 demonstrates that

$$A(e^{j\omega t} + e^{-j\omega t}) = 2A\cos\omega t \qquad \cos\omega t = \frac{e^{j\omega t} + e^{-j\omega t}}{2}$$

and that

$$A(e^{j\omega t} - e^{-j\omega t}) = 2jA\sin\omega t \qquad \sin\omega t = \frac{e^{j\omega t} - e^{-j\omega t}}{2j}$$

Caution must be observed in the use of complex numbers to represent real physical quantities; for example, acoustic power may be computed as the product of a radiation resistance, R a real number, by the velocity squared. Here velocity may be expressed as the real part of $\mathbf{A}e^{j\omega t}$ or as $v = A\cos(\omega t + \varphi)$. The real part of $R(\mathbf{A}e^{j\omega t})^2$ is not equal to $RA^2\cos^2(\omega t + \varphi)$.

1.4 THE THIN RING AS A HARMONIC OSCILLATOR

Neither the mass and spring nor the pendulum, frequently mentioned as examples of the harmonic oscillator, are used as underwater acoustic devices. A practical instrument and an excellent subject for demonstrating the behavior of an oscillator in water is the thin-walled ring, so the discussion continues

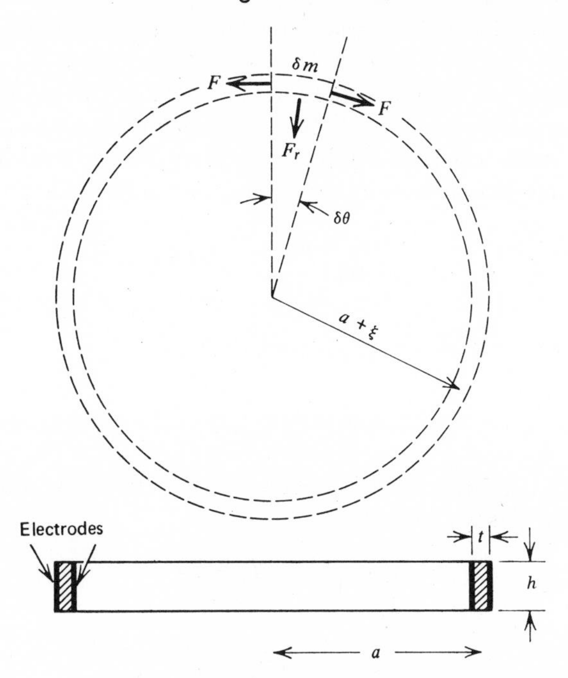

FIG. 1.3 Radially vibrating thin ring.

with one such as that shown in Fig. 1.3. In practice this ring may be of a material that can be excited to motion by an electric field, or a material that may be similarly excited by a magnetic field. With t the wall thickness and h the wall height both small compared with the ring radius a, all parts of the ring will move in phase with the same amplitude at the radial which is the fundamental mode of vibration. The only strain of consequence in the body is that due to a change in the circumferential length. It is assumed that all material properties of the ring are quite uniform and that an internally applied force has caused the ring to expand uniformly giving the radius an increment ξ. The expansion has developed a strain S, defined as the increase in length per unit length, in the wall:

$$S = \frac{2\pi(a + \xi) - 2\pi a}{2\pi a} = \frac{\xi}{a} \tag{1.14}$$

and applying Hooke's law,

$$\frac{F}{ht} = \frac{Y\xi}{a} \tag{1.15}$$

where F is the force and Y is an elastic modulus for the material. The value of this modulus may vary considerably depending on the field applied in developing the strain. In many cases it will differ only slightly from Young's modulus.

The resultant F_r of the tangential forces acts radially to return the ring to its equilibrium position at $\xi = 0$. If the expanding force is instantaneously released, an increment of the ring is subjected to the restoring force F_r. Using Newton's second law—rate of change of momentum equals applied force,

$$\rho hta\delta\theta \frac{d^2\xi}{dt^2} = \frac{-Yht\delta\theta\xi}{a} \tag{1.16}$$

where ρ is the density of the ring material. Then

$$\frac{d^2\xi}{dt^2} = -\frac{Y\xi}{\rho a^2} \tag{1.17}$$

Equation 1.17 is similar to (1.2) defining simple harmonic motion. If the exponential $\mathbf{A}e^{j\omega t}$ is a solution,

$$-\omega^2\mathbf{A}e^{j\omega t} = -\left(\frac{Y}{\rho a^2}\right)\mathbf{A}e^{j\omega t} \tag{1.18}$$

and

$$\omega^2 = \frac{Y}{\rho a^2} = \omega_0^{\ 2} \tag{1.19}$$

Since $\omega_0 = 2\pi f_0 = (1/a)\sqrt{Y/\rho} = c_m/a$,

$$f_0 = \frac{c_m}{2\pi a} = \frac{c_m}{\lambda_m} \tag{1.20}$$

when c_m is the velocity of sound in the material and λ_m, which is also the mean circumference of the ring, is the wavelength of sound in the material at frequency f_0. Although the latter relationship implies a wave motion, the ring is vibrating as a lumped component system. By replacing the angle increment in (1.16) by the total angle 2π the constants for the equation

$$m\frac{d^2\xi}{dt^2} + s\xi = 0$$

are

$$m = 2\pi aht\rho \quad \text{and} \quad s = \frac{2\pi htY}{a} \tag{1.21}$$

1.4.1 The ring with damping

Because of the contact of the inner and outer surfaces with the medium, a vibrating ring radiates sound at an energy rate proportional to the radial velocity of the ring wall. Consequently, a damping force must be introduced into the equation of motion, which becomes

$$m\frac{d^2\xi}{dt^2} + R_m\frac{d\xi}{dt} + s\xi = 0 \tag{1.22}$$

Since the product of R_m by velocity yields a force, R_m has the nature of a mechanical impedance with units of kilograms per second.

The damping effect absorbs more or less quickly the energy originally put into the system. Therefore the solution to (1.22) does not describe a simple harmonic motion. It is logical to assume that a decaying vibratory motion could be of the form

$$\xi = \mathbf{A}e^{-\alpha t}e^{j\omega t} = \mathbf{A}e^{\gamma t}$$

Substituting this into (1.22),

$$(m\gamma^2 + R_m\gamma + s)\mathbf{A}e^{\gamma t} = 0 \tag{1.23}$$

Solving the quadratic for γ,

$$\gamma = -\frac{R_m}{2m} \pm \left[\left(\frac{R_m}{2m}\right)^2 - \frac{s}{m}\right]^{1/2} \tag{1.24}$$

Making the following substitutions,

$$\alpha = \frac{R_m}{2m} \qquad \frac{s}{m} = \omega_0^2$$

(1.24) may be rewritten as

$$\gamma = -\alpha \pm \sqrt{\alpha^2 - \omega_0^2} \tag{1.25}$$

Of the three possible cases: $\alpha^2 > \omega_0^2$, $\alpha^2 = \omega_0^2$, and $\alpha^2 < \omega_0^2$, only the last is oscillatory and therefore capable of producing sound. For this condition γ is a complex number expressed as

$$\gamma = -\alpha \pm j\omega_d$$

where

$$\omega_d = \sqrt{\omega_0^2 - \alpha^2}$$

is the angular velocity of the freely vibrating ring with damping.

Using the two values of γ from (1.25), we get as a complete solution to (1.22)

$$\xi = e^{-\alpha t}(\mathbf{A}_1e^{j\omega_d t} + \mathbf{A}_2e^{-j\omega_d t})$$

which is equivalent to

$$\xi = Ae^{-\alpha t}\cos(\omega_d t + \varphi) \tag{1.26}$$

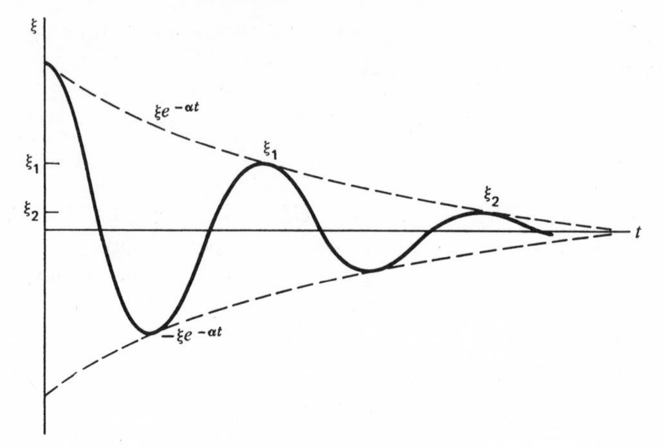

FIG. 1.4 Damped oscillations.

Figure 1.4 shows the displacement, ξ, from the equilibrium position of the ring wall as a function of time following an initial release at $t = 0$ and $\xi = \xi_0$. Although the period of this kind of motion does not change with time, the amplitude does change so the motion is not periodic. It is possible to gather some useful information about this vibrator by studying an oscillogram of the motion similar to Figure 1.4; for example, looking at the displacements shown on the figure as ξ_1 and ξ_2, which occur at times t, and at $t_1 + T$ one period later, we have

$$\xi_1 = Ae^{-\alpha t_1} \cos(\omega_d t_1 + \varphi)$$

$$\xi_2 = Ae^{-\alpha(t_1+T)} \cos[\omega_d(t_1 + T) + \varphi]$$

Dividing ξ_1 by ξ_2 and taking advantage of the fact that

$$\cos(\omega_d t_1 + \varphi) = \cos[\omega_d(t_1 + T) + \varphi]$$

the result is

$$\frac{\xi_1}{\xi_2} = e^{\alpha T}$$

Defining

$$Q_m = \frac{\omega_0 m}{R_m} = \frac{2\pi f_0 m}{R_m} = \frac{\pi}{\alpha T} \tag{1.27}$$

we get

$$\frac{\xi_1}{\xi_2} = e^{\pi/Q_m} \tag{1.28}$$

The magnitude of Q_m can be determined by taking the logarithm of (1.27) and solving

$$Q_m = \frac{0.4343\pi}{\log(\xi_1/\xi_2)} \tag{1.29}$$

A more accurate result may be obtained by measuring the decay of ξ over n periods to give

$$Q_m = \frac{0.4343n\pi}{\log \xi_1/\xi_n} \tag{1.30}$$

The definition of Q_m above is only one of three that are in common use. It is a valuable measure of the quality of a vibrator before the load is applied, an indicator of the type of load after it is applied, and a limiting factor in the band pass behavior of the system.

1.5 FORCED OSCILLATIONS

The ring discussed in Section 1.4 may serve as a sound source when driven by an electric field if piezoelectric or by a magnetic field if magnetostrictive. On the other hand, it may serve as a sound receiver when the driving force is developed by a sound pressure field in the surrounding medium. The driving force may be applied in pulses or continuously. It may be periodic of a single frequency or of a number of harmonically related frequencies or it may be as complex as white noise. In this discussion it is assumed that any of these can be expressed as a Fourier series and that the driving force in the following equation may be one term of such a series. Then (1.22) becomes

$$m\frac{d^2\xi}{dt^2} + R_m\frac{d\xi}{dt} + s\xi = Fe^{j\omega t} \tag{1.31}$$

A complete solution to (1.31) must satisfy the special case where F is zero. For this condition the solution is that found for (1.22), which is

$$\xi_t = Ae^{-\alpha t}\cos(\omega_d t + \varphi)$$

The subscript is attached to ξ_t to identify it as a transient part of the total displacement. That remaining after ξ_t has subsided will be identified at ξ_s corresponding to a *steady-state* condition. It is to be observed at this point that the two arbitrary constants required to satisfy the initial conditions are contained in the transient solution, so none are permissible in the steady-state solution. It is sometimes referred to as "the motion that has forgotten how it started."

If the solution to (1.31) is

$$\xi = \xi_t + \xi_s \tag{1.32}$$

simple reasoning leads us to find ξ_s. Substituting $\xi_t + \xi_s$ for ξ in the equation yields

$$m\frac{d^2\xi_s}{dt^2} + R_m\frac{d\xi_s}{dt} + s\xi_s = Fe^{j\omega t} \tag{1.33}$$

since ξ_t is the solution with F equated to zero. It is logical to assume that the ring will vibrate at the frequency of the driving force and with some amplitude A dependent on the amplitude of the driving force and impedance to motion of the ring. So the assumption that the solution ξ_s is the real part of $Ae^{j\omega t}$ gives

$$(-\omega^2 m + j\omega R_m + s)Ae^{j\omega t} = Fe^{j\omega t} \tag{1.34}$$

which determines the amplitude

$$A = \frac{F}{-\omega^2 m + j\omega R_m + s} = \frac{F}{j\omega[R_m + j(\omega m - s/\omega)]} \tag{1.35}$$

The denominator of (1.35) is so written to show the analogy to the electric series RLC circuit impedance. In keeping with this analogy we shall define a mechanical impedance

$$Z_m = R_m + j\left(\omega m - \frac{s}{\omega}\right) = R_m + jX_m = |Z_m|\, e^{j\varphi} \tag{1.36}$$

where

$$\tan^{-1}\theta = \frac{X_m}{R_m} \tag{1.37}$$

and

$$|Z_m| = \sqrt{R_m^{\ 2} + X_m^{\ 2}} \tag{1.38}$$

With these substitutions the steady-state displacement may be written as

$$\xi_s = \frac{Fe^{j(\omega t - \theta)}}{j\omega\,|Z_m|} \tag{1.39}$$

the real part being

$$\xi_s = \frac{F\sin(\omega t - \theta)}{\omega\,|Z_m|} \tag{1.40}$$

The total displacement from its rest position of the ring wall is

$$\xi = \xi_t + \xi_s = Ae^{-\alpha t}\cos(\omega_d t + \varphi) + \frac{F\sin(\omega t - \theta)}{\omega\,|Z_m|} \tag{1.41}$$

Transient phenomena are of considerable importance in underwater acoustics for several reasons. First, they may serve as a useful signal and often are so employed in limited range depth sounders. Bursts of energy may be thrown into the system through a switching mechanism and the resulting decaying oscillation shaped as desired by the addition of other components to the circuit. Second, most sonar is pulsed and in such operations short pulses have an advantage. It is obvious that a continuous constant frequency signal cannot determine range. A very short pulse has an advantage in identification of the target and has less interference from reverberation. Pulse excitation permits the development of much higher power levels without cavitation, and because of its short duty cycle requires less energy.

When the driving signal is first applied to the oscillator, the two responses, transient and steady state, could differ considerably in frequency. In most cases, however, the oscillator is serving as a source and except for extremely short pulses, the transient and steady-state responses will be quite close together in frequency. Initial conditions are likely to be zero displacement and velocity. The response of transducers to short pulses are shown in Figure 1.5*a* and 1.5*b*. The first shows the performance of a unit having a high Q_m. The pulse was not long enough to produce a steady-state condition. The second response is that of transducer of low Q_m requiring about three periods to reach the steady-state condition. Naturally the second transducer can transmit or receive information at a much faster rate than the first.

One may observe from the figures that the decay time of the transient occurs both at the beginning and the end of the pulse. One measure of this time, termed the decay modulus τ, is that required for the transient displacement to decay to $1/e$ of its initial value. From (1.26) the decay of ξ is fixed by the factor $e^{-\alpha t}$ and will be reduced to $1/e$ of its initial value at

$$t = \tau = \frac{1}{\alpha} = \frac{Q_m T}{\pi} \tag{1.42}$$

which shows the decay modulus to be directly proportional to Q_m. Therefore an oscillator with a relatively high Q_m must go through a number of cycles before the steady state, which usually contains the useful information, is established. In the meantime the front edge of the pulse has traveled some distance through the medium. It is also apparent that the transient limits the pulse-repetition rate. Perhaps the most serious handicap of a high Q_m is its limitation of projector bandwidth. Very short pulses and shaped pulses require a broad band for transmission. In our consideration of the advantages of a low Q_m we are assuming that R_m contained therein is the result of useful energy radiation, and we must not lose sight of the fact that the power capability and efficiency of many sound sources require a high Q_m.

1.5.1 The driven oscillator as a sound source

From (1.33) we see that the damping force involves velocity. Since this force supplies the energy going out as acoustic radiation, the velocity amplitude developed by the driving force is of more fundamental interest than

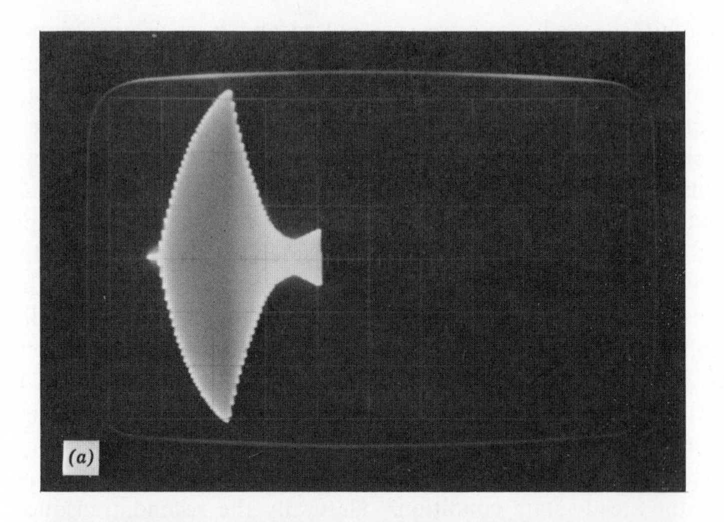

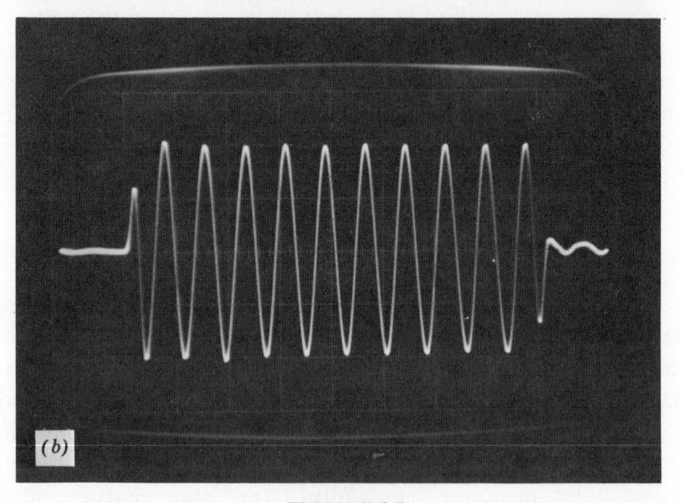

FIGURE 1.5

displacement. If the apparatus is to endure, however, definite limits on displacement must be observed to avoid destructive strains. By and large a discussion of power sources is concerned with the steady-state solution to (1.33) and particularly with its first derivative with respect to time, which is the radial velocity of the ring wall. From (1.39)

$$\xi_s = \frac{Fe^{j(\omega t-\theta)}}{j\omega Z_m}$$

and, taking the derivative,

$$\frac{d}{dt}(\xi_s) = \frac{Fe^{j(\omega t-\theta)}}{|Z_m|} \tag{1.43}$$

the real part of which is

$$v_s = \frac{F}{|Z_m|}\cos(\omega t - \theta) \tag{1.44}$$

The rate of energy dissipation varies through a period as a harmonic function of time so the average power represented must be found by an integration process. After the transient has decayed the motion does not change from cycle to cycle. Therefore averaging through one period is sufficient to establish the general average. Instantaneous power is the product of the driving force by the velocity it produces. So the average power will be found by integrating this product over one period and dividing by the time:

$$\begin{aligned} P &= \frac{1}{T}\int_0^T F\cos\omega t\left[\frac{F}{|Z_m|}\cos(\omega t - \theta)\right] dt \\ &= \frac{F^2}{|Z_m|\,T}\int_0^T (\cos\theta\cos^2\omega t + \sin\theta\sin\omega t\cos\omega t)\,dt \\ &= \frac{F^2}{2\,|Z_m|}\cos\theta = R_m\frac{F^2}{2\,|Z_m|^2} \end{aligned} \tag{1.45}$$

Here the cosine of the phase angle between force and velocity, analogous to the phase angle between voltage and current, plays the role of a power factor. Later we shall use rms values for amplitudes such as F, thereby eliminating the factor of 2 in the denominator. The average power consumed by the damping force of (1.33) is the time average obtained from

$$\begin{aligned} \frac{1}{T}\int_0^T R_m\left(\frac{d\xi}{dt}\right)^2 dt &= \frac{1}{T}\int_0^T R_m\frac{F^2\cos^2(\omega t - \theta)}{|Z_m|^2}\,dt \\ &= R_m\frac{F^2}{2\,|Z_m|^2} = \left(\frac{F^2}{2\,|Z_m|}\right)\left(\frac{R_m}{|Z_m|}\right) \\ &= \frac{F^2}{2\,|Z_m|}\cos\theta \end{aligned} \tag{1.46}$$

The equality of the results from (1.45) and (1.46) demonstrates another analogy with the series RLC electric circuit where all of the average power is consumed by R.

At this point a useful comment can be made about the impedance of the medium to the motion of the vibrating surfaces. Though a possible reactive component of that impedance was ignored in setting up the equation of motion, that component is by no means negligible in most cases. With reference to the ring, the dimension h of the radiating surface is quite small compared with the wavelength of sound in the medium. Because of this fact, a component of the impedance equivalent to increased mass is present. This has the effect of lowering the resonant frequency and increasing Q_m. It is possible to improve this difficulty and still retain the desirable features of the ring by making a coaxial assembly of a number of rings, thereby forming a cylinder of some length. Unless insulated acoustically from the walls of the vibrating ring, the medium inside will also have a contribution to the radiation impedance that could vary quite radically with frequency.

1.6 RESONANCE OF OSCILLATORY SYSTEMS

If the expression for mechanical impedance

$$Z_m = R_m + j\left(\omega m - \frac{s}{\omega}\right)$$

is examined for dependence on frequency, it is easily done graphically by using a plot in the complex plane. In Figure 1.6, it is seen that the imaginary or reactive term moves from a very large negative value through zero to a very large positive value as the frequency increases from zero. The mechanical impedances passes through a minimum at ω_0, where

$$X_m = \omega_0 m - \frac{s}{\omega_0} = 0 \qquad (1.47)$$

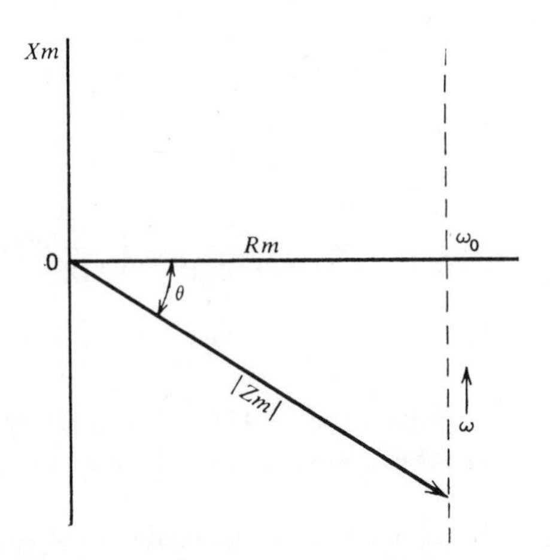

FIG. 1.6 Impedance locus as a function of frequency.

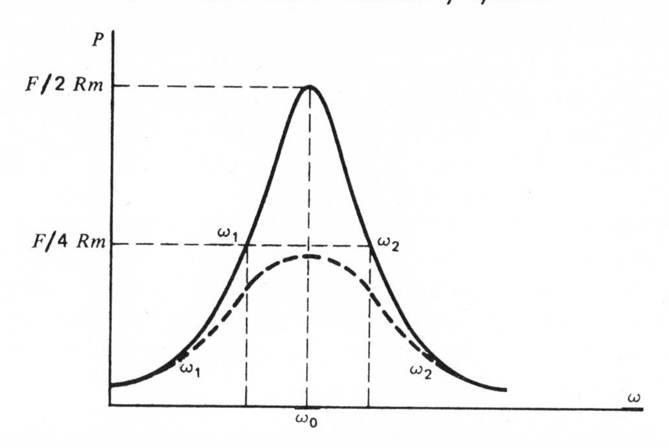

FIG. 1.7 Resonance loci for simple oscillators with constant driving force.

or

$$\omega_0 = \left(\frac{s}{m}\right)^{1/2} \qquad f_0 = \frac{1}{2\pi}\left(\frac{s}{m}\right)^{1/2} \tag{1.48}$$

The condition termed series resonance of the mechanical system exists when the reactive component of impedance vanishes. The corresponding frequency f_0 is also the natural system of vibration of the undamped oscillator.

The quantities s and m that determine the resonant frequency are properties peculiar to a system. For the simple oscillator under discussion, m is the total mass of the vibrating system and s is the total force exerted on this mass per unit displacement to restore it to the equilibrium position. To repeat (1.21)

$$m = 2\pi aht\rho \quad \text{and} \quad s = \frac{2\pi htY}{a}$$

In using these relationships to compute frequencies, we will become aware of the fact that the elastic modulus Y may have several values determined by the nature of the driving field.

A critical property of a sound projector is its behavior as a function of frequency particularly in the neighborhood of resonance. Figure 1.7 demonstrates the variation as a function of frequency of the power developed under the application of a constant driving force. The curves are loci of (1.45), repeated here for convenience.

$$P = R_m \frac{F^2}{2\,|Z_m|^2} = \frac{F^2}{2R_m} \quad \text{at} \quad \omega = \omega_0$$

for two different values of R_m. We may write the equation for power as

$$P = \frac{F^2}{2R_m[1 + Q_m^2(\omega/\omega_0 - \omega_0/\omega)]^2} \tag{1.49}$$

where

$$Q_m = \frac{\omega_0 m}{R_m}$$

Three angular frequencies are marked on the figure as having special significance. They are ω_0 at the peak power point, and the two points ω_1 and ω_2. The latter are the half-power points. Referring to (1.49), at $\omega = \omega_0$ the power delivered by the sound source is

$$P = \frac{F^2}{2R_m} \tag{1.50}$$

At the half-power points the term in brackets has the value 2 or

$$Q_m\left(\frac{\omega}{\omega_0} - \frac{\omega_0}{\omega}\right) = \pm 1 \tag{1.51}$$

If ω_1 and ω_2 are the angular frequencies at the lower and upper half-power points, respectively, Q_m may be found in terms of the three frequencies. Substituting into (1.51)

$$Q_m(\omega_2^2 - \omega_0^2) = \omega_2\omega_0$$
$$Q_m(\omega_1^2 - \omega_0^2) = \omega_1\omega_0$$

Subtracting and factoring

$$Q_m = \frac{\omega_0}{\omega_2 - \omega_1} \tag{1.52}$$

which enables us to predict the bandwidth of a sound source from its physical constants. Also from (1.51)

$$\frac{\omega_2^2 - \omega_0^2}{\omega_2\omega_0} = -\frac{\omega_1^2 - \omega_0^2}{\omega_1\omega_0} \qquad \omega_2\omega_1(\omega_2 + \omega_1) = \omega_0^2(\omega_2 + \omega_1)$$

giving

$$\omega_0 = \sqrt{\omega_2\omega_1} \tag{1.53}$$

It is common practice to refer to the difference between the half-power point frequencies as the system bandwidth. From (1.52) we see that the bandwidth is inversely proportional to Q_m. Referring to Q_m as a quality factor is a carryover from electric circuit theory where a high Q means high quality. In sound sources it more likely means the opposite.

1.7 PHASE PROBLEMS

The difference in phase between the driving force and the response of the driven mechanism is of vital importance because, in most cases, a sound source consists of an assembly or array of a number of smaller sources, each of which must conform within limits to a preassigned phase pattern. Among the responses to the driving force—displacement, velocity, or acceleration—we choose to deal with the velocity because it is more simply connected with measurement techniques.

Figure 1.8 shows how the phase angle between velocity and driving force behaves with change in frequency. We may take an analytical look at this change in a narrow region including resonance by finding the derivative of θ with respect to angular velocity. The phase angle is given by

$$\tan\theta = \frac{X_m}{R_m}$$

and taking the derivative

$$\sec^2\theta\,\frac{d\theta}{d\omega} = \frac{1}{R_m}\frac{dX_m}{d\omega}$$

Near resonance this equation becomes

$$d\theta = 2Q_m\,d\left(\frac{\omega - \omega_0}{\omega_0}\right) = 2Q_m\frac{d\omega}{\omega_0} \tag{1.54}$$

which says that the change in phase between the driving force and the response with frequency change is proportional to Q_m. As an illustration of the effect

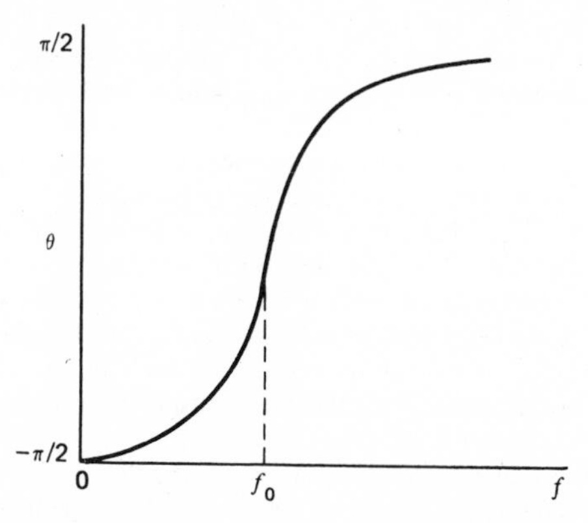

FIG. 1.8 Phase lag of velocity versus driving force.

that a difference in resonance frequency of the elements in an array may have on its performance, consider the following situation: two sound sources are operating side by side with the same driving force of frequency ω_1. One source has an ω_0, 5% above ω_1, and the other an ω_0, 5% below ω_1. The Q_m for each source is five. At a distant point on a line normal to the line joining these two sources, the signals from them would be in phase if there were no difference in their resonant frequencies. The actual phase difference between the two signals is

$$d\theta = 2Q_m \frac{2(\omega_1 - \omega_0)}{\omega_0} = 10 \times 0.1 = 1 \text{ rad}$$

This difference in phase is sufficient to cause serious malfunctioning of the array.

So far the discussion of phase has involved driving force and velocity. Looking into the electrical terminals, we deal with the phase between current and voltage. Impedance as determined at these terminals is the resultant of several terms, one of which is that due to motion. Phase changes in this term in the region of mechanical resonance will not be revealed in their proper perspective by the input impedance at the electrical terminal without a thorough analysis of the equivalent circuit.

PROBLEMS

1.1 Given $\log_b n$, prove that $\log_a n = \log_a b \log_b n$ and therefore $\log_e n = 2.306 \log_{10} n$.

1.2 A polar plot of the sound pressure radiated by a certain transducer as a function of angle ϕ from the normal is described by $p/p_0 = \sin \phi/\phi$. Find the pressure level in dB with reference to p_0 at $\phi \neq 172°$.

1.3 Given springs in tandem of stiffnesses s_1 and s_2, extended distances x_1 and x_2 by a common force, and also capacitors in parallel of elastances S_1 and S_2 charged with q_1 and q_2 by a common voltage, show the analogy between these two systems by computing the stored energy in terms of s_1, s_2, and x_1 for the springs and S_1, S_2, and q_1 for the capacitors.

1.4 If x, a real number, is to be represented by the sum of two complex numbers $\mathbf{A}_1 e^{j\omega t} + \mathbf{A}_2 e^{-j\omega t}$, show that $\mathbf{A}_1$ and $\mathbf{A}_2$ must be complex conjugates.

1.5 A spring and mass system has a period of 2 sec. A force of 3 N is required to increase the spring length by 1 m. Find the mass on the spring.

1.6 The instantaneous power delivered to a system by a driving force $Fe^{j\omega t}$ is the product of the real parts of the force and the resulting velocity $Ae^{j\omega t}$, where A is a real number. Show that the product of the real parts is not the same as the real part of the product of the two complex numbers. Find the average value over a cycle for each of the two products.

1.7 One of the ceramic materials used in manufacturing thin-walled rings has the physical constants $\rho = 5700$ kg/m^3, $Y = 11 \times 10^{10}$ N/m^2. Find the mean diameter of a ring having an undamped fundamental resonance in the radial mode of 2 kHz. The ring is 0.13 m in height and weighs 97.8 kg. Calculate the wall thickness.

1.8 If the ring of problem 1.7 is so mounted that when immersed its inside wall is not in contact with the water, the damping resistance, *Rm* of (1.31) is given approximately by the product $\rho_0 c_0 S$ where S is the area of the outside cylinder wall and ρ_0 and c_0 are the density and propagation velocity of sound in water. Under these conditions find a general expression for the mechanical *Qm* of vibrating rings as functions of the ρc product of ring material and of water and of ring dimensions. Find *Qm* for the ring of problem 1.7.

1.9 The ring of problem 1.8 is driven at its resonant frequency with a sinusoidal force of 4.6×10^4 N. Find the velocity and displacement amplitudes.

1.10 Four rings are assembled coaxially to form a water-tight cylinder. Each ring has a mean diameter of 0.3 m, a wall thickness of 0.03 m, and a wall height of 0.1 m. The method of mounting results in a frictional damping resistance equivalent to 30% of that due to radiation. Assume that the latter is $\rho_0 c_0 S$ as defined in problem 1.8. The physical constants of the material are $\rho = 7500$ kg/m^3, $Y = 8.11 \times 10^{10}$ N/m^2. Find the undamped resonant frequency f_0, the total mass m, the stiffness s, and *Qm*. What driving force F is needed to develop a velocity amplitude of 0.03 m/sec?

1.11 The ceramic rings of problem 1.10 will fail under a tension exceeding 2000 psi. Compute the maximum allowable displacement ξ, the acoustic power being radiated, and the total power consumed.

1.12 A ring-style transducer, resonant at 2 kHz, has a *Qm* of 4 and a mass of 36 kg. Find the constants *Rm* and s for the equation: $m(d^2\xi/dt^2) + Rm(d\xi/dt) + s = F \cos \omega t$. Also find the half-power frequencies f_2 and f_1 on the power–resonance curve. Assume that the transducer is driven at 1.9 kHz. How far down in decibels is the power output from that produced by the same driving force at resonance? Find the phase angle between the driving force and velocity.

CHAPTER 2

Distributed Systems

2.1 INTRODUCTION

Most of the vibrating structures used in underwater acoustics do not fall into the category of simple harmonic oscillators. In dealing with one of the latter, consisting of a mass and a springlike component, we assume that all parts of the mass experience the same displacement simultaneously and that all sections of the spring are stretched equal amounts. However, if we could take a look at any given instant at all parts of an extended body in vibration, we might find some parts under compression, other parts under extension, and still others under no strain at all. In addition, some sections of the body could be moving forward while others are moving backward and still others not in motion.

Sound waves develop motion of this nature when moving through either a limited or an extensive medium. Although we experience sound waves as longitudinal waves of vibratory motion whose displacement is in the direction of, rather than transverse to, the direction of propagation, mechanical devices producing sound in water or air may be vibrating longitudinally, transversely, radially, or in some combination. Before taking up the problems of sound propagation in a virtually unlimited medium, we shall discuss its action in the common types of structures used in projectors and hydrophones where it is appropriate to treat the action as wave motion. These structures are modified rods, cylinders, pistons, and plates. Membranes and very thin plates, so useful in air acoustics, have little application in water.

2.2 THE UNIFORM BAR

When a bar such as that shown in Figure 2.1 is put under tension by forces along its length, it will not only extend in length but also contract in cross section, so that a vibratory disturbance moving along the bar will set up a transverse motion. However, if the bar is very long and is small in cross

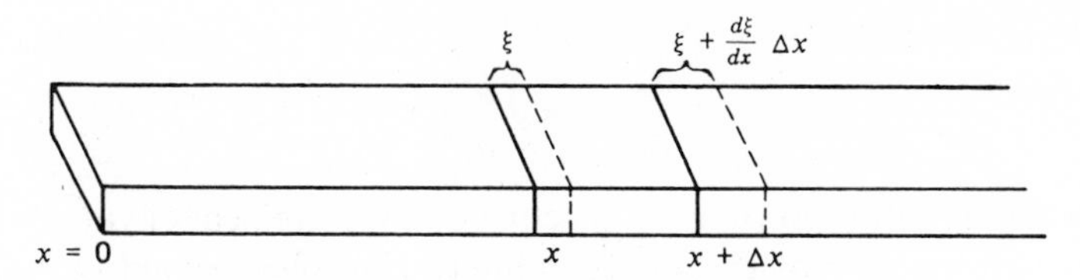

FIG. 2.1 A uniform bar under static stress.

section, we will find the transverse motion of no interest. Let us concentrate our attention on the highly magnified section of the bar where the solid lines at x and $x + \Delta x$ mark its boundaries when no stress is present. Certain properties of the bar are to be noted: its composition, of density ρ, is uniform, its cross-sectional area is A, and the elastic modulus applicable under these conditions is Y. If a static tensile stress T is applied to each end of the bar, the boundary at x will move to $x + \xi$, and the boundary at $x + \Delta x$ will move up to $x + \Delta x + (d\xi/dx)\,\Delta x$. In the static case, the derivative of ξ is not a partial and the change in length of the section is exactly $(d\xi/dx)\,\Delta x$ if the material obeys Hooke's law. By definition the strain S is the elongation divided by the length of the section, or

$$S = \frac{(d\xi/dx)\,\Delta x}{\Delta x} = \frac{d\xi}{dx} \tag{2.1}$$

The situation is quite different when a sound wave is moving down the bar. Tensile forces across sections are not only a function of time but also of position along the bar. In this case we must refer to the strain at the point x and not across the section Δx. Mathematically speaking, the strain at x is

$$S_x = \mathscr{L}_{\Delta x \to 0} \frac{(\partial\xi/\partial x)\,\Delta x}{\Delta x} = \left(\frac{\partial \xi}{\partial x}\right)_x \tag{2.2}$$

Also, there is at $x + \Delta x$ a strain given approximately by

$$S_{x+\Delta x} = \left(\frac{\partial \xi}{\partial x}\right)_x + \frac{\partial}{\partial x}\left(\frac{\partial \xi}{\partial x}\right)\Delta x \tag{2.3}$$

Corresponding to the strains S_x and $S_{x+\Delta x}$ are forces acting across the boundaries of Δx given by

$$F_x = AYS_x \quad \text{and} \quad F_{x+\Delta x} = AYS_{x+\Delta x} \tag{2.4}$$

Equation (2.4) indicates an unbalanced force acting on the section given by the limit of

$$\Delta F = \frac{\partial F}{\partial x}\Delta x = AY(S_{x+\Delta x} - S_x) \tag{2.5}$$

and from (2.2) and (2.3)

$$\frac{\partial F}{\partial x} = AY\frac{\partial^2 \xi}{\partial x^2} \tag{2.6}$$

The unbalanced force shown by (2.5) acting on the increment of volume $A\,\Delta x$ provides a change of momentum according to Newton's Second Law of

$$\rho A\,\Delta x\frac{\partial^2 \xi}{\partial t^2} = AY\frac{\partial^2 \xi}{\partial x^2}\Delta x \tag{2.7}$$

from which comes the equation describing the motion in the bar along the direction of its length:

$$\frac{\partial^2 \xi}{\partial t^2} = c^2\frac{\partial^2 \xi}{\partial x^2} \tag{2.8}$$

where

$$c^2 = Y/\rho \tag{2.9}$$

We have here a partial differential equation with the displacement ξ as a function of the two independent variables: time t and distance along the bar x. As the motion is pictured by Figure 2.1 and the foregoing equations, ξ is the displacement from its rest position of a plane section normal to the bar's length. The time function will describe the plane's motion. The fact that ξ differs from plane to plane means that strains in the material are created by the motion. If these strains exceed those conforming to Hooke's law, the preceding discussion is invalid.

In looking for a solution to (2.8) one is reminded that solutions to partial differential equations are general functions. With no more information than that supplied by the equation, we can only fix the argument of these functions. Physically, this means that many kinds of motion can satisfy the given conditions. We shall first take a look at the most general solution and then apply restrictions that lead to specific solutions.

Testing $\xi = f_1(x - ct)$,

$$\frac{\partial \xi}{\partial x} = f_1'(x - ct) \quad\text{and}\quad \frac{\partial^2 \xi}{\partial x^2} = f_1''(x - ct)$$

where f_1' and f_1'' are the first and second derivatives of the function f_1 with respect to the argument $(x - ct)$. Similarly,

$$\frac{\partial \xi}{\partial t} = -cf_1'(x - ct) \quad\text{and}\quad \frac{\partial^2 \xi}{\partial t^2} = c^2(x - ct)$$

Insertion of these derivatives shows that the general function $f_1(x - ct)$ is a solution to (2.8). In like manner one can show that another function f_2 of another argument $(x + ct)$ is also a solution. The two furnish the required

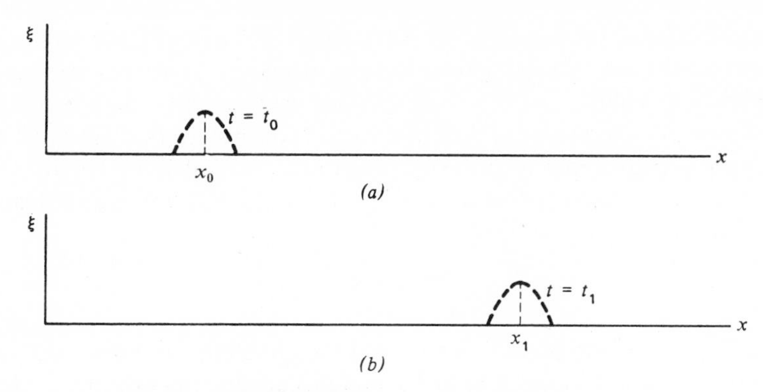

FIG. 2.2 Longitudinal wave moving along a bar.

number of functions to be a complete solution of a second-order partial differential equation.

Two important properties of the kind of wave motion that conforms to the specifications just listed may be deduced from the general solution

$$\xi = f_1(x - ct) + f_2(x + ct) \tag{2.10}$$

Referring to Figure 2.2*a*, we see an arbitrary function $\xi = f_1(x - ct)$ plotted at a time t_0 in the limited region containing the point x_0. In other words, the dotted curve is the locus of $\xi = f_1(x - ct_0)$ while allowing x to vary from $x_0 - \Delta x$ to $x_0 + \Delta x$. Figure 2.2*b* shows the locus of the function $f_1(x - ct_1)$ plotted about the point x_1 while allowing x to vary from $x_1 - \Delta x$ to $x_1 + \Delta x$. There are certainly values of x and t for which the following equality holds:

$$x_1 - ct_1 = x_0 - ct_0 \tag{2.11}$$

and therefore

$$x_1 + \Delta x - ct_1 = x_0 + \Delta x - ct_1 \tag{2.12}$$

Equation 2.12 shows that as Δx varies over the limited region, the shapes of the curves in the two cases are identical. Also from (2.11):

$$\frac{x_1 - x_0}{t_1 - t_0} = c \tag{2.13}$$

According to (2.13), c is the velocity with which this particular shape moves down the bar. It is not to be confused with $\partial\xi/\partial t$, the oscillatory velocity of the motion of a plane about its equilibrium position. To repeat, for this type of motion, a wave moves along the bar, unchanging in shape, at a constant velocity c.

A similar argument will show that $\xi = f_2(x + ct)$ represents a wave moving in the opposite or negative direction. Recalling that $c^2 = Y/\rho$, properties of the bar only, we see that c, called the velocity of propagation, does not depend on direction or wave shape. Frequently the two waves result from the same incident and are alike in shape. However, they may be completely unrelated, and of course it is not necessary that both be present.

Following is a summary of definitions and relationships that have just been developed. They apply to plane waves in a bar of indefinite length. First there are the relations between force F, tension T, pressure p, and strain S, given by

$$T = \frac{F}{A} \qquad T = YS \qquad p = -YS$$

where S is elongation per unit length and therefore dimensionless, and Y, in Newtons per square meter is Young's modulus of elasticity or some modification thereof. We have shown that $S = \partial\xi/\partial x$ and from (2.9) that $Y = \rho c^2$. Therefore

$$p = -\rho c^2 \frac{\partial \xi}{\partial x} \tag{2.14}$$

Also

$$\frac{\partial \xi}{\partial t} = -cf_1' \qquad \frac{\partial \xi}{\partial x} = f_1'$$

so

$$\frac{\partial \xi}{\partial t} = -c\frac{\partial \xi}{\partial x} \qquad p = \rho c \frac{\partial \xi}{\partial t} \tag{2.15a}$$

for the wave moving in the positive direction and

$$p = -\rho c \frac{\partial \xi}{\partial t} \tag{2.15b}$$

for the wave moving in the negative direction.

2.2.1 The uniform bar with boundary conditions

To approach a less general solution to the problem of wave motion in a uniform bar let us assume a solution

$$\xi = X(x)\,T(t) \tag{2.16}$$

which is a product of a function of x by a function of t. Substituting from (2.16) into (2.8), we get

$$X\frac{d^2T}{dt^2} = c^2T\frac{d^2X}{d^2x} \tag{2.17}$$

and separating the variables

$$\frac{d^2X}{X\,dx^2} = \frac{1}{c^2}\frac{d^2T}{T\,dt^2} \tag{2.18}$$

Since in (2.18) we have a function of x only, equal to a function of time only, each must be equal to the same constant which will arbitrarily be called $-k^2$. This yields the two separate equations

$$\frac{d^2X}{dx^2} + k^2X = 0 \tag{2.19a}$$

$$\frac{d^2T}{dt^2} + \omega^2T = 0 \tag{2.19b}$$

where $\omega^2 = k^2c^2$, which are recognized as similar to the equation of simple harmonic motion. The displacement therefore is the real part of

$$\xi = Ae^{j(\omega t - kx)} \quad \text{or} \quad \xi = Ae^{j(\omega t + kx)} \tag{2.20}$$

or the sum of the two, where A is an arbitrary complex constant. Equation 2.20 is still a quite general solution until we restrict the value of k.

In our study of the simple oscillator we noticed that there was only one way or mode in which the system could vibrate. The amplitude and phase are determined by initial conditions. Theoretically, distributed systems may have an infinite number of modes, but the imposition of boundary conditions allows only certain ones. For purposes of illustration let us assume that a wave is induced in the bar that conforms to

$$\xi = (A \cos kx + B \sin kx)e^{j\omega t} \tag{2.21}$$

where A and B are arbitrary constants, and that this bar of length l is clamped at one end and free at the other. The boundary conditions at $x = 0$ are $\xi = 0$ and at $x = l$, $\partial\xi/\partial x = 0$. These conditions require that

$$\xi_0 = Ae^{j\omega t} = 0 \qquad A = 0 \tag{2.22a}$$

$$\left(\frac{\partial\xi}{\partial x}\right)_l = Bk \cos kle^{j\omega t} = 0 \qquad \cos kl = 0 \tag{2.22b}$$

$$\xi = B \sin \frac{n\pi x}{2l} e^{j\omega t} \tag{2.23}$$

where n is an odd integer. Equation 2.22*b* restricts values of k to odd multiples of $\pi/2l$. The freely vibrating bar develops only odd harmonics. Energy losses have been ignored as small so presumably the wave once started will move to and fro on the bar for a long time. Incident waves and reflected waves will add up along the bar to nodes and loops that are fixed in position. Actually as applied to transducers, n will be limited to unity, as was true of the quarter wave longitudinal vibrator once very popular for the construction of piezoelectric sound sources.

Another set of boundary conditions leading to useful applications are

$$\frac{\partial \xi}{\partial x} = 0 \quad \text{at} \quad x = 0 \qquad \xi = 0 \quad \text{at} \quad x = \frac{l}{2} \qquad \frac{\partial \xi}{\partial x} = 0 \quad \text{at} \quad x = l \tag{2.24}$$

Again using the form

$$\xi = (A \cos kx + B \sin kx)e^{j\omega t} \tag{2.21}$$

$$\frac{\partial \xi}{\partial x} = k(-A \sin kx + B \cos kx)e^{j\omega t} \tag{2.25}$$

The use of the boundary condition at $x = 0$ yields $B = 0$ and its use at $x = l$ restricts values of k to $n\pi/l$, where n is an integer. Therefore

$$\xi = Ae^{j\omega t} \cos \frac{n\pi}{l} x \tag{2.26}$$

To determine the effect of the boundary conditions at $x = l/2$, insert this value into (2.26) to find

$$\xi_{l/2} = Ae^{j\omega t} \cos \frac{n\pi}{2} \tag{2.27}$$

Only odd values of n are allowed and in use n will most likely be unity.

The longitudinal vibrator just discussed is known as the half-wave vibrator with a nodal point at the center. It and its modifications are the basic elements of many types of underwater sound sources.

2.2.2 Energy flow of plane waves

Much of the phenomena usually discussed under the heading of acoustic plane waves are also common to the propagation of longitudinal waves in slender rods. Therefore we shall continue with them here rather than in a separate chapter. The small losses that accompany wave propagation will be ignored because they are not a ruling factor in transducer design. The engineer who happens to be working with propagation in materials with heavy losses should examine special discussions of this problem.[1,2] If the motion in the bar has the form of harmonic waves, a solution to (2.8) is

$$\xi = Ae^{j(\omega t - kx)} + Be^{j(\omega t + kx)} \tag{2.28}$$

We recognize that (2.28) expresses the total displacement ξ as the sum of displacements $\xi_+ = Ae^{j(\omega t - kx)}$, a wave moving to the right, and $\xi_- = Be^{j(\omega t + kx)}$, a wave moving to the left.

[1] W. G. Cady, *Piezoelectricity*, McGraw-Hill, New York, 1946.

[2] W. P. Mason, *Physical Acoustics and Properties of Solids*, Van Nostrand, Princeton, New Jersey, 1958.

In terms of these functions, then, we have from (2.14) the pressure

$$p = -\rho c^2 \frac{\partial \xi}{\partial x} = j\rho c\omega(\xi_+ - \xi_-) \tag{2.29}$$

The strain is

$$S = \frac{\partial \xi}{\partial x} = -jk(\xi_+ - \xi_-) \tag{2.30}$$

and the particle velocity

$$v = \frac{\partial \xi}{\partial t} = j\omega(\xi_+ + \xi_-) \tag{2.31}$$

A matter of fundamental interest to the acoustic engineer is the rate of energy flow. It has been defined as the acoustic intensity and is measured in watts per square meter, a quantity readily determined from the energy density and rate of propagation. Referring back to the uniform bar of Figure 2.1, we find the energy to be kinetic, due to particle motion, and potential, due to strain or change in volume. As the thickness Δx of the elementary volume approaches zero, all particles contained therein are moving with the same velocity v so the kinetic energy of the element is

$$\Delta W_k = \tfrac{1}{2}\Delta m v^2 = \tfrac{1}{2}(\rho A\, \Delta x)v^2 \tag{2.32a}$$

The potential energy is that stored in the element $A\,\Delta x$ by compression. The volume V of this element is $A\,\Delta x(1 + \partial\xi/\partial x)$ where $\partial\xi/\partial x$ ranges from zero to a maximum value. The energy stored then is the integral

$$\Delta W_p = -\int p \times d\left[A\,\Delta x\left(1 + \frac{\partial \xi}{\partial x}\right)\right] \tag{2.32b}$$

To change the integrand into a function of one variable

$$dV = d\left[A\,\Delta x\left(1 + \frac{\partial \xi}{\partial x}\right)\right] = A\,\Delta x \frac{\partial^2 \xi}{\partial x^2}\, dx \tag{2.33}$$

since $A\,\Delta x$ is not a function of x. From (2.29), $\partial\xi/\partial x = -p/\rho c^2$, so

$$\frac{d}{dx}\left(\frac{\partial \xi}{\partial x}\right) dx = d\,\frac{-p}{\rho c^2} \tag{2.34}$$

and (2.32) becomes

$$\Delta W_p = \frac{A\,\Delta x}{\rho c^2}\int_0^p p\, dp = \frac{A\,\Delta x}{2\rho c^2}\, p^2 \tag{2.35}$$

The total energy ΔW in the volume element $A\,\Delta x$ is the sum $\Delta W_k + \Delta W_p$ or

$$\Delta W = \tfrac{1}{2}\rho\left(\frac{v^2 + p^2}{\rho^2 c^2}\right) A\,\Delta x \tag{2.36a}$$

and the total energy per unit volume W in joules per cubic meter is

$$W = \frac{\Delta W}{A\,\Delta x} = \tfrac{1}{2}\rho\left(\frac{v^2 + p^2}{\rho^2 c^2}\right) \tag{2.36b}$$

The variables of (2.36) are functions of space and time. We are going to be concerned with the average or rms values since these correspond to the accepted mode for naming quantities of energy delivered to or received from acoustic systems. We may find the values by an integration over a complete period T at a fixed point x_0, or by integration over a full wavelength at a fixed time t_0. The results may be expressed as functions of displacement, velocity, or pressure amplitudes. It is perhaps most straightforward to develop the expression first as functions of displacement. Dealing with the wave moving to the right, from (2.28), (2.29), and (2.31) let us use A_1 as the magnitude of the displacement amplitude and substitute into (2.36) the fact shown by (2.29) and (2.31) that $p_+ = \rho c v_+$. Then

$$W_+ = \rho v_+^2 \tag{2.37}$$

The average of W_+ at a fixed point over one period is

$$(\overline{W}_+)_x = \frac{1}{T}\int^T \rho v_+^2 \, dt$$

if

$$\xi_+ = A_1 \cos(\omega t - kx + \varphi)$$

$$v_+ = \omega A_1 \sin(\omega t - kx + \varphi)$$

and

$$(\overline{W}_+)_x = \frac{\rho\omega^2 A_1^{\,2}}{T}\int_0^T \sin^2(\omega t - kx_0 + \varphi)$$

$$= \frac{\rho\omega^2 A_1^{\,2}}{T}\left[\int_0^T \tfrac{1}{2}\,dt + \frac{1}{2\omega}\int_0^T \cos 2(\omega t - kx_0 + \varphi)2\omega\,dt\right] \tag{2.38}$$

$$(\overline{W}_+)_x = \frac{\rho\omega^2 A_1^{\,2}}{2} + 0 \tag{2.39}$$

To find the corresponding quantity $(\overline{W}_+)_x$ in terms of a velocity amplitude U_x, substitute it for $-\omega A_1$ in (2.39) to obtain

$$(\overline{W}_+)_x = \frac{\rho U^2}{2} \tag{2.40}$$

or in terms of pressure substitute $p_+/\rho c$ for $-\omega A_1$ in (2.39):

$$(\overline{W}_+)_x = \frac{p_+^2}{2\rho c^2} \tag{2.41}$$

The average of W_+ at a fixed time over a wavelength is the integral

$$(\overline{W}_+)_{t_0} = \frac{1}{\lambda}\int_0^\lambda \rho v_+^2\, dx$$

or

$$(\overline{W}_+)_{t_0} = \frac{\rho\omega^2 {A_1}^2}{\lambda}\int_0^\lambda \sin^2(\omega t_0 - kx)\, dx = \frac{\rho\omega^2 {A_1}^2}{2}$$

which is identical with the average over a period.

With acoustic sound sources, we are dealing generally with the power being developed so that acoustic intensity is the more useful concept. For a plane wave the intensity is the average rate of flow of energy through a cross section normal to the direction of flow. A very simple way of arriving at an expression for this quantity is to say that in one period, T, a section of length λ, containing average energy $\overline{W}c$, where c is the velocity of propagation, would flow through unit cross-section. Since the sound field could be fluctuating rapidly in both space and time, our definition must be the limit of this average as the volume and time approach a zero limit. However, as long as c is constant the average rate of flow is

$$I = \overline{W}c = \frac{\rho c\omega^2 {A_1}^2}{2} = \frac{p^2}{2\rho c} \tag{2.42a}$$

As a matter of convenience and to conform to conventional engineering practice, amplitudes of displacement, velocity, and pressure are expressed in rms values, or by peak amplitude divided by the square root of two. This removes the two from the denominators of the expressions for energy density and intensity, leaving

$$I = \frac{p^2}{\rho c} \tag{2.42b}$$

Here p is called the effective pressure.

2.3 TRANSMISSION PROBLEMS

The progressing acoustic wave coming to a boundary where there is a change in the characteristics of the conducting medium may be reflected partially or completely, or it may be refracted, or both reflected and refracted. The general problem belongs to our discussion of sound transmission in the sea. For the moment, we shall see how these phenomena affect the design of the longitudinal vibrator type of transducer which may consist of one or more parts, each with its own set of physical properties. Our problem is to determine what happens to the uniform bar when terminated with resistive, reactive, or complex loads. Most of these vibrators in current use have at least three sections, usually referred to as head, active section, and tail. These sections may be held together by a fourth component. Some are so simple as to simulate a mass and spring system while others are quite complex distributed systems. The discussion will begin with the behavior of plane acoustic waves restricted by boundary layers.

2.3.1 Reflection at normal incidence by a boundary layer

We approach in a simple manner the problem of a uniform half-wave bar immersed in a sound field. Some assumptions not always true in a characteristic situation are made in order to get started. One is that this bar, though quite small in cross section, presents an area normal to the direction of wave propagation that appears to the sound wave as an element of a large, continuous plane surface. This effect can be approached by the assembly of a large number of bars. Another assumption, more reasonable, is that the bar is not losing energy due to supporting fixtures, internal friction, or coupling to an external system.

Consider the situation represented by Figure 2.3 in which the incident wave may be represented by

$$p_i = A_1 e^{j(\omega t - k_1 x)} \tag{2.43}$$

and a wave reflected at the boundary $x = 0$

$$p_r = B_1 e^{j(\omega t + k_1 x)} \tag{2.44}$$

and a wave transmitted through the boundary layer into medium 2

$$p_t = A_2 e^{j(\omega t - k_2 x)} \tag{2.45}$$

It is to be observed that medium 1 and medium 2 are indefinitely extended, and also that the frequency is not changed by change of medium. Since

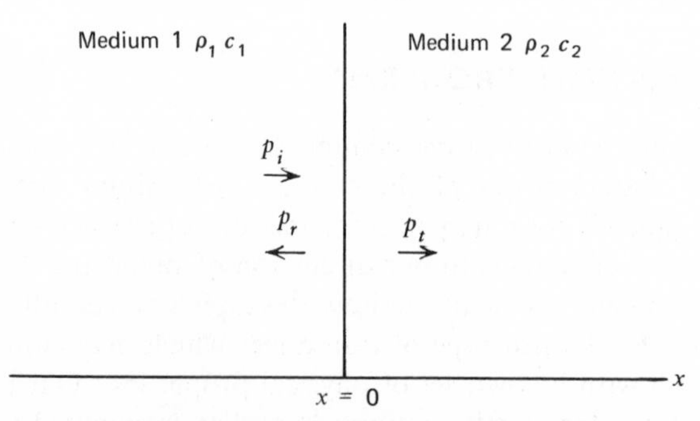

FIG. 2.3 Normal incidence of a plane wave at a boundary.

$f\lambda = c$, the wavelength increases directly as c, the velocity of propagation. Therefore, the wave number $k = 2\pi/\lambda$ will change with change of medium.

The two boundary conditions to be satisfied at $x = 0$ by this set of equations are continuity of pressure across and continuity of particle velocity of the boundary. From (2.15a)

$$p_i = \rho_1 c_1 v_i \qquad p_r = -\rho_1 c_1 v_r \qquad p_t = \rho_2 c_2 v_t$$

To continue our analogy with electric circuits we compare the ratio of pressure to velocity with the ratio of voltage to current, the ratio in each case being an impedance. There are several impedances defined in acoustics and it is essential to identify each carefully. In general the ratio of effective pressure to effective velocity at a point in a sound field is a complex quantity. It is termed the specific acoustic impedance. In a plane progressive wave, the ratio, a real number is the product of the density ρ of the medium by the velocity c of propagation of sound in the medium. The product ρc is called the characteristic impedance and is a most critical property of the medium.

The discussion of sound transmission in this and following sections is greatly simplified by the assumption that we are dealing with characteristic impedances—real numbers. As a result, the phase difference between pressure and velocity is either zero or 180°, and only the real parts of expressions such as (2.43) are needed. Thus in Figure 2.3

$$z_1 = \rho_1 c_1 \qquad z_2 = \rho_2 c_2 \quad p_i = z_1 v_i \qquad p_r = -z_1 v_r \qquad p_t = z_2 v_t \tag{2.46}$$

The boundary conditions at $x = 0$ for pressure and particle velocity are

$$A_1 + B_1 = A_2 \tag{2.47}$$
$$z_2(A_1 - B_1) = z_1 A_1$$

From (2.47) we get the amplitude of the transmitted wave as

$$A_2 = \frac{2z_2}{z_1 + z_2} A_1 \tag{2.48}$$

and that of the reflected wave as

$$B_1 = \frac{z_2 - z_1}{z_2 + z_1} A_1 \tag{2.49}$$

The significance of these results for wave transmission through the boundary is obvious. If we envision medium 2 as a receiving hydrophone, there are two effects of special interest. First, the existence of the reflected wave means that the introduction of the hydrophone into the sound field has altered the field. Whether this is of serious consequence may be determined. Second, A_2, which is the total pressure amplitude at the boundary, may vary from $2A_1$ or twice the pressure of the incident sound field, to a very low value. The high value occurs when $z_2 \gg z_1$ and the low value when the reverse condition exists. If the second medium is really a hydrophone, z_2 would be of the nature of a mechanical impedance and more a characteristic of the structure rather than of the material from which it is made; for example, if the area presented to the sound wave departs widely from our first assumption about the area, z_2 will be small. A pressure-sensitive hydrophone is very poor under these conditions. On the other hand, if we were working with a velocity-sensitive hydrophone, we would take advantage of the fact that the total velocity amplitude at the boundary is

$$v_T = \frac{p_T}{z_2} = \frac{2}{z_1 + z_2} A_1 \tag{2.50}$$

which is twice the velocity amplitude of the incident sound field if $z_2 \ll z_1$.

2.3.2 The resonant bar

There are several procedures used in the determination of design parameters for the resonator which we call the equivalent half-wave vibrator. The half-wave uniform bar is a special case. A method of general applicability is to treat the bar as an acoustic transmission line.[3] The following treats the bar as a mechanical device, which it is, and is general enough for our purpose. Given a bar of length l, cross-sectional area S, and characteristic impedance ρc, the bar has a characteristic mechanical impedance z_m of $\rho c S$. In addition, let us terminate the bar at $x = l$ with an impedance which may be a mass, another bar, or a radiation load. An initial acoustic wave moving in the positive x direction will be partially reflected at $x = l$ resulting in waves moving in both directions along the bar. In order to employ the concept of mechanical impedance defined as F/v or the ratio of force to velocity, we shall describe the motion in terms of compressive forces rather than pressures. In the bar section from $x = 0$ to $x = l$, incident and reflected waves may be represented by

$$\mathbf{F}_i = \mathbf{A}e^{j(\omega t - kx)} \tag{2.51a}$$

and

$$\mathbf{F}_r = \mathbf{B}e^{j(\omega t + kx)} \tag{2.51b}$$

Correspondingly, particle velocities in terms of these forces are

$$\mathbf{v}_i = \frac{\mathbf{F}_i}{z_m} \qquad \mathbf{v}_r = -\frac{\mathbf{F}_r}{z_m} \tag{2.51c}$$

[3] W. P. Mason, *Vibration and Sound*, Chapter 6, McGraw-Hill, New York, 1948.

The mechanical impedance at any point in the bar will be

$$\mathbf{z} = \frac{\mathbf{F}_i + \mathbf{F}_r}{\mathbf{v}_i + \mathbf{v}_r} = z_m \frac{\mathbf{F}_i + \mathbf{F}_r}{\mathbf{F}_i - \mathbf{F}_r} \tag{2.52a}$$

or

$$\mathbf{z}_x = z_m \frac{\mathbf{A}e^{-jkx} + \mathbf{B}e^{jkx}}{\mathbf{A}e^{-jkx} - \mathbf{B}e^{jkx}} \tag{2.52b}$$

If there is a harmonic driving force of angular frequency ω applied to the bar at $x = 0$, the bar will present to this force an impedance given by setting $x = 0$ in (2.52a).

$$\mathbf{z}_0 = z_m \frac{\mathbf{A} + \mathbf{B}}{\mathbf{A} - \mathbf{B}} \tag{2.53}$$

In like manner at $x = l$

$$\mathbf{z}l = z_m \frac{\mathbf{A}e^{-jkl} + \mathbf{B}e^{jkl}}{\mathbf{A}e^{-jkl} - \mathbf{B}e^{jkl}} \tag{2.54}$$

At l, the point of reflection, there is continuity of force and of velocity across the boundary. Therefore there is also continuity of the ratio of the two, the mechanical impedance. In other words, the impedance used to terminate the bar at $x = l$ matches the impedance x_l given by (2.54). Eliminating the amplitudes **A** and **B** between (2.53) and (2.54) thereby yielding the input impedance of the bar $\mathbf{z}_0$ as a function of the terminating impedance and length of the line,

$$\mathbf{z}_0 = z_m \frac{\mathbf{z}_l + jz_m \tan kl}{z_m + jz_l \tan kl} \tag{2.55}$$

To illustrate the utility of (2.55) let us solve several examples that may be taken from Figure 2.4. This figure shows two bars firmly attached at $x = l$. The bar of length l has a density ρ_1, cross-sectional area S_1, and velocity of sound propagation c_1, or as previously defined, a characteristic mechanical impedance $z_{m1} = \rho_1 c_1 S_1$. Similarly, the bar of length a has a characteristic mechanical impedance $z_{m2} = \rho_2 c_2 S_2$. From the definition of z_m we can see

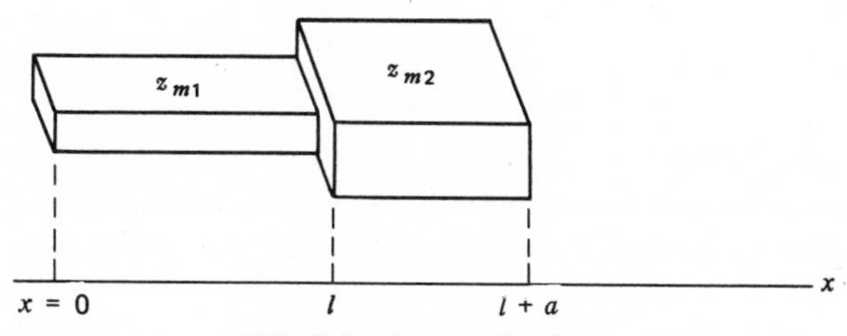

FIG. 2.4 Composite bar.

that if any one or all its factors change at the boundary an incident acoustic wave will be partially or totally reflected.

The driving force will be applied at $x = 0$, the bar marked 2 is to be considered the terminating load, and the mechanical impedance of the loaded bar computed at $x = 0$, is the input impedance z_0 facing the driver. We must first determine the terminating impedance z_l which is that of a bar of length a with a terminating impedance of zero at $z = l + a$. Accordingly, its input impedance at $x = l$ is found from (2.55) by the substitution of zero for z_l. The result, of course, is the input impedance of an unloaded uniform bar for which

$$z_0 = jz_{m2} \tan k_2 a \tag{2.56a}$$

In finding equivalent electrical circuits for oscillating mechanical systems we employ the analogy between the input impedance of an RLC circuit in the vicinity of resonance to that given by (2.56a) as a approaches a half-wave length. This analogy is demonstrated as follows: rewriting the equation for a bar of length l and characteristic mechanical impedance z_m,

$$z_0 = jz_m \tan kl = jz_m \tan\left(\frac{\pi\omega}{\omega_0}\right) \tag{2.56b}$$

since $k = \omega/c$ and $l/c = \pi/\omega_0$ for the half-wave resonant frequency f_0.

In the vicinity of resonance $\pi\omega/\omega_0 \to \pi$ and

$$jz_m \tan\left(\pi\frac{\omega}{\omega_0}\right) = -jz_m \tan \pi\left(\frac{\omega - \omega_0}{\omega}\right) \approx jz_m \pi\left(\frac{\omega - \omega_0}{\omega_0}\right) \tag{2.57}$$

Now consider the electrical impedance

$$z_e = j\left(L\omega - \frac{1}{C_e\omega}\right) = jL\omega_0\left(\frac{\omega}{\omega_0} - \frac{\omega_0}{\omega}\right)$$

$$= jL\omega_0\left[\frac{(\omega - \omega_0)(\omega + \omega_0)}{\omega\omega_0}\right] \approx j2L\omega_0\left(\frac{\omega - \omega_0}{\omega_0}\right) \tag{2.58}$$

A comparison of (2.57) and (2.58) indicates that in the vicinity of resonance the input impedance of the uniform half-wave bar is

$$z_0 = j\left(M\omega - \frac{K}{\omega}\right) \tag{2.59}$$

where $M = \rho Sl/2$ and $K = \pi^2 YS/2l$. Equation (2.59) also shows that the bar behaves as a lumped mass and stiffness system having one-half the mass of the bar. The stiffness of the bar is $\pi^2/2$ times that of the same bar under static stress.

To proceed with the problem of the composite bar of Figure 2.4 we insert the impedance of bar 2 as seen at $x = l$ into (2.55). The input impedance is

$$\mathbf{z}_0 = z_{m1} \cdot \frac{jz_{m2} \tan k_2 a + jz_{m1} \tan k_1 l}{z_{m1} - z_{m2} \tan k_2 a \tan k_1 l} \tag{2.60}$$

Equation 2.60 is utilized mostly to determine dimensions l and a for a desired resonant frequency. Resonance is that condition in which the reactive component of the impedance disappears. Since the denominator of (2.60) is a real number, a resonant state exists when

$$z_{m1} \tan k_1 l = -z_{m2} \tan k_2 a \tag{2.61}$$

The possible solutions to this equation are limited by several practical requirements. As a transducer Section 1 will be the active driving section related to the electrical impedance and power capability. Section 2 is likely to be an inert material acting as a radiating diaphragm. The one requirement it must meet is one of stiffness to meet the boundary condition at $x = l$ which says, in effect, that plane waves are transmitted into bar 2 with no bending of the anterior surface of that bar. This is a separate problem to be discussed later. The relative dimensions of a and l are also related to the overall weight and bandwidth. There will therefore be a best possible solution to (2.61).

If section a is so short that $k_2 a$ is small, we may replace $z_{m2} \tan k_2 a$ by $z_{m2} k_2 a$, thereby getting the more readily handled expression

$$z_{m1} \tan k_1 l = -z_{m2} k_2 a = -\rho_2 S_2 a \omega = -M_2 \omega$$

or

$$\tan k_1 l = -\frac{M_2 \omega}{\rho_0 c_1 S_1} \tag{2.62}$$

In some cases, however, this approximation should be used only as a preliminary search for a more accurate answer.

2.3.3 The equivalent half-wave composite resonator

The composite bar of Figure 2.4 was in common use as a sound projector for sonar prior to and during World War II. The projector was an assembly of several hundred thin-walled nickel tubes silver-braised into circular slots cut into the back side of a thick-walled steel disc which served as the radiating face. The tubes were driven by an alternating magnetic field supplied by current-carrying coils about each tube. The assembly was sealed against sea water. Each tube with its associated section of the disc behaved as the mass loaded bar just described. As we may deduce by looking at (2.62), the length of a thin nickel tube loaded by a heavy mass will exceed a quarter-wave only slightly. This instrument has two unquestionable virtues—it is rugged and it gets a lot of power out of a small amount of nickel. Its one serious fault is discussed subsequently.

Setting the numerator of (2.60) to zero determines the dimension of an equivalent half-wave bar. Setting the denominator to zero determines the dimension of a bar having a very high input impedance at $x = 0$. Therefore if

$$z_{m1} - z_{m2} \tan k_2 a \tan k_1 l = 0$$

$$\tan k_1 l \tan k_2 a = z_{m1}/z_{m2} \tag{2.63}$$

This equation may be used for designing a quarter-wave resonator. To understand how (2.63) is used in the design of a half-wave resonator, let us consider the structure shown in Figure 2.5. The section from $x = -q$ to $x = a$ is a uniform bar of characteristic mechanical impedance $z_{m2} = \rho_2 c_2 S_2$. The other two sections are uniform bars of mechanical properties as shown. The plane cross section at $x = 0$ is as yet undetermined except by the condition that when the bar is vibrating as a half-wave resonator, the mechanical impedance z_0 as given by (2.60) is very large. This leads to the two design equations:

$$\begin{aligned} \tan k_2 a \tan k_3 b &= \frac{z_{m2}}{z_{m3}} \\ \tan k_2 q \tan k_1 L &= \frac{z_{m2}}{z_{m1}} \end{aligned} \tag{2.64}$$

Figure 2.5 is diagrammatic but it is usually a simple matter to convert actual configurations to equivalent rectangular sections. If the active material of the transducer is a piezoceramic, it will make up the central section and will most likely be tubular. The right-hand section will supply the radiating surface and the left-hand section a loading mass for bandwidth control. Equation 2.64 appears to contain even more arbitrary parameters that does (2.63). Fortunately these are limited by design specifications and afford a greater opportunity to meet these specifications.

We consider first the distribution of strain along the bar. Ceramics have excellent compressive strength characteristics, but are poor in tensile strength.

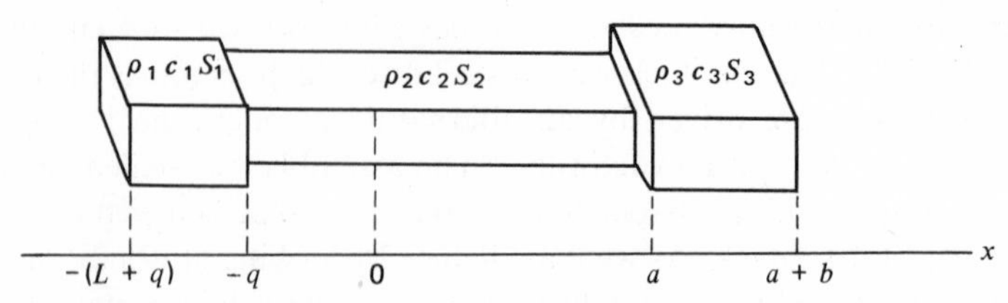

FIG. 2.5 An equivalent half-wave composite resonator.

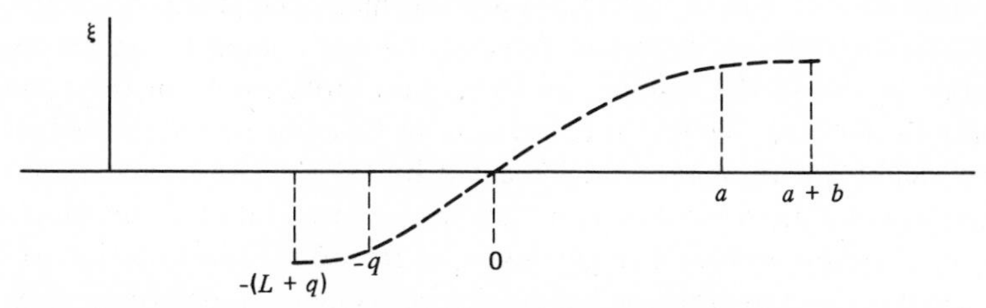

FIG. 2.6 Particle displacement along vibrating bar. $-q \leq x \leq a$, $\xi = A \sin kx$; $a \leq x \leq a + b$, $\xi = A(\sin k_1a/\cos k_2b) \cos k_2(a + b - x)$; $-(L + q) \leq x \leq -q$, $\xi = A(\sin k_1q/\cos k_3L) \cos k_3(q + L + x)$.

For this reason they are used in a prestressed condition. We must know the strain distribution in order to determine the amount of prestress needed. The power to be delivered per unit area determines the displacement amplitude at $x = a + b$ in Figure 2.6, where the time factor $e^{j\omega t}$ is included in the complex amplitude **A**. If the displacement amplitude at this point is B, maximum stress which occurs at $x = 0$ will be

$$T = Y\left(\frac{\partial \xi}{\partial x}\right)_{x=0} = \rho_1 c_1 \omega \frac{\cos k_2 b}{\sin k_1 a} B \tag{2.65}$$

The dimensional choices to be made in the design of the composite resonator bar are lengths and cross section areas. Frequently these bars are assembled horizontally as staves which are shaped as a sector of a right circular cylindrical shell; for example, the complete cylinder may contain 48 staves, the distance between radiating surfaces of each stave being one-half the wavelength of sound developed in the medium. In other words, the outer circumference of the cylinder would be 24 times the wavelength of sound in water. Since it is usually desirable to pack closely the cylindrical surface, the cross-sectional areas of head and tail are pretty well determined by the choice of operating frequency.

A little experience in design soon teaches one how length and weight of the bar varies with change in frequency. The item likely to be decisive is the effect of dimensions on the usable frequency band, or the mechanical quality factor Q_m. We may recall from our previous discussion of the simple harmonic vibrator two definitions of Q_m, namely

$$Q_m = \frac{\omega_0 M}{R_m} \tag{1.27}$$

and

$$Q_m = \frac{\omega_0}{\omega_2 - \omega_1} \tag{1.52}$$

The constants M and R_m of (1.27) are the mass and damping factor of a system having a single degree of freedom. Their related forces are exerted at single points in the system, which is quite different from the condition existing in the bar. We find it convenient to take advantage of the fact that in the vicinity of resonance, as indicated by (2.56), (2.57), and (2.58), the parameters of a distributed system may be converted into the constants of an equivalent system having a single degree of freedom. The selection of these equivalent constants for the single-degree system is determined by the condition that at every instant the kinetic energies of the two systems must be the same. It is possible to choose an arbitrary point in the distributed system as the location of the hypothetical equivalent mass and damping force. The damping force of interest to us arises from the radiation impedance and we choose the coordinate of the radiating surface as that point at which the equivalent mass $\bar{M}$ of the single-degree system will have kinetic energy at every instant equal to the kinetic energy possessed by the entire bar. To make the condition clearer we assume that $\bar{M}$ has a velocity amplitude identical with that of the radiating surface and that R_m is a radiation impedance which under ideal conditions is the product of the radiating surface by the characteristic impedance of the medium.

To become more familiar with the method we shall try it out on the uniform half-wave bar and on the composite bar of Figure 2.4. At every instant the kinetic energy of the equivalent mass due to its motion will be $\frac{1}{2}\bar{M}v^2$, where v is the velocity amplitude to be associated with the equivalent mass. Given a uniform bar of length l and characteristic mechanical impedance z_m, we may write an equation for the forced vibration of the bar in its fundamental mode as

$$\xi = \mathbf{A} \cos kx \tag{2.66}$$

where $\mathbf{A}$ contains the factor $e^{j\omega t}$. Then

$$\mathbf{v} = \frac{\partial \xi}{\partial t} = j\omega \mathbf{A} \cos kx \tag{2.67a}$$

The end of the bar, $x = l$, has been chosen as the point having the velocity to be ascribed to the equivalent mass, and this at resonance is

$$(\mathbf{v})_{x=l} = j\omega \mathbf{A} \tag{2.67b}$$

The total kinetic energy possessed by the bar is determined by integration. Consider at a point x along the bar, an increment of mass $\Delta\rho\; S\; \Delta x$ having a velocity given by (2.67a). Its peak instantaneous kinetic energy is

$$\Delta \mathrm{kE} = \rho S\, \Delta x\, \omega^2 A^2 \tag{2.68}$$

where A is the real value for the displacement amplitude. To equate the kinetic energy of the equivalent mass to that of the bar we have

$$\tfrac{1}{2}\bar{M}\omega^2A^2 = \tfrac{1}{2}\rho S\omega^2A^2\int_0^l \cos^2 kx\,dx \tag{2.69}$$

$$\tfrac{1}{2}\bar{M}\omega^2A^2 = \tfrac{1}{4}\rho S\omega^2R^2l \tag{2.70a}$$

$$\bar{M} = \tfrac{1}{2}\rho Sl \tag{2.70b}$$

showing that the equivalent mass is one-half the total mass of the bar, a result also given by (2.59). To find the quality factor Q_m we use the equivalent mass in (1.27):

$$Q_m = \frac{\rho Sl\omega_0}{2\rho_0c_0S_0}$$

where $\rho_0c_0S_0$ is the mechanical impedance of the medium. Since $\omega_0 = \pi c/l$

$$Q_m = \frac{\pi\rho c}{2\rho_0c_0} \tag{2.71}$$

If we were to use uniform solid rods of a typical piezoelectric ceramic in an array, the quality factor would be something like

$$Q_m = \frac{\pi \times 5.6 \times 4.5 \times 10^6}{2 \times 1.5 \times 10^6} \approx 26$$

a quantity far too high for most purposes. The use of the composite structures takes care of this problem.

The composite bar of Figure 2.4 radiates sound energy from the face at $x = l + a$, so our equivalent mass will be given the velocity amplitude of that surface. Figure 2.7 is a reproduction of Figure 2.4 including the boundary conditions, and particle displacement as a function of x. Equating the kinetic energy of the equivalent mass having a velocity amplitude of

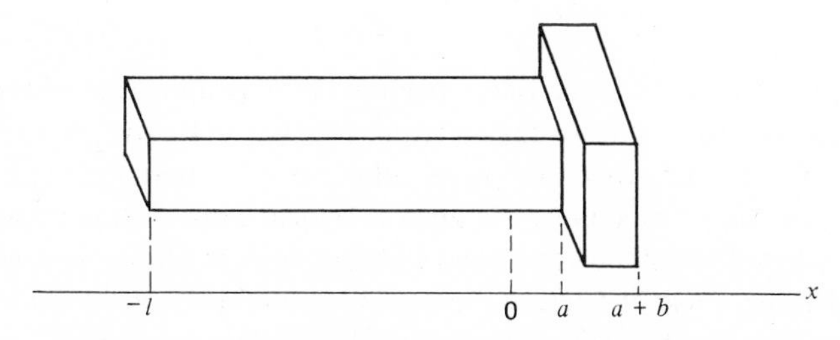

FIG. 2.7 Composite bar with particle displacement as a function of x along the bar. $-l \leq x \leq a$, $\xi = A \sin kx$; $a \leq x \leq a + b$, $\xi = A(\sin k_1a/\cos k_2b) \cos k_2(b + a - x)$.

$\omega A(\sin k_1a/\cos k_2b)$ with the total kinetic energy of the bar gives

$$\tfrac{1}{2}\bar{M}\omega^2 A^2 \frac{\sin^2 k_1a}{\cos^2 k_2b} = \tfrac{1}{2}\rho_1 S_1 A^2 \int_{-l}^{a} \sin^2 k_1 x \, dx + \cdots$$

$$+ \tfrac{1}{2}\rho_2 S_2 A^2 \frac{\sin^2 k_1a}{\cos^2 k_2b} \int_{a}^{a+b} \cos^2 k_2(a + b - x) \, dx \quad (2.72)$$

$$\bar{M} \frac{\sin^2 k_1a}{\cos^2 k_2b} = \tfrac{1}{2}\rho_1 S_1(a + l) + \tfrac{1}{2}\rho_2 S_2 \frac{\sin^2 k_1a}{\cos^2 k_2b} b$$

$$- \frac{1}{4k_1} \rho_1 S_1 \sin 2k_1a + \frac{1}{4k_2} \rho_2 S_2 \frac{\sin^2 k_1a}{\cos^2 k_2b} \sin 2k_2b \quad (2.73)$$

By employing the boundary condition at $x = a$ that

$$\tan k_1a \tan k_2b = \frac{z_{m1}}{z_{m2}} \quad (2.74)$$

we find that the last two terms of (2.73) cancel each other, leaving

$$\bar{M} = \tfrac{1}{2}M_1 + \tfrac{1}{2}M_2 \frac{\cos^2 k_2b}{\sin^2 k_1a} \quad (2.75)$$

where M_1 and M_2 are the masses of the two components of the bar.

2.3.4 Quality factor of composite longitudinal vibrators

Many designs permit the use of shortcuts which simplify the computation involved in arriving at a value for the equivalent mass of a distributed system; for example, in Figure 2.4 the dimension a of Section 2 occurs in (2.61) which determines the length of Section 1. If a in that equation is replaced by $\lambda/8$, the result is

$$\tan k_1 l = -\frac{z_{m2}}{z_{m1}}$$

and if the ratio of these characteristic mechanical impedances is of the order of six or so, a variation due to the difference in $\tan \pi/4$ and $\pi/4$ would not seriously change the length l. However, the use of a material with a low characteristic acoustic impedance requires a critical look at this approximation. To use it in finding the equivalent mass of the bar shown in Figure 2.4 let us consider the mass of Section 2 as concentrated at $x = l$. If the bar is vibrating, we may write an equation for the displacement as a harmonic function of x and time

$$\xi = A \cos k_1 x e^{j\omega t}$$

and a corresponding velocity

$$v = j\omega A \cos k_1 x e^{j\omega t}$$

then at $x = 0$, $v_0 = \omega A e^{j\omega t}$ and at $x = l$, $v_l = \omega A \cos kle^{j\omega t}$ which gives the ratio

$$\frac{v_l}{v_0} = \cos kl \tag{2.76}$$

Considering the equivalent mass $\bar{M}$ also concentrated at $x = l$, we shall set up an expression for the maximum instantaneous kinetic energy of the system and equate it to that of the hypothetical equivalent mass thus

$$\tfrac{1}{2}\bar{M}(v_l)^2 = \tfrac{1}{2}\rho_1 S_1 v_0 \int_0^l \cos^2 k_1 x \, dx + \tfrac{1}{2}M_2(v_l)^2 \tag{2.77}$$

or

$$M = M_2 + \frac{\rho_1 S_1}{2k_1} \sec^2 k_1 l(k_1 l + \sin k_1 l \cos k_1 l) \tag{2.78}$$

The foregoing method may be applied to the complex resonator of Figure 2.5, where the displacement node is at $x = 0$; M_1 is concentrated at $x = -q$ and M_3 at $x = a$. Considering the radiating surface to be at a, the equivalent mass should be given the velocity amplitude at a. The solution is subject to the following boundary condition:
for

$$-q \leq x \leq a \qquad \xi = A \sin k_2 x e^{j\omega t} \qquad v = j\omega A \sin k_2 x e^{j\omega t}$$

the velocity amplitude of M_1 is $v_1 = A \sin k_2 q$ and the velocity amplitude of M_3 is $v_3 = A \sin k_2 a$.

Equating the maximum instantaneous kinetic energy possessed by the equivalent mass $\bar{M}$ to that of the vibrating system, we have

$$\tfrac{1}{2}\bar{M}(v_3)^2 = \tfrac{1}{2}M_1(v_1)^2 + \tfrac{1}{2}M_3(v_3)^2 + \frac{\rho_2 S_2 A^2}{2}\int_{-q}^{a} \sin^2 k_2 x \, dx \tag{2.79}$$

or

$$\bar{M} = M_3 + M_1 \frac{\sin^2 k_2 q}{\sin^2 k_2 a} + \frac{\rho_2 S_2}{4k_2 \sin^2 k_2 a}[2k_2(a + q) - (\sin 2k_2 a + \sin 2k_2 q)] \tag{2.80}$$

A simple form of this type of resonator is the dumbbell where $M_1 = M_3 = M$ and therefore $q = a$. The equivalent mass is

$$\bar{M} = 2M + \frac{\rho_2 S_2 a}{\sin^2 ka} - \frac{\rho_2 S_2}{k_2} \cot k_2 a \tag{2.81}$$

PROBLEMS

2.1 A slender ADP crystal bar 0.09 m long is vibrating with one end free and the other radiating into the characteristic impedance of water. The longitudinal velocity of sound in ADP is 3280 m/sec and its density is 1800 kg/m³. Find the fundamental resonant frequency and Qm.

2.2 A slender barium titanate bar 0.2 m long is vibrating as a half-wave with both ends radiating into the characteristic impedance of water. The longitudinal velocity of sound in barium titanate is 4400 m/sec and its density is 5700 kg/m³. Find the frequency of the half-power points on the power resonance curve.

2.3 Figure 2.8 shows a mass-loaded tube with linear dimensions given in meters. It is free at the ends and supported at the nodal point. The right section is a mass of 0.443 kg. The tube is barium titanate having a longitudinal velocity of sound of 4400 m/sec and a density of 5700 kg/m³.

(*a*) Find the length of the bar.
(*b*) Locate the nodal point for a fundamental resonance of 10 kHz.

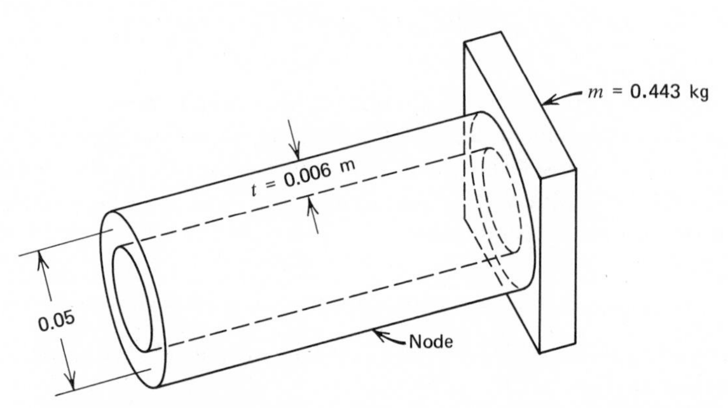

FIG. 2.8 Mass-loaded tube. Tube mean diameter = 0.05 m; wall thickness = 0.006 m.

2.4 Figure 2.9 shows a longitudinal vibrator consisting of masses m_1 and m_3 joined by a thin uniform bar of wave number k_2. Show that when the bar is supported at 0 and free at both ends:

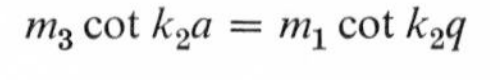

$$m_3 \cot k_2 a = m_1 \cot k_2 q$$

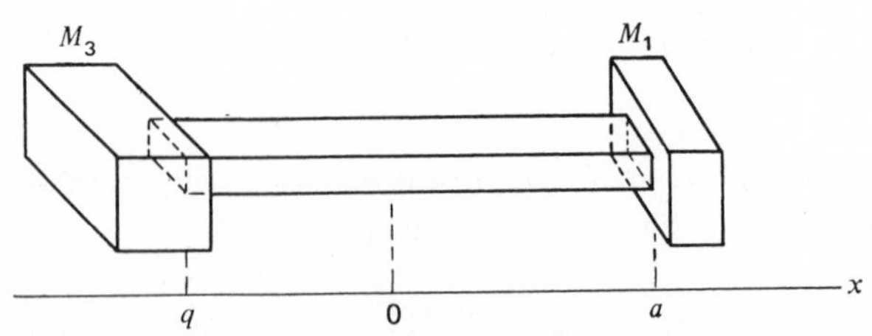

FIG. 2.9 Mass-loaded bars.

2.5 The longitudinal vibrator of Figure 2.8 is altered by attaching a mass of 1 kg to the left end while shortening the center section to maintain a resonance frequency of 10 kHz. Find the length of the center section.

2.6 The composite resonator of Figure 2.9 is an assembly of $m_1 = 8.5$ kg, $m_3 = 1.7$ kg, and a ceramic cylinder of length $q + a$, having a cross sectional area $S_2 = 0.00144$ m^2, a density ρ_2 of 5700 kg/m^3, and a propagation velocity c_2 of 4400 m/sec.

(*a*) Find the length $q + a$ for a fundamental resonance of 5 kHz.
(*b*) For a peak amplitude of displacement at $x = a$ of 2×10^{-5} m, find the peak strain at $x = 0$. If the radiating area is 0.0144 m^2 working into the full characteristic impedance of water, what is Q_m?

2.7 Determine the answers to problem 6 if S_2, the cross-sectional area of the ceramic tube is increased to 0.00288 m^2. If section S_2 is bonded to mass m_1 by a cement having a maximum bond strength of 500 lb/in.^{2}, will it hold at the stress developed by the amplitude of problem 6?

CHAPTER 3

Wave Acoustics

There are two general methods which are often used to solve problems that arise in the field of underwater acoustics. The first approach is based on wave motion calculations and includes information concerning both amplitude and phase of the sound. The second approach, which utilizes optical ray techniques, is, in reality, an approximation of the wave acoustical approach, although this technique is often considered separately. In this form phase information concerning wave propagation is usually discarded. The first method is introduced in this chapter and ray acoustics in the following but it must be remembered that wave acoustics will solve all problems, provided that the difficulty of a calculation is not a factor to be considered.

3.1 METHODS OF DESCRIBING PROPERTIES OF PARTICLES

There are two general systems for describing particle properties such as position, pressure, velocity, or density. The first, the Lagrangian system, is usually referred to as the "material" system. The second, the Eulerian system, is known as the "spatial" system.

3.1.1 Lagrangian description

The Lagrangian description is best suited for describing *identified* particles. The system may best be described by the following statement. "If you tell me *which* particle to consider and *when* you need the information, I'll describe the properties of the particle at that time." Thus, if F is some property of a particle, then

$$F = F(x_0, y_0, z_0, t - t_0)$$

where the position of the particle to be identified is at x_0, y_0, z_0 at time, t_0, and we desire information at time t. Keeping account of identified particles is not the most efficient method of describing acoustical properties and therefore this book will be based on the alternate description.

3.1.2 Eulerian description

The Eulerian description is better suited for describing "fields" of particles where it is not necessary to identify any single particle and hence is ideal for acoustical or fluid mechanical problems. To summarize: "if you tell me *where* and *when* to look, I'll describe the properties of the particles which happen to be there at that time." Thus

$$F = F(x, y, z, t)$$

It can be seen that the location and time coordinates are independent variables and therefore the time derivative of the property F may not be simple.

3.1.3 Time derivative of a particle property in the Eulerian description

Consider some property, F, of a group of particles described by the Eulerian description, then

$$F = F(x, y, z, t) \tag{3.1}$$

If F becomes $F + dF$ in time dt, it may be represented by the first- and second-order terms of a Taylor's series

$$F + dF = F(x, y, z, t) + \frac{\partial F}{\partial x}\,dx + \frac{\partial F}{\partial y}\,dy + \frac{\partial F}{\partial z}\,dz + \frac{\partial F}{\partial t}\,dt \tag{3.2}$$

It follows that

$$\frac{dF}{dt} = u_x\frac{\partial F}{\partial x} + u_y\frac{\partial F}{\partial y} + u_z\frac{\partial F}{\partial z} + \frac{\partial F}{\partial t} \tag{3.3}$$

where

$$u_x = \frac{dx}{dt} \qquad u_y = \frac{dy}{dt} \qquad u_z = \frac{dz}{dt}$$

The particle property derivative in the Eulerian system is called the "total" or "hydrodynamic" derivative and is usually represented by the operator D/Dt. If "grad" is defined in Cartesian coordinates as $\hat{i}(\partial/\partial x) + \hat{j}(\partial/\partial y) + \hat{k}(\partial/\partial z)$ and particle velocity $\mathbf{u}$ as $\hat{i}u_x + \hat{j}u_y + \hat{k}u_z$, then the hydrodynamic operator becomes

$$\frac{D}{Dt} = \mathbf{u} \cdot \text{grad} + \frac{\partial}{\partial t} \tag{3.4}$$

The term $\mathbf{u} \cdot \text{grad}$ describes the change in the property with position *at a given instant of time* while the term $\partial/\partial t$ describes the change with time *at the given position* x, y, z.

The wave equation may now be derived in Cartesian coordinates using the Eulerian description.

3.2 WAVE EQUATION IN THREE DIMENSIONS

A particle system must be compatible with the physical laws of our universe. There are three such laws which play an important part in the description of the particles of a fluid system.

The first law which cannot be violated is the law of the conservation of mass. This law, in mathematical terms, is called the "equation of continuity." It may be stated in three ways:

1. We cannot create or destroy matter.
2. Matter that goes into a box and does not come out must still be there.
3. Matter that is in a sealed box must always be there.

The third form will be used to derive the equation of continuity.

The second law which must not be violated is Newton's second law of motion—the force necessary to move matter must be equal to the rate of change of its momentum.

The third law is the existence of an equation of state, that is, a thermodynamic relation between sets of variables such as pressure and density.

If a system obeys these three laws it will be shown that it may be described by a single equation called the "wave equation."

3.2.1 Equation of continuity

Consider a *sealed* box of finite volume $Q = dx\,dy\,dz$. The volume must be large enough to contain a sufficient number of particles so that they may be considered to be a part of a continuous medium. Let ρ be the density of the particles (i.e., mass/unit volume). The total mass of the particles in the sealed box is ρQ. If the mass is to remain invariant with time then

$$\frac{D}{Dt}(\rho Q) = 0 \tag{3.5}$$

Differentiating,

$$\rho\frac{DQ}{Dt} + Q\frac{D\rho}{Dt} = 0 \tag{3.6}$$

where D/Dt is defined by (3.4) and DQ/Dt may be interpreted in the following manner.

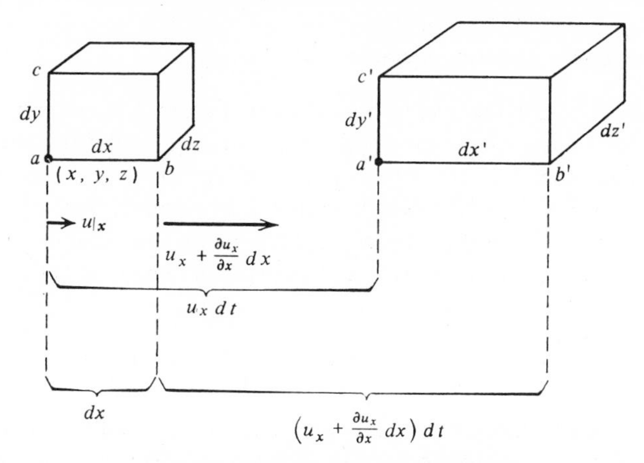

FIG. 3.1 Movement of "sealed box."

If the volume Q of the sealed box is changed during time dt, its new volume Q' will be given by

$$Q' = Q + \frac{DQ}{Dt}\,dt$$

or

$$\frac{DQ}{Dt} = \frac{\text{new volume} - \text{old volume}}{dt} \tag{3.7}$$

Consider Figure 3.1. The x component of the velocity of the particle located at point a is u_x. At point b the particle velocity must be $[u_x + (\partial u_x/\partial x)\,dx]$. During time dt the particle at a moves to a', a distance of $u_x\,dt$ and the particle at b moves to b', a distance of $[u_x + (\partial u/\partial x)\,dx]\,dt$. Thus the new length $dx' = dx + (\partial u_x/\partial x)\,dt\,dx$. Similarly $dy' = dy + (\partial u_x/\partial y)\,dt\,dy$ and $dz' = dz + (\partial u_z/\partial z)\,dt\,dz$. The new volume Q', keeping only linear terms, is

$$Q' = Q + \left(\frac{\partial u_x}{\partial x} + \frac{\partial u_y}{\partial y} + \frac{\partial u_z}{\partial z}\right) Q\,dt$$

so that

$$\frac{DQ}{Dt} = \frac{Q' - Q}{dt} = \left(\frac{\partial u_x}{\partial x} + \frac{\partial u_y}{\partial y} + \frac{\partial u_z}{\partial z}\right) Q = (\text{div}\,\mathbf{u})Q \tag{3.8}$$

Combining (3.8), (3.6), and (3.4) the equation of continuity is

$$\text{div}\,(\rho\mathbf{u}) + \frac{\partial \rho}{\partial t} = 0 \tag{3.9}$$

Thus the equation of continuity is a relationship between four parameters—the three components of particle velocity and the density.

3.2.2 Newton's second equation of motion

Once again consider the *sealed* box Q as in Figure 3.2. If pressure P is applied to side $dy\,dz$, the force applied to that side is $P\,dy\,dz$, whereas a distance dx away the force must be $-[P + (\partial P/\partial x)\,dx]\,dy\,dz$. The net force in the x direction is thus $-(\partial P/\partial x)Q$ and this must equal the rate of change of momentum in that direction. Since the mass of the particles in the box is ρQ, the x momentum must be $\rho Q u_x$ and the rate of change is $(D/Dt)(\rho Q u_x)$. Combining,

$$-\frac{\partial P}{\partial x}Q = \frac{D}{Dt}(\rho Q u_x) = \rho Q\frac{Du_x}{Dt} + u_x\frac{D(\rho Q)}{Dt} \tag{3.10}$$

It can be seen that the last term on the right is identically zero from the definition of the equation of continuity. Equation 3.10 with (3.4) yields

$$-\frac{\partial P}{\partial x} = \rho\left(\mathbf{u}\cdot\text{grad}\,u_x + \frac{\partial u_x}{\partial t}\right)$$

If we consider the forces and momenta in the y and z directions, Newton's second equation of motion takes the form

$$\frac{1}{\rho}\,\text{grad}\,P + (\mathbf{u}\cdot\text{grad})\mathbf{u} + \frac{\partial\mathbf{u}}{\partial t} = 0 \tag{3.11}$$

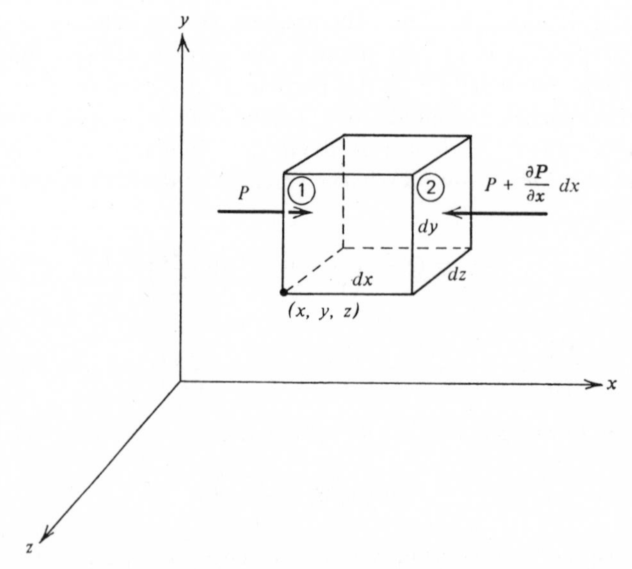

FIG. 3.2 Pressure forces in x direction.

It should be noted that shear forces have been ignored since this book is primarily concerned with sound propagation in fluids. The number of parameters have been increased to five (u_x, u_y, u_z, ρ, P) but only four equations have been derived (equation of continuity plus three equations of motion). One more independent equation is necessary to determine uniquely any parameter. In order to simplify the calculations the "acoustic approximation" will be considered.

3.2.3 Acoustic approximation

If the condensation s is defined by the equation

$$s \equiv \frac{\rho - \rho_a}{\rho_a} \tag{3.12}$$

where ρ_a is the ambient density (the density which exists in the absence of the acoustic wave) then (3.9) and (3.11) become, respectively,

$$\begin{gathered} \rho_a(1 + s) \operatorname{div} \mathbf{u} + \mathbf{u} \cdot \operatorname{grad} [\rho_a(1 + s)] + \frac{\partial}{\partial t} [\rho_a(1 + s)] = 0 \\ \frac{1}{\rho_a(1 + s)} \operatorname{grad} P + (\mathbf{u} \cdot \operatorname{grad})\mathbf{u} + \frac{\partial \mathbf{u}}{\partial t} = 0 \end{gathered} \tag{3.13}$$

It can be seen that the condensation is a dimensionless quantity and in most acoustical problems is much less than unity. The fact that the deviations in the density are small compared to the ambient density and realizing further that the ambient density is independent of position and time, (3.13) becomes

$$\begin{gathered} \operatorname{div} \mathbf{u} + \mathbf{u} \cdot \operatorname{grad} s + \frac{\partial s}{\partial t} = 0 \\ \frac{1}{\rho_a} \operatorname{grad} P + (\mathbf{u} \cdot \operatorname{grad})\mathbf{u} + \frac{\partial \mathbf{u}}{\partial t} = 0 \end{gathered} \tag{3.14}$$

Another consequence of the acoustic approximation is that the terms $\mathbf{u} \cdot \operatorname{grad} s$ and $(\mathbf{u} \cdot \operatorname{grad})\mathbf{u}$ are of second order compared to the other terms in the equation (see problem 1) and may be disregarded. Equations 3.14 are now

$$\operatorname{div} \mathbf{u} + \frac{\partial s}{\partial t} = 0 \tag{3.15}$$

$$\frac{1}{\rho_a} \operatorname{grad} P + \frac{\partial \mathbf{u}}{\partial t} = 0 \tag{3.16}$$

It is apparent that by taking the difference of the partial derivative with respect to time of (3.15) and the divergence of (3.16) it is possible to eliminate the particle velocity **u** with the result

$$\frac{1}{\rho_a}\nabla^2 P = \frac{\partial^2 s}{\partial t^2} \tag{3.17}$$

where the operator ∇^2 means div grad and in Cartesian coordinates is $(\partial^2/\partial x^2) + (\partial^2/\partial y^2) + (\partial^2/\partial z^2)$. To determine uniquely either P or s one more relation must exist between these two quantities.

3.2.4 Relation between pressure and condensation

If it is assumed that a relation exists between pressure and density then $P = P(\rho)$ and therefore grad $P = (dP/d\rho)$ grad ρ. In terms of the condensation grad $P = \rho_a c^2$ grad s, where c^2 is defined as $(dP/d\rho)$. If c^2 is a constant, which may not always be true, then the divergence of grad P is

$$\nabla^2 P = \rho_a c^2 \nabla^2 s \tag{3.18}$$

3.2.5 Wave equation for condensation and pressure fluctuations

By substituting (3.18) into (3.17) the wave equation for condensation fluctuations is obtained:

$$\boxed{\nabla^2 s = \frac{1}{c^2}\frac{\partial^2 s}{\partial t^2}} \tag{3.19}$$

In a similar manner to that described in Section 3.2.4 it is possible to obtain the relation

$$\frac{\partial^2 P}{\partial t^2} = \rho_a c^2 \frac{\partial^2 s}{\partial t^2} \tag{3.20}$$

By combining (3.20) with (3.17) and further by defining the acoustic pressure p by the relationship $P = P_0 + p$ (where P_0 is the ambient pressure), the wave equation for pressure fluctuations is obtained:

$$\boxed{\nabla^2 p = \frac{1}{c^2}\frac{\partial^2 p}{\partial t^2}} \tag{3.21}$$

3.2.6 Wave equation for velocity fluctuations

From Section 3.2.4 grad $P = \rho_a c^2$ grad s. If this relation is combined with (3.16) and, as before, (3.15) is considered, then

$$\text{div}\,\mathbf{u} + \frac{\partial s}{\partial t} = 0 \tag{3.22}$$

$$c^2\,\text{grad}\,s + \frac{\partial \mathbf{u}}{\partial t} = 0 \tag{3.23}$$

It can be seen that the condensation may be eliminated by taking the difference between the gradient of (3.22) and the partial derivative with respect to time of (3.23); thus

$$\text{grad div}\,\mathbf{u} = \frac{1}{c^2}\frac{\partial^2 \mathbf{u}}{\partial t^2}$$

Expanding gives

$$\text{curl curl}\,\mathbf{u} + \nabla^2 \mathbf{u} = \frac{1}{c^2}\frac{\partial^2 \mathbf{u}}{\partial t^2} \tag{3.24}$$

It is apparent that this equation does not have the same form as (3.19) and (3.21). If, however, only a class of solutions are considered such that their velocity field is irrotational (curl $\mathbf{u} = 0$), (3.24) reduces to the following equation which describes the velocity fluctuations in a fluid medium.

$$\boxed{\nabla^2 \mathbf{u} = \frac{1}{c^2}\frac{\partial^2 \mathbf{u}}{\partial t^2}} \tag{3.25}$$

3.2.7 Summary

It is apparent that the form of the equations which satisfy three physical laws (conservation of mass, Newton's second law of motion, and existence of an equation of state) is the same for pressure, particle velocity, and density fluctuations, provided that only linear terms are kept, condensation is small compared to unity, only irrotational motion is considered, and the ratio $dP/d\rho$ is a constant.

The wave equations obtained under these conditions are

$$\nabla^2 \begin{Bmatrix} p \\ \mathbf{u} \\ s \end{Bmatrix} = \frac{1}{c^2}\frac{\partial^2}{\partial t^2}\begin{Bmatrix} p \\ \mathbf{u} \\ s \end{Bmatrix}$$

3.3 VELOCITY POTENTIAL

Since the form of the equations describing pressure, particle velocity, and density variation are the same, it seems possible that one equation and one parameter might exist from which all other quantities could be derived. A parameter that satisfies this requirement is the scalar velocity potential φ. If the particle motion is irrotational, it can be shown that the particle velocity $\mathbf{u}$ can be derived from the gradient of a velocity potential or

$$\mathbf{u} = -\text{grad}\ \varphi \tag{3.26}$$

It should be noted that curl grad φ is identically zero. Combining (3.26) with (3.16) and integrating leads to the following relation between the acoustic pressure and the velocity potential

$$p = \rho_a \frac{\partial \varphi}{\partial t} \tag{3.27}$$

If (3.27) is used in conjunction with the equation of state, grad $p = \rho_a c^2$ grad s, the following relation may be obtained

$$s = \frac{1}{c^2} \frac{\partial \varphi}{\partial t} \tag{3.28}$$

Finally, if (3.26) and (3.28) are combined with (3.15), a wave equation is obtained for the velocity potential with identical form to those obtained earlier. Thus one equation with a scalar parameter is obtainable from which all acoustic variables are derivable.

$$\nabla^2 \varphi = \frac{1}{c^2} \frac{\partial^2 \varphi}{\partial t^2} \tag{3.29}$$

and where

$$\mathbf{u} = -\text{grad}\ \varphi$$

$$p = \rho_a \frac{\partial \varphi}{\partial t}$$

$$s = \frac{1}{c^2} \frac{\partial \varphi}{\partial t}$$

3.4 PLANE WAVE SOLUTION OF THE WAVE EQUATION

Consider the wave equation for a three-dimensional acoustic wave whose parameters are only a function of x.

$$\frac{\partial^2 \varphi}{\partial x^2} = \frac{1}{c^2} \frac{\partial^2 \varphi}{\partial t^2} \tag{3.30}$$

where $\varphi = \varphi(x, t)$ and $u_x = -\partial\varphi/\partial x$, $u_y = u_z = 0$, $p = \rho_a(\partial\varphi/\partial t)$, $s = (1/c^2)(\partial\varphi/\partial t)$. A solution of this equation was found by D'Alembert to be $\varphi(x, t) = \varphi(t \pm x/c)$. Although any function of $(t \pm x/c)$ will satisfy the wave equation, care must be taken such that the solution does not violate any assumptions used in deriving the equation. [Does tan $(t \pm x/c)$ satisfy the wave equation? Does it comply with all assumptions?]

3.4.1 Physical meaning of the plane wave solution

Consider $\varphi(t - x/c)$ to represent a disturbance which is at x_1 at t_1, thus $\varphi(x_1 t_1) = \varphi(t_1 - x_1/c)$. At time t_2 this *particular* disturbance will be at x_2; therefore $\varphi(x_1, t_1) = \varphi(x_2, t_2)$ or

$$\varphi\left(t_1 - \frac{x_1}{c}\right) = \varphi\left(t_2 - \frac{x_2}{c}\right)$$

Regardless of the form of the function, this condition is always satisfied if $t_1 - x_1/c = t_2 - x_2/c$ which gives rise to the relation

$$x_2 = x_1 + c(t_2 - t_1)$$

It can immediately be seen that the constant c has the units of length/time and corresponds to the velocity of the disturbance (wave velocity). Thus $\varphi(t - x/c)$ represents a disturbance which propagates (travels) in the positive x direction while $\varphi(t + x/c)$ represents a disturbance propagating in the negative x direction.

3.4.2 Examples of solutions and some special relations

An often used solution to the plane wave equation is

$$\varphi(x, t) = A \sin \omega\left(t - \frac{x}{c}\right) \tag{3.31}$$

where ω is an unspecified constant and A is called the amplitude. If this function is plotted at a particular time, t_0, it may have the appearance as shown in Figure 3.3. From the periodic form of the sine function it is known that the value of the function and its derivative repeat every 2π radians.

$$A \sin\left(\omega t_0 - \frac{\omega x_1}{c}\right) = A \sin\left(\omega t_0 - \frac{\omega x_2}{c} + 2\pi\right)$$

This condition is satisfied if $-\omega x_1/c = -\omega x_2/c + 2\pi$ or

$$\frac{\omega}{c}(x_2 - x_1) = 2\pi$$

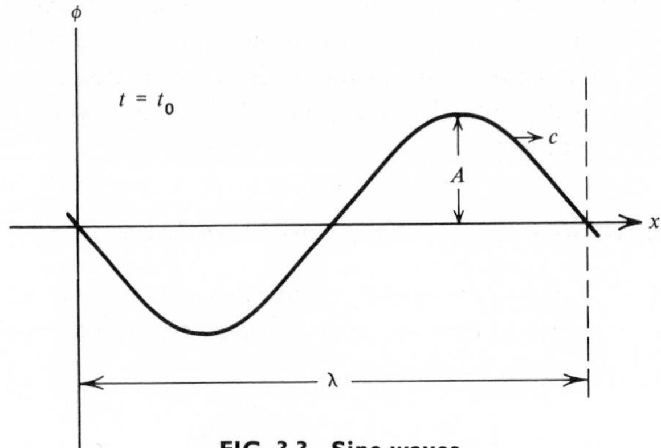

FIG. 3.3 Sine waves.

If the distance $x_2 - x_1$, the distance between the same values of the function, is called the wavelength λ, then

$$\frac{\omega}{c} = \frac{2\pi}{\lambda} \equiv k \tag{3.32}$$

where k is called the wave number. In a similar manner consider the time variation at position x_0, thus A $\sin(\omega t_1 - \omega x_0/c + 2\pi) = A \sin(\omega t_2 - \omega x_0/c)$ or

$$\omega(t_2 - t_1) = 2\pi$$

If the time between consecutive variations of the sine function is called the period T, then

$$\omega = \frac{2\pi}{T} = \omega\pi f \tag{3.33}$$

where f is called the frequency and is defined as $1/T$ (f is the number of times that the function repeats per second). The units of f are Hertz (Hz). Equations 3.32 and 3.33 may be combined to form the following relation:

$$f\lambda = c \tag{3.34}$$

Other possible solutions to the wave equation are

(1) $\varphi = A \cos(\omega t - kx)$
(2) $\varphi = \text{Re}\{Be^{i(\omega t - kx)}\}$

where B is allowed to be complex. Some useful relations are (a) $(\partial/\partial t) = i\omega$; (b) $(\partial/\partial x) = -ik$ for a wave propagating in the positive x direction; (c) $(\partial/\partial x) = ik$ for a wave propagating in the negative x direction.

3.5 SPECIFIC ACOUSTIC IMPEDANCE

The ratio of the acoustic pressure to the particle velocity is called the specific acoustic impedance or

$$Z_{\text{sp.ac.}} \equiv \frac{p}{\mathbf{u}} \tag{3.35}$$

For a plane cosinusoidal wave traveling in the positive x direction where $p = \rho_a(\partial\varphi/\partial t)$ and $u_x = -\partial\varphi/\partial x$ the specific acoustic impedance equals the product of the ambient density and the wave velocity $\rho_a c$. (For the negative x direction $Z_{\text{sp.ac.}} = -\rho_a c$.) It may be simply shown that the ratio of the particle velocity to the wave velocity is equal to the condensation

$$\frac{u_x}{c} = s \tag{3.36}$$

3.6 WAVE VELOCITY IN FLUIDS

From the definition of the wave velocity $c = (dP/d\rho)^{1/2}$ it can be seen that some relationship must exist between pressure and density to define the velocity. For liquids this relation can be found from a knowledge of the compressibility K.

$$K = -\frac{1}{V}\frac{dV}{dP}$$

where V is the specific volume and is equal to $1/\rho$. The wave velocity is thus

$$c = \left(\frac{1}{\rho_a K}\right)^{1/2}$$

which, for pure water, is 1.43×10^3 m/sec. The velocity of sound in sea-water will be studied in Chapter 4.

3.7 REFLECTION AND TRANSMISSION OF ACOUSTIC PLANE WAVES AT BOUNDARIES

One of the most important problems that may be solved through the use of wave acoustics is that of the propagation of an acoustic wave which impinges on the interface between two media. Information concerning the amount of signal transmitted through and reflected by the boundary is extremely useful in underwater acoustics.

3.7.1 Normal incidence—two media

Consider the problem represented by Figure 3.4, that of two semi-infinite media whose total acoustic properties may be described by a knowledge of their densities and compressibilities (thus their wave velocities). Medium one has density ρ_1 and wave velocity c_1, whereas medium two is characterized by ρ_2 and c_2. Let the boundary between the two media be established at $x = 0$. If a source of acoustic plane waves is located in medium one and, further, the plane waves impinge normally on the boundary, some of the energy may be reflected at the boundary and some may be transmitted into medium two. Therefore in medium one there must be an incident wave, p_i, and possibly a reflected wave, p_r,

$$p_i = p_{0i}e^{i(\omega t - k_1 x)}$$

$$p_r = p_{0r}e^{i(\omega t + k_1 x)}$$

where $k_1 = \omega/c_1$ and where p_{0r} is yet to be determined. Associated with each wave is the particle velocity

$$u_i = u_{0i}e^{i(\omega t - k_1 x)}$$

$$u_r = u_{0r}e^{i(\omega t + k_1 x)}$$

In medium two only a transmitted wave can exist;

$$p_t = p_{0t}e^{i(\omega t - k_2 x)}$$

$$u_t = u_{0t}e^{i(\omega t - k_2 x)}$$

At the interface two conditions must be satisfied

1. The pressure across the boundary must be continuous, since the pressure is a scalar, single-valued point function.

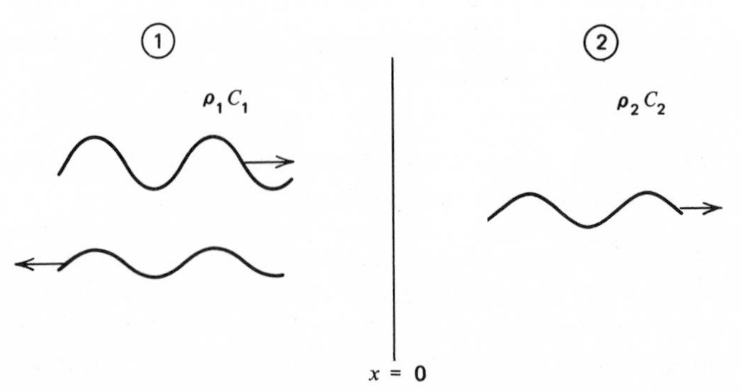

FIG. 3.4 Normal incidence—two media.

2. The normal component of the particle velocity must also be continuous. (If it were not, medium two might pull away from medium one, leaving a void.)

Therefore at $x = 0$

$$p_i + p_r = p_t$$

$$u_i + u_r = u_t$$

From (3.35) $u_i = p_i/\rho_1 c_1$, $u_r = -p_r/\rho_1 c_1$, $u_t = p_t/\rho_2 c_2$. Combining all equations

$$p_{0i} + p_{0r} = p_{0t}$$

$$\frac{p_{0i}}{\rho_1 c_1} - \frac{p_{0r}}{\rho_1 c_1} = \frac{p_{0t}}{\rho_2 c_2}$$

Solving for the reflected wave amplitude in terms of the incident wave amplitude

$$p_{0r} = p_{0i}\left(\frac{\rho_2 c_2 - \rho_1 c_1}{\rho_2 c_2 + \rho_1 c_1}\right) \tag{3.37}$$

If $\rho_2 c_2 \gg \rho_1 c_1$ (i.e., $z_2 = p_{0t}/u_{0t}$ is large when u_{0t} is small, hence medium two approximates a rigid wall), then $p_{0r} = p_{0i}$. The reflected signal has the same amplitude and phase as the incident signal. For $\rho_2 c_2 \ll \rho_1 c_1$ (or p_{0t} is small, as in the case of water–air interface), which is commonly called "pressure release," the reflected signal has the same amplitude but is π radians opposite in phase to the incident signal.

The *sound power reflection coefficient* α_r is defined by the relation

$$\alpha_r = \frac{I_r}{I_i}$$

where I_r is intensity of the reflected wave and I_i is the intensity of the incident wave. (The intensity is the amount of energy passing through a unit area per unit time.) For the case of a plane wave it can be shown that $I = \frac{1}{2}p^2/\rho c$. Thus α_r is given by

$$\alpha_r = \left(\frac{\rho_2 c_2 - \rho_1 c_1}{\rho_2 c_2 + \rho_1 c_1}\right)^2 \tag{3.38}$$

Since energy must be conserved, α_t, the *sound power transmission coefficient* must be given as

$$\alpha_t = 1 - \alpha_r = \frac{4\rho_2 c_2 \rho_1 c_1}{(\rho_2 c_2 + \rho_1 c_1)^2} \tag{3.39}$$

Note the symmetry of α_t. The amount of energy crossing the interface does not depend on which medium contains the source of acoustic energy.

3.7.2 Normal incidence—three media

Consider Figure 3.5. Once again all acoustic properties of each medium are defined by its density and compressibility.

As in Section 3.7.1, it can be assumed that there may be transmitted and reflected waves in media one and two but only a transmitted wave in medium three. The boundary conditions at $x = 0$ and $x = l$ are also as indicated. The sound power transmission coefficient is

$$\alpha_t = \frac{4\rho_3 c_3 \rho_1 c_1}{(\rho_3 c_3 + \rho_1 c_1)^2 \cos^2 k_2 l + (\rho_2 c_2 + \rho_3 c_3 \rho_1 c_1/\rho_2 c_2)^2 \sin^2 k_2 l} \tag{3.40}$$

This equation may be further simplified for special cases; for example, if $\rho_1 c_1 = \rho_3 c_3$, then

$$\alpha_t = \frac{4}{4\cos^2 k_2 l + (\rho_2 c_2/\rho_1 c_1 + \rho_1 c_1/\rho_2 c_2)^2 \sin^2 k_2 l} \tag{3.41}$$

The presence of an intermediate layer between two media whose acoustic properties differ from each other can often be used to advantage. Consider (3.40) under the following conditions: $k_2 l = [(2n - 1)\pi]/2$ and $\rho_2 c_2 = (\rho_1 c_1 \rho_3 c_3)^{1/2}$. The first condition implies that the layer l may be a quarter wavelength in thickness whereas the second condition states that the acoustical properties of the material is just the geometric mean of the two outer media. Equation 3.40 becomes unity and it is seen that the inner layer acts as a transformer in order to match the acoustic impedance of medium one to that of medium two, thereby causing full transmission of the signal.

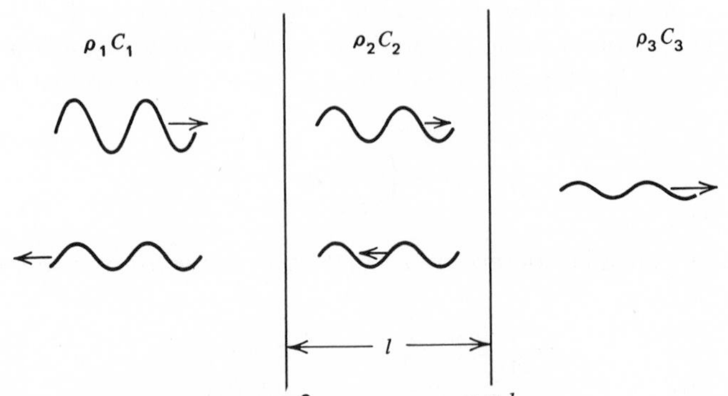

FIG. 3.5 Normal incidence—three media.

3.7.3 Oblique incidence—two media

Consider Figure 3.6. A plane wave which propagates in an arbitrary positive x and y direction in the $z = 0$ plane is given by the relation

$$p = p_0 \exp \{i[\omega t - k(x \cos \theta + y \sin \theta)]\}$$

whereas for a negative x direction and a positive y direction

$$p = p_0 \exp \{i[\omega t - k(-x \cos \theta + y \sin \theta)]\}$$

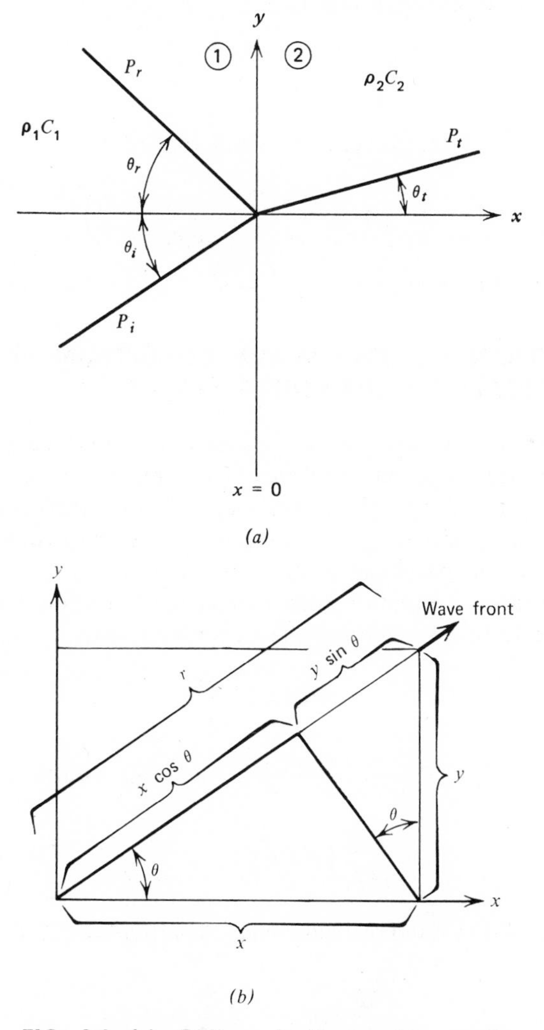

FIG. 3.6 (*a*) Oblique incidence—two media. (*b*) Path of ray.

Thus the acoustic pressure equations in the two media are given by

$$p_i = p_{0i} \exp \{i[\omega t - k_1(x \cos \theta_i + y \sin \theta_i)]\}$$
$$p_r = p_{0r} \exp \{i[\omega t - k_1(-x \cos \theta_r + y \sin \theta_r)]\}$$
$$p_t = p_{0t} \exp \{i[\omega t - k_2(x \cos \theta_t + y \sin \theta_t)]\}$$

where θ_i is the angle of incidence, θ_r is the angle of reflection, and θ_t is the angle of refraction. The particle velocities are once again given by (3.35). It can be shown that Snell's law must hold (law of refraction) and the law of reflection must be satisfied; that is

$$\frac{\sin \theta_i}{\sin \theta_t} = \frac{k_2}{k_1} \quad \text{and} \quad \theta_r = \theta_i$$

If the boundary conditions are once again applied, then

$$\alpha_r = \left(\frac{\rho_2 c_2 \cos \theta_i - \rho_1 c_1 \cos \theta_t}{\rho_2 c_2 \cos \theta_i + \rho_1 c_1 \cos \theta_t}\right)^2$$

which reduces to (3.38) when $\theta_i = \theta_t = 0°$.

3.8 SOLUTION OF THE WAVE EQUATION FOR SPHERICALLY EXPANDING WAVES

In the following sections some basic sources of acoustic energy are investigated. Since the output characteristic of each source may be computed from the sum of individual point sources (superposition theorem), every point source emitting a spherically expanding wave, the wave equation should be investigated for this particular form of the solution.

Consider the wave equation with a sinusoidally varying time dependence and restricted to be a function of the r-coordinate only,

$$\nabla^2 \varphi = \frac{1}{c^2} \frac{\partial^2 \varphi}{\partial t^2} \tag{3.42}$$

where $\varphi = \psi(r)e^{i\omega t}$. Expanding $\nabla^2 \varphi$ in spherical coordinates leads to the equation

$$\frac{1}{r^2} \frac{\partial}{\partial r}\left(r^2 \frac{\partial \psi}{\partial r}\right) = -\frac{\omega^2}{c^2} \psi \tag{3.43}$$

Multiplying (3.43) by r and realizing that $[\partial^2(\psi r)]/\partial r^2 = (1/r)(\partial/\partial r)[r^2(\partial \psi/\partial r)]$ gives

$$\frac{\partial^2}{\partial r^2}(\psi r) = -\frac{\omega^2}{c^2}(\psi r)$$

whose general solution may be written as $\psi r = Ae^{\pm ikr}$; the final solution for the wave equation is thus

$$\varphi = \frac{\varphi_0}{r} e^{i(\omega t - kr)} \tag{3.44}$$

In contrast to the plane wave solution the spherically expanding wave contains a $1/r$ dependence.

Since the specific acoustic impedance is defined as the ratio of the acoustic pressure to the particle velocity, it can be seen from the relations $p = \rho_a(\partial\varphi/\partial t)$ and $u_r = -\partial\varphi/\partial r$ (u_r being the radial particle velocity) that the impedance is complex.

$$z = \frac{\rho_a(\partial\varphi/\partial t)}{-\partial\varphi/\partial r} = \rho_a c\left(\frac{k^2r^2}{1 + k^2r^2} + i\frac{kr}{1 + k^2r^2}\right) \tag{3.45}$$

For small values of kr both the real and imaginary components of the impedance approach zero, whereas for large values of kr the impedance approaches $\rho_a c$, the impedance of a plane wave.

3.9 SIMPLE SOURCES

As mentioned in the preceding section, it is possible to calculate the basic characteristics of a complex source by considering it to be formed of simple point sources, each radiating with an appropriate amplitude and phase. The fact that this procedure can be done arises from the principle of superposition, which is applicable to linear equations.

3.9.1 Source strength

The source strength Q is defined in the following manner

$$Q \equiv \int_S \mathbf{u}_s \cdot \hat{n}\, ds \tag{3.46}$$

where u_s is the velocity vector on the radiating surface of the source, $\hat{n}$ is a unit vector normal to the surface, and the equation is integrated over the surface. It can be seen that Q represents the volume velocity of the source. If the basic source is considered to be a monopole, that is a sphere whose total surface periodically expands and contracts radially with the same phase, then the source strength Q is $4\pi r_0^2 u_0$, where r_0 is the nominal radius of the source and u_0 the amplitude of vibration.

3.9.2 Radiation from a monopole source

The potential function φ for a spherically expanding periodic wave was found to be $\varphi = (\varphi_0/r)e^{i(\omega t - kr)}$. In order to evaluate φ for a monopole source, it is necessary to apply the boundary condition that at the surface of the source $u = u_0 e^{i\omega t}$. The radial velocity of the wave is given by $u = -\partial\varphi/\partial r$, which, from (3.44), leads to

$$u = (1 + ikr)\frac{\varphi_0}{r^2}\, e^{i(\omega t - kr)} = (1 + k^2r^2)^{1/2}\frac{\varphi_0}{r^2}\, e^{i(\omega t - kr + \tan^{-1} kr)}$$

However, this equation under the condition that $kr \ll 1$ (and therefore $\tan^{-1} kr \cong kr$) reduces to

$$u = \frac{\varphi_0}{r^2}\, e^{i\omega t}$$

This condition is met in the region in which the distance from the source is small compared with a wavelength. Therefore at the surface of the monopole, $r = r_0$,

$$u = \frac{\varphi_0}{r_0^{\,2}}\, e^{i\omega t} = u_0 e^{i\omega t}$$

or $\varphi_0 = u_0 r_0^2$. In terms of the source strength, $Q = 4\pi r_0^2 u_0$, the potential field radiated by a monopole is given as

$$\varphi = \frac{Q}{4\pi r}\, e^{i(\omega t - kr)} \tag{3.47}$$

3.9.3 Radiation from a dipole source

A dipole consists of two monopoles of equal strength whose phase differs from one another by π radians. If they are separated by a distance d and the point P under consideration is located a distance r from the midpoint between the two sources, the potential at P is given by

$$\varphi_P = \frac{Q}{4\pi r_1}\, e^{i(\omega t - kr_1)} - \frac{Q}{4\pi r_2}\, e^{i(\omega t - kr_2)}$$

As shown in Figure 3.7, r_1 and r_2 expressed in the same coordinate system are

$$r_1 = r - \frac{d}{2}\cos\theta + \text{other higher order terms}$$

$$r_2 = r + \frac{d}{2}\cos\theta + \text{other higher order terms}$$

Therefore the potential at P becomes

$$\varphi_P = \frac{Q}{4\pi r}\, e^{i(\omega t - kr)}\left[\frac{e^{i(kd/2)\cos\theta}}{1 - (d/2r)\cos\theta} - \frac{e^{-i(kd/2)\cos\theta}}{1 + (d/2r)\cos\theta}\right]$$

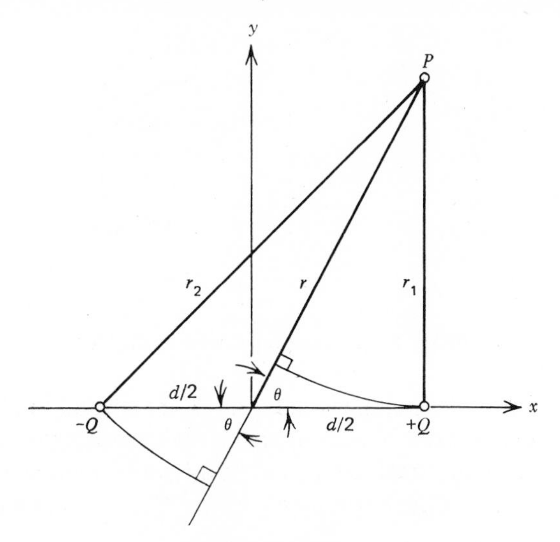

FIG. 3.7 Dipole.

If $d/r \ll 1$ (far field conditions), which implies that the point P is far away compared to the spacing between the sources, this equation reduces to the form

$$\varphi_P = \frac{Q}{4\pi r} e^{i(\omega t - kr)}[e^{i(kd/2)\cos\theta} - e^{-i(kd/2)\cos\theta}]$$

The definition of the sine function is, however,

$$\sin x = \frac{e^{ix} - e^{-ix}}{2i}$$

which further simplifies the equation for the potential φ_P to

$$\varphi_P = \frac{Q}{4\pi r} e^{i(\omega t - kr)}\left[2i \sin\left(\frac{kd}{2}\cos\theta\right)\right]$$

If $kd \ll 1$, implying that the spacing between the sources is small compared to the wavelength, then

$$\varphi_P = i\frac{Qkd\cos\theta}{4\pi r} e^{i(\omega t - kr)}$$

and therefore

$$p = -\frac{\rho_a \omega^2 d\cos\theta}{4\pi rc} e^{i(\omega t - kr)} \tag{3.48}$$

The acoustic pressure is seen to be proportional to the square of the frequency and in addition is modified by an angular dependence given by $\cos\theta$ as shown

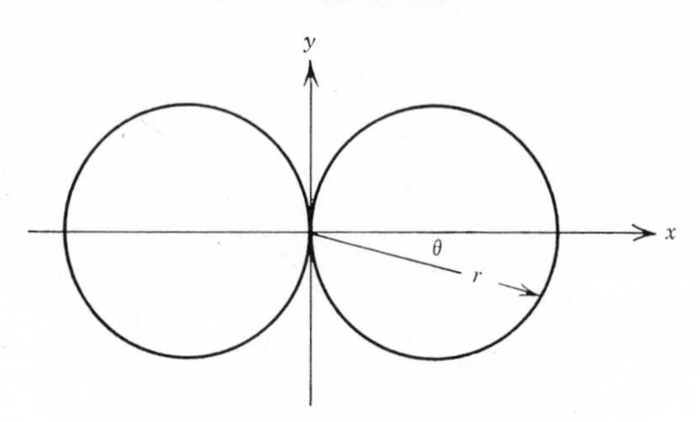

FIG. 3.8 Radiation pattern of dipole.

in Figure 3.8. It must be emphasized that (3.48) is valid only when (*a*) the radius of the sources are small compared to the wavelength, (*b*) the spacing between the sources are small compared to the wavelength, and (*c*) the point under observation is distant compared to the spacing between the sources.

3.9.4 Radiation from a piston source in an infinite baffle

The radiation pattern from a circular piston source can be calculated in much the same manner as the radiation from a dipole source. Consider the piston shown in Figure 3.9*a*. The potential at point *P* which arises from

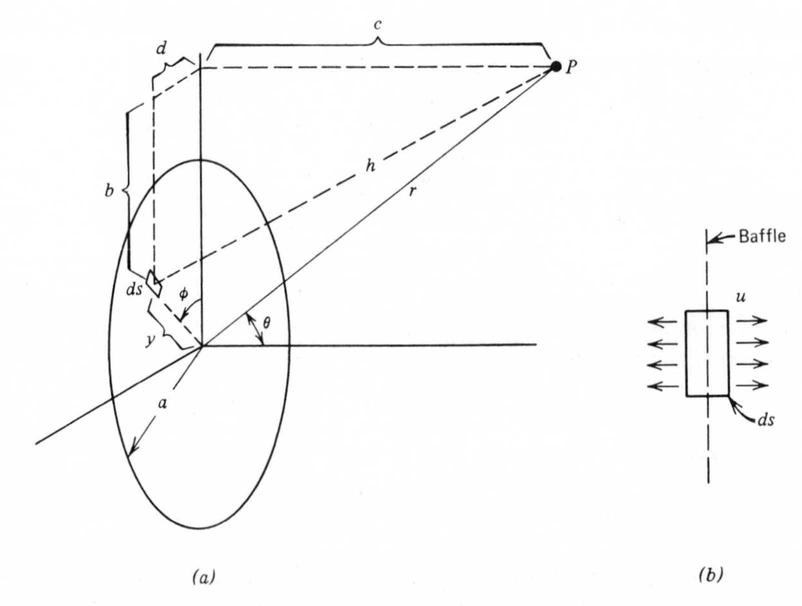

FIG. 3.9 (*a*) Piston source. (*b*) Detail of piston source.

the small element of radiating area ds is given by

$$d\varphi = \frac{dQ}{4\pi h} e^{i(\omega t - kh)} \tag{3.49}$$

It can be shown that a monopole shaped like a "pill box," as in Figure 3.9*b*, expanding and contracting normal to its flat sides but without the baffle, will produce the same potential at P as the piston in an infinite baffle. It will also satisfy the boundary condition that the particle velocity normal to the baffle is zero. Since the source strength Q is defined as $\int \mathbf{u} \cdot \hat{n}\, ds$, the source strength dQ for the pill box is $2u_0\, ds$, where u_0 is the amplitude of oscillation. It is convenient to establish a coordinate system at the center of the piston and it is seen that

$$h = r\left(1 - \frac{2y}{r} \sin\theta \cos\phi + \frac{y^2}{r^2}\right)^{1/2}$$

since $c = r\cos\theta$, $d = y\sin\phi$, and $b = r\sin\theta - y\cos\phi$. If only points satisfying the condition $y/r \ll 1$ are considered ("far field" calculations), h may be approximated by $(1 - 2y/r \sin\theta\cos\phi)^{1/2}$. Under the same condition, a further approximation may be applied; that is, only the first two terms of the binomial expansion of h will be retained, thus

$$h = 1 - \frac{y}{r} \sin\theta \cos\phi \tag{3.50}$$

Combining (3.49) and (3.50) and the value of the source strength, the following equation is obtained for the far field potential at point P due to an element of surface $ds = y\, d\phi\, dy$ on the piston.

$$d\varphi = \frac{u_0 y\, d\phi\, dy}{2\pi r[1 - (y/r)\sin\theta\cos\phi]} \exp\left[i(\omega t - kr + ky\sin\theta\cos\phi)\right] \tag{3.51}$$

Since y/r is small compared with unity, it is possible to ignore the contribution of the term $(y/r)\sin\theta\cos\phi$ in the denominator of (3.51) because it will only slightly affect the amplitude of the potential; however, the term $ky\sin\theta\cos\phi$ cannot be similarly ignored, since it affects the phase of the potential.

The potential of P is thus

$$\varphi_P = \frac{u_0}{2\pi r} e^{i(\omega t - kr)} \int_0^{2\pi} \int_0^a e^{iky\sin\theta\cos\phi} y\, dy\, d\phi \tag{3.52}$$

The simplest method of evaluating (3.52) is to make the following change of variables. Set $z = (k\sin\theta)y$, where k and θ are constant for a given location P. Equation 3.52 takes on the form

$$\varphi_P = \frac{u_0}{2\pi r} e^{i(\omega t - kr)} \frac{1}{k^2 \sin^2\theta} \int_0^{ak\sin\theta} z\, dz \int_0^{2\pi} e^{iz\cos\phi}\, d\phi \tag{3.53}$$

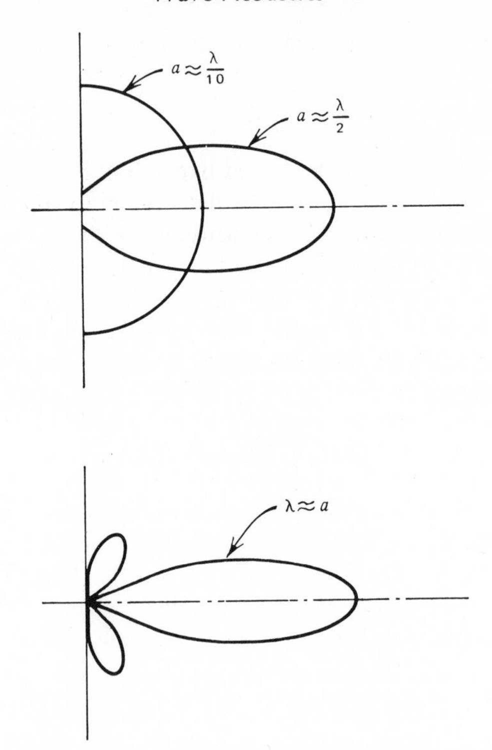

FIG. 3.10 Radiation pattern of piston source.

However, the definition of the Bessel function of the zeroth kind is

$$J_0(z) = \frac{1}{2\pi} \int_0^{2\pi} e^{iz \cos \phi} \, d\phi$$

and further

$$\int zJ_0(z) \, dz = zJ_1(z)$$

where $J_1(z)$ is the Bessel function of the first kind. By substituting into (3.53) the following equation is obtained:

$$\varphi_P = \frac{u_0 a^2}{r} e^{i(\omega t - kr)} \left[\frac{J_1(ak \sin \theta)}{ak \sin \theta} \right] \tag{3.54}$$

A polar plot of the function $J_1(ak \sin \theta)/(ak \sin \theta)$ is given in Figure 3.10 for various values of λ/a. Since $J_1(ak \sin \theta)$ has its first zero when $ka \sin \theta = 1.2\pi$, the angle θ_0 which indicates the spread in the main lobe of the beam can be found from the relation

$$\theta_0 = \sin^{-1} \frac{0.6\lambda}{a} \tag{3.55}$$

3.9.5 Radiation impedance of a piston source

The radiation impedance of a source is defined as the ratio of the force necessary to obtain a particular particle velocity at the surface of the source to that particle velocity; thus, if the force is given for a sinusoidally varying piston as $F = F_0 e^{i\omega t}$, the radiation impedance is given by

$$Z_r = \frac{F_0 e^{i\omega t}}{u_0 e^{i\omega t}} = \frac{F_0}{u_0} \tag{3.56}$$

To obtain the radiation impedance of a circular piston source in an infinite baffle it is necessary to calculate the total force on a small area ds of the piston arising from the rest of the piston and then to sum up all the contributions from each element ds. Consider the construction in Figure 3.11. The pressure on ds caused by the movement of ds' can be found from (3.49) and (3.27). Thus

$$dp = \frac{i\omega\rho_a}{4\pi h}(2u_0 y\, d\psi\, dy)e^{i(\omega t - kh)}$$

where h is given by the relation $h^2 = x^2 + y^2 - 2ry \cos(\psi - \theta)$. The pressure on ds is then the sum of all the contributions from each ds' on the entire piston:

$$p = \frac{i\omega\rho_a u_0}{2\pi} e^{i\omega t} \int_0^a y\, dy \int_0^{2\pi} \frac{e^{-ikh}}{h}\, d\psi$$

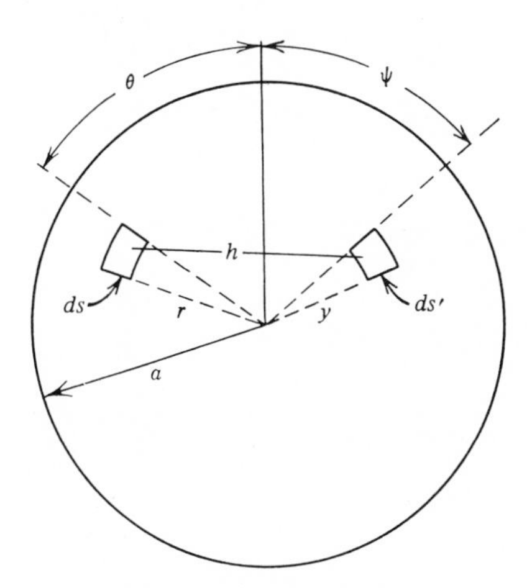

FIG. 3.11 Piston source.

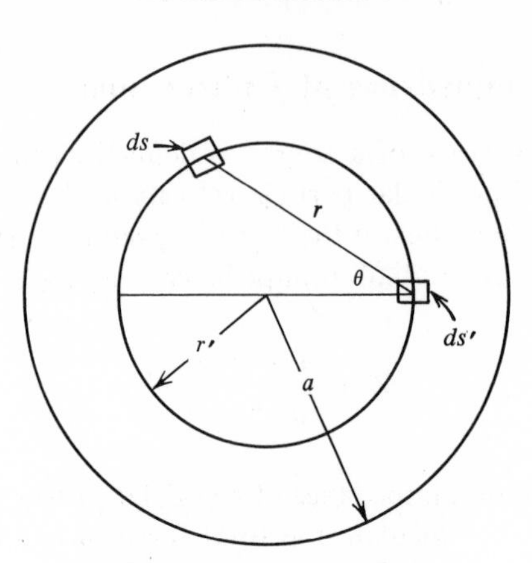

FIG. 3.12 Piston source.

Since the force on the area *ds* is given by *p ds*, the total force on the piston is

$$F = \frac{i\omega\rho_a u_0}{2\pi} e^{i\omega t} \int_0^a r\,dr \int_0^{2\pi} d\theta \int_0^a y\,dy$$

$$\times \int_0^{2\pi} \frac{\exp\{-ik[r^2 + y^2 - 2ry\cos(\psi - \theta)]^{1/2}\}}{[r^2 + y^2 - 2ry\cos(\psi - \theta)]^{1/2}}\,d\psi \quad (3.57)$$

The evaluation of this integral is extremely complex and is usually somewhat simplified in the following manner. It can be seen from (3.57) that each element of area on the piston is summed twice. In the calculation that follows each element is considered only once and the result multiplied by two. A new diagram, as in Figure 3.12, is utilized in which the pressure on ds' is obtained only from within a disk of radius r'. It follows that

$$p_{ds'} = \frac{i\omega\rho_a u_0}{2\pi} e^{i\omega t} \int_0^{\pi} d\theta \int_0^{2r'\sin\theta} e^{-ikr}\,dr$$

Performing this double integration and realizing that the definition of the Bessel function $J_0(z)$ is given by

$$J_0(z) = \frac{1}{\pi}\int_0^{\pi} \cos(z\sin\theta)\,d\theta$$

the following equation is obtained.

$$p_{ds'} = \frac{\omega\rho_a u_0}{2k} e^{i\omega t}\left[1 - J_0(2kr') + \frac{i}{\pi}\int_0^{\pi} \sin(2kr'\sin\theta)\,d\theta\right]$$

Therefore, one-half the total force on the piston is given by

$$\tfrac{1}{2}F = \frac{\omega\rho_a u_0}{2k}\, e^{i\omega t}\int_0^{2\pi} d\psi \int_0^a r'\, dr'\left[1 - J_0(2kr') + \frac{i}{\pi}\int_0^{\pi} \sin\,(2kr'\sin\theta)\, d\theta\right]$$

Carrying out the integrations indicated and also realizing that $\int zJ_0(z)\, dz = zJ_1(z)$ gives the following relation:

$$F = \frac{\omega\rho_a u_0 \pi a^2}{k}\, e^{i\omega t}\left[1 - \frac{J_1(2ka)}{ka} + \frac{i2}{\pi a^2}\int_0^a r'\, dr' \int_0^{\pi} \sin\,(2kr'\sin\theta)\, d\theta\right]$$

There remains only the simplification of the terms still left under the double integral. This term is usually evaluated in the following manner: effect the transformation $z = 2kr'$. It can also be seen from the form of the integrand that the value of the integral between the limits of zero and π are just twice that of the integral between the limits zero and $\pi/2$. Thus

$$\frac{i2}{\pi a^2}\int_0^a r'\, dr' \int_0^{\pi} \sin\,(2kr'\sin\theta)\, d\theta = \frac{i}{\pi a^2 k^2}\int_0^{2ka} z\, dz \int_0^{\pi/2} \sin\,(z\sin\theta)\, d\theta \qquad (3.58)$$

Let $K = \int_0^{\pi/2} \sin\,(z\sin\theta)\, d\theta$. It can be shown that

$$\frac{1}{z}\frac{d}{dz}\left(z\frac{dK}{dz}\right) = K + \frac{1}{z} \qquad (3.59)$$

by using the relation $\sin^2\theta = 1 - \cos^2\theta$ and by integrating by parts. Multiplying (3.59) by $z\, dz$ and integrating between the limits zero and $2ka$ leads to the relation

$$\int_0^{2ka} zK\, dz = 2ka - 2ka\frac{dK}{dz}$$

Therefore (3.58) is seen to be

$$\int_0^{2ka} z\, dz \int_0^{\pi/2} \sin\,(z\sin\theta)\, d\theta = 2ka\left[1 - \int_0^{\pi/2} \cos\,(2ka\sin\theta)\sin\theta\, d\theta\right]$$

Using the relation $\cos x = 1 - 2\sin^2 x/2$, the following equation is obtained:

$$\int_0^{2ka} z\, dz \int_0^{\pi/2} \sin\,(z\sin\theta)\, d\theta = 4ka\int_0^{\pi/2} \sin^2(ka\sin\theta)\sin\theta\, d\theta$$

Since $k = \omega/c$, the total force on the piston is now given by

$$F = \rho_a c\pi a^2 u_0 e^{i\omega t}\left[1 - \frac{J_1(2ka)}{ka} + \frac{i4}{\pi ak}\int_0^{\pi/2} \sin^2(ka\sin\theta)\sin\theta\, d\theta\right]$$

$$= \rho_a c\pi a^2 u_0 e^{i\omega t}\{R_1(2ka) + iX_1(2ka)\} \qquad (3.60)$$

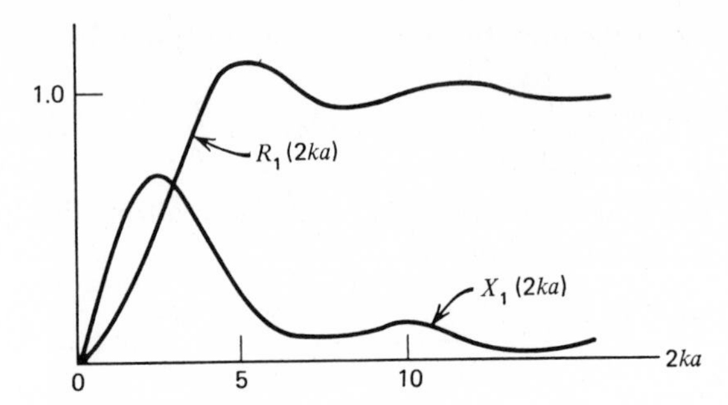

FIG. 3.13 Real and imaginary parts of a piston source impedance.

where

$$R_1(2ka) = 1 - \frac{J_1(2ka)}{ka} = \frac{(2ka)^2}{2 \cdot 4} - \frac{(2ka)^4}{2 \cdot 4^2 \cdot 6} + \frac{(2ka)^6}{2 \cdot 4^2 \cdot 6^2 \cdot 8} - \cdots$$

$$X_1(2ka) = \frac{4}{\pi a k} \int_0^{\pi/2} \sin^2(ka \sin \theta) \sin \theta \, d\theta$$

$$= \frac{4}{\pi}\left[\frac{(2ka)}{3} - \frac{(2ka)^3}{3^2 \cdot 5} + \frac{(2ka)^5}{3^2 \cdot 5^2 \cdot 7} - \cdots\right]$$

The impedance of the piston from (3.56) is then

$$z_r = \rho_0 c \pi a^2 [R_1(2ka) + iX_1(2ka)]$$

The functions R_1 and X_1 are plotted as a function of $2ka$ in Figure 3.13.

PROBLEMS

3.1 Show that $\mathbf{u} \cdot \text{grad}\, s$ and $(\mathbf{u} \cdot \text{grad})\, \mathbf{u}$ are of second order in (3.14). (*Hint:* Show that these terms are each smaller than every other term in their respective equations, providing $s < 1$.)

3.2 Consider a plane wave of acoustic pressure amplitude 0.0002 dyn/cm^2 at 1000 Hz (threshold of hearing). If this plane wave propagates in air, obtain (*a*) velocity potential, (*b*) condensation amplitude, (*c*) velocity amplitude, and (*d*) particle displacement amplitude. Repeat for an acoustic pressure of 160 dB re 0.0002 μbars (noise of jet engine)

3.3 Obtain the sound power transmission coefficient of a plane wave of frequency 1000 Hz, propagating through a steel panel 1 in. thick, which separates two tanks of water. (Density of steel, 7.7 g/cm; velocity of sound in steel, 6.1×10^5 cm/sec.)

3.4 What is the sound power transmission coefficient for a plane wave impinging normally on the surface of the sea?

3.5 A monopole of strength Q is placed a short distance below the surface of the sea at depth d, thus forming a dipole. Obtain an expression for the pressure amplitude as a function of horizontal distance, r, from the source, for families of constant depth, d.

3.6 A sonar radiator, designed in the shape of a piston source, is used both for transmitting and receiving; it has a diameter of 1 m. As a receiver it is capable of detecting a signal of 60 dB less energy than the transmitted signal. At what minimum frequency should it be operating in order that the reflection from a vertical cliff is just detected? The cliff is 10^4 m distant and reflects only one-tenth of the energy striking it. Assume no viscous attenuation by the water.

CHAPTER 4

Basic Ray Acoustics

4.1 INTRODUCTION

In Chapter 3 a method is developed whereby essentially most problems in fundamental acoustics can be solved. It consists of obtaining a solution to the linearized wave equation for the velocity potential which is compatible with boundary and initial conditions. From the velocity potential all acoustic parameters may be obtained. An "ideal" underwater acoustical problem would consist of a directional source placed in a homogeneous, isotropic medium such as pure water; the bottom surface would be flat and rigid (therefore having zero particle velocity) while the upper surface would also be flat but would present a pressure-release boundary (zero pressure at the surface).

It is unfortunate that very few of these conditions are ever met. Seawater is neither homogeneous nor isotropic but contains many impurities (including bubbles and fish). The velocity of sound varies with temperature, pressure, and salinity and therefore with position and time. The bottom is seldom flat (if it is possible even to define the bottom), while the surface is never flat, and is always a function of time. To find a solution to the wave equation compatible with these conditions is usually impossible. It is possible, however, to solve in an approximate manner quite difficult acoustical problems by the use of ray (or optical) techniques wherein the phase information contained in the acoustic wave is discarded. This chapter, then, deals with elementary concepts in ray acoustical techniques.

4.2 PATH OF A SOUND RAY

The theory that permits the calculation of the path of an acoustical signal as if the path were that of a light ray is based on Huygens method. This method is a statement of the assumption that ray paths are everywhere normal to their wave front. Note that this does not imply that all ray paths are straight lines. Spherically expanding waves, however, implicitly satisfy Huygens method but all acoustic waves will do so if certain conditions are met. Under these conditions the path of the sound is governed by the "eikonal equation." These conditions are (a) if the path of the sound ray is curving due to a change in the acoustical index of refraction, the radius of curvature must be large compared to the wavelength of the sound, (b) the acoustical index of refraction should not change appreciably over a wavelength, and (c) the percentage change in the amplitude of the signal should be small over a wavelength. It is apparent that the conditions for ray acoustics are generally satisfied for very high frequencies. It serves the purpose of this chapter to consider the case of a sound ray propagating in a medium whose velocity of sound is a linear function of the depth since other functions may be approximated by a series of linear steps. Since the complexities of obtaining the path of a sound ray in a general medium is staggering, the method of stepwise linearizing the velocity profile is often used, frequently with the aid of a computer.

4.2.1 Equation of a circular sound path

The equation of a circle can be described by the equation $x^2 + y^2 = r^2 =$ constant. It is possible, however, to describe the locus of points on a circle in other ways. Consider the arc of a circle as shown in Figure 4.1. The depth d can be seen to satisfy the relation

$$d = R(1 - \cos \theta) \tag{4.1}$$

where R is the radius of curvature, a constant, and θ is the angle formed by the tangent to the circle and the horizon. If two points are considered on the circle, their difference in depth $\Delta d = d_2 - d_1$ is given by

$$\Delta d = R(\cos \theta_1 - \cos \theta_2) \tag{4.2}$$

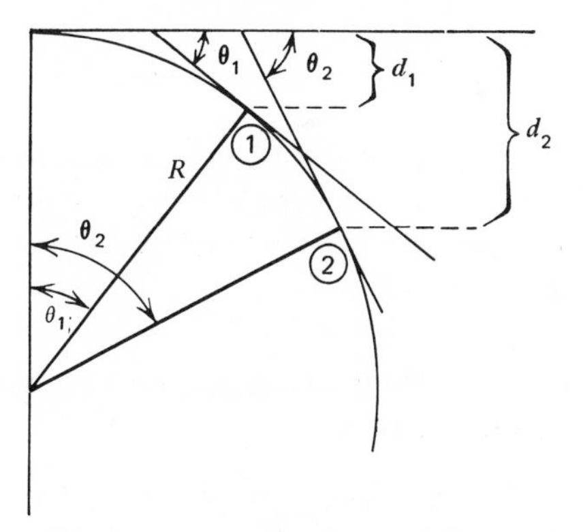

FIG. 4.1 Geometry of a circle.

Equation 4.2 is thus the equation of a circle with radius of curvature R. It can be stated that if a ray path were to satisfy this relation, the path described would be a circle.

4.2.2 Relation between acoustic path and velocity gradient

If the velocity profile in a medium is considered to be linear, the velocity can be described by a relation of the type

$$c = A + gd$$

where c is the velocity at depth d, g is the gradient of the velocity, and A is a constant to be determined. The difference in depth $\Delta d = d_2 - d_1$ at two points of the ray path is given by

$$\Delta d = \frac{c_2 - c_1}{g} \tag{4.3}$$

As in the case of optical ray paths, Snell's law is applicable.

$$\frac{\sin \varphi_i}{\sin \varphi_r} = \frac{\cos \theta_i}{\cos \theta_r} = \frac{c_i}{c_r} \tag{4.4}$$

where φ_i is the angle between the incident path and the normal to the interface, θ_i is the angle between the incident path and the interface, c_i is the velocity of sound in the medium containing the incident sound ray and similarly for the refracted wave. If c_0 is defined as that velocity in which the ray path is horizontal, then the velocity at any depth can be given by the general equation

$$c = c_0 \cos \theta \tag{4.5}$$

Combining (4.3) and (4.5), the relation describing the path of a sound ray in a medium with a constant velocity gradient is

$$\Delta d = -\frac{c_0}{g}(\cos \theta_1 - \cos \theta_2) \tag{4.6}$$

If this equation is compared with (4.2), it can immediately be seen that the path of the sound ray is an arc of a circle with a radius of curvature equal to $-c_0/g$. It is therefore possible to plot "piecemeal" the path of a sound ray for "linearized" velocity profiles.

4.2.3 Geometrical method of obtaining ray paths for constant velocity gradient profiles

Consider Figure 4.2. Let a source be placed at depth d_1 and consider an arbitrary ray emanating from the source that forms an angle ϕ with the

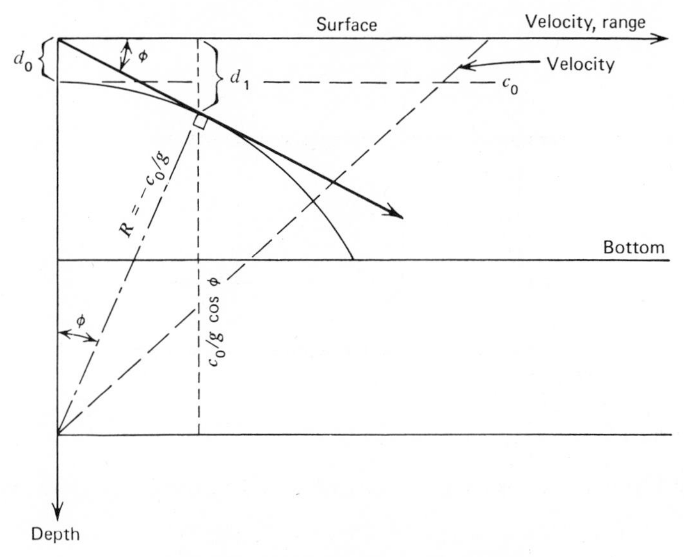

FIG. 4.2 Ray path for linear velocity.

horizon. At some depth d_0 the ray will be horizontal and will have a velocity c_0. The equation describing the velocity of sound can then be given by

$$c = c_0 + g(d - d_0)$$

The radius of curvature for this particular ray must be $-c_0/g$ and must lie in a direction normal to the ray. From Figure 4.2 it can be seen that the origin of the radius of curvature must lie at some depth given by the expression $d_1 + R \cos \varphi$. The velocity at that depth is then

$$c = c_0 + g(d_1 + R \cos \varphi - d_0)$$

or

$$\frac{c}{g} = -R(1 - \cos \varphi) + (d_1 - d_0)$$

but $R(1 - \cos \varphi) = (d_1 - d_0)$ thus the velocity at the origin of the radius of curvature would be exactly zero if the velocity profile were to extend indefinitely. This fact makes it possible to determine geometrically the path of a sound ray. A line is drawn horizontally at the depth corresponding to a zero sound velocity. In addition, a line is constructed normal to the direction of a projected ray. The intersection of these two lines is the origin of the radius of curvature. An example is shown in Figure 4.3.

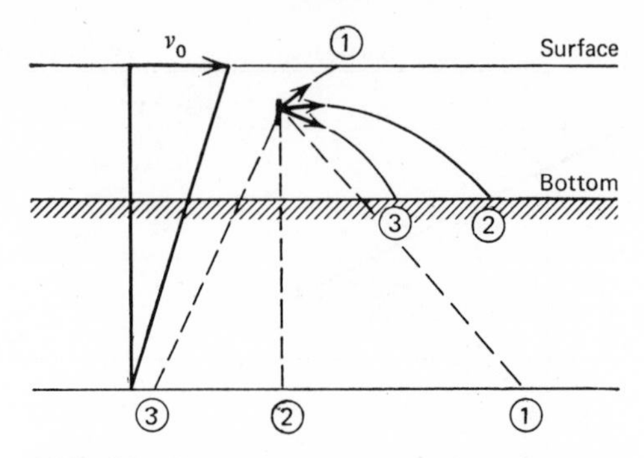

FIG. 4.3 Geometrical method of obtaining ray paths.

4.2.4 Sound velocity in the sea

A simplified equation for the velocity of sound in the sea is given by

$$c = 1449 + 4.6T - 0.055T^2 + 0.0003T^3 + (1.39 - 0.012T)(s - 35) + 0.017d \tag{4.8}$$

where c is the velocity of sound in meters per second, T is the temperature in degrees centigrade, s is the salinity in parts per thousand, and d is the depth in meters. It can be seen that the effect of temperature changes on the sound velocity are far greater than that of changes in depth. However, as the temperature stabilizes, changes in depth become important. Thus the sound velocity may be seen to decrease with temperature and finally to increase with depth. Figure 4.4 illustrates a typical temperature and velocity profile.

4.2.5 Some examples of ray paths

When particular velocity gradients are present in the sea, it is possible to create rather unique and unexpected situations. Consider Figure 4.5a, that

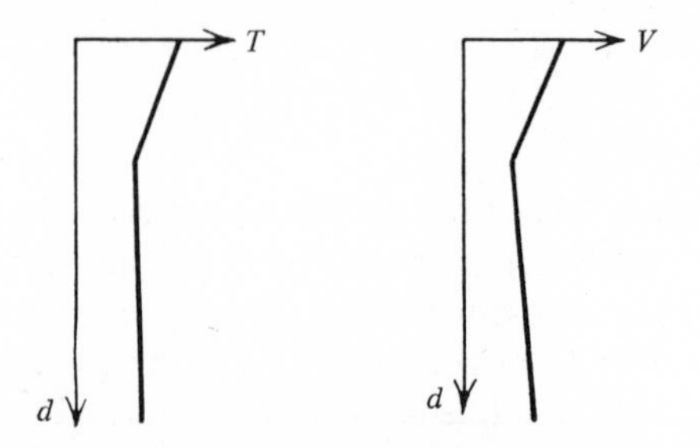

FIG. 4.4 Typical temperature and velocity profile.

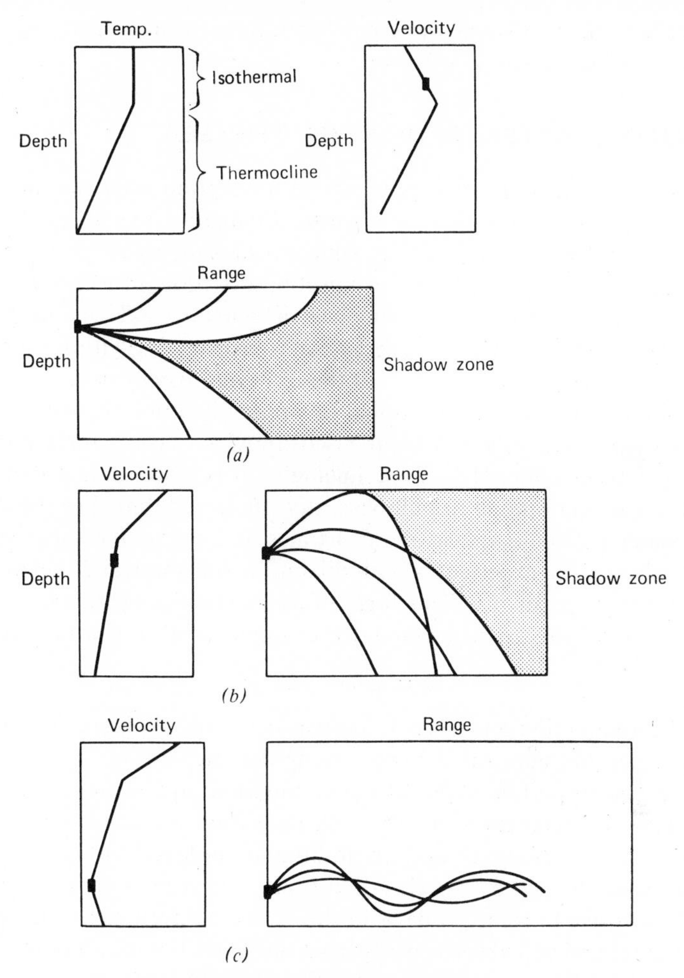

FIG. 4.5 Special ray paths.

of a region of constant temperature (increasing sound velocity with depth) followed by a steep thermocline (strong decreasing sound velocity with depth). If a transducer is placed in the position indicated, it can be seen that a region devoid of sound rays is formed (assuming no reflections). This situation is repeated in Figure 4.5*b* under slightly different conditions. These regions are called shadow zones.

A quite different sound structure can be seen in Figure 4.5*c*. The path of any sound ray leaving the source will oscillate about the axis of the source in such a way that the sound is trapped. This phenomenon is known as a sound channel. Sounds trapped in such a manner have been known to travel thousands of miles. Much work has been done to investigate the possible uses of these channels.

4.3 SOUND ATTENUATION IN THE SEA

Of the many possible sources of sound absorption only two play a predominant role in determining the acoustic attenuation in the sea.

Because of the relative motion which occurs between portions of the medium during compression and expansion, a frictional form of energy loss occurs. As in all forms of friction, heat is generated. This energy can be obtained only from the acoustic energy that is propagating within the medium. It can be assumed, as Stokes did, that these viscous losses can be resolved into a bulk viscosity which is proportional to a pure dilatation (or volume change without shape change) and a shear viscosity which occurs for a pure shear motion (shape change only). Experimentally, it has been found that in pure water the attenuation due to bulk viscosity can be attributed to the structure of the water molecule and is about 3 times as large as the shear viscosity. (Even a plane wave experiences both dilatation and shear and therefore loses energy through viscosity.) This form of attenuation process is proportional to the square of the frequency and, for example, at 41°F can be given by

$$\alpha = 2.9 \times 10^{-7} f^2 \tag{4.9}$$

where the attenuation α is given in decibels per yard. At higher temperatures the coefficient becomes smaller, decreasing the attenuation.

The second important cause of the attenuation present in seawater is due to one type of "relaxation process." A relaxation process is an interaction between the acoustic wave and an additional energy-absorbing mechanism which is characterized by a "relaxation time." That is to say that the mechanism is sensitive to some property of the acoustic wave (such as temperature and pressure) and will absorb energy from the wave, but the amount absorbed is a function of time (as in the charging and discharging of a capacitor through a resistor). Among the many types of relaxation process, chemical relaxation is important in seawater. The amount of dissociation of the salt $MgSO_4$ which is present in seawater will vary with the presence of the acoustic wave. It can be shown that at about 41°F the attenuation due to this phenomenon can be given by the equation

$$\alpha = \frac{0.033 f^2}{f^2 + 3600} \tag{4.10}$$

where α is the attenuation in decibels per yard and f is the frequency of the acoustic wave. Thus the total attenuation at about 41°F is the sum of the attenuations calculated in (4.9) and (4.10).

4.4 SIMPLIFIED THEORY OF SCATTERING BY A SINGLE PARTICLE

The theory of scattering of sound waves by particles is extremely complicated and only a very simplified version is presented here. It will be assumed that the object ensonified is large compared with the wavelength of the sound so that it is still possible to consider the acoustic wave as an optic ray. It is also assumed that when the sound wave strikes the particle all the energy that is reradiated is scattered uniformly in all directions.

Consider Figure 4.6. If the intensity I (power transmitted per unit area) of a plane wave intercepts a target of projected area A, the power intercepted and reradiated must be given by

$$W = IkA$$

where W is the power intercepted, I is the intensity, A is the projected area of the target, and k is a factor which allows the possibility that the target may "appear" larger or smaller to the acoustic wave than would be concluded from its physical dimensions; for example, a spherical bubble of volume $\frac{4}{3}\pi r_0^3$ has a projected area of πr_0^2, whereas in fact the bubble under certain conditions may appear to the sound wave almost 1000 times larger than this amount. The combination kA is called the "target area," σ thus

$$W = I\sigma \tag{4.11}$$

Since this power is now to be uniformly distributed, the intensity I_s at some point due to the reradiation is

$$I_s = \frac{I\sigma}{4\pi r^2}$$

where r is the distance from the point of observation to the scatterer. If the incident energy is derived from a source whose intensity at a unit distance is I_0, then $I = I_0/r'^2$, where r' is the distance between the source and the scatterer. The intensity of the scattered sound is $I_0\sigma/(4\pi r^2 r'^2)$, and if the

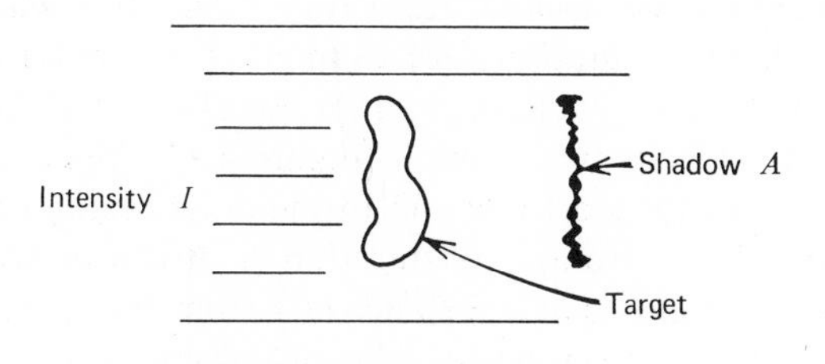

FIG. 4.6 Intensity intercepted by a target.

source is at the same location as the receiver (most usual case) then

$$I_s = \frac{I_0 \sigma}{4\pi r^4} \tag{4.12}$$

It should be especially noted that the sound intensity decreases as the fourth power of the separation.

A more convenient set of units is found by considering the definition of the echo level

$$E = 10 \log I_s$$

where the echo level E is given in decibels. Equation 4.12 becomes

$$E = 10 \log I_s = 10 \log I_0 + 10 \log \frac{\sigma}{4\pi} - 40 \log r \tag{4.13}$$

The quantity $10 \log \sigma/4\pi$ is further defined as the target strength, T.

4.5 UNDERWATER "REVERBERATION"

The problem of underwater reverberation is quite different from that which is commonly encountered in architectural acoustics. Although the sounds perceived are similar in nature to that of room reverberation, the causes are not the same. They arise principally from multiple scattering (a) within the volume of the sea and (b) from the sea surface and bottom.

Before it is possible to consider the problem of reverberation, it is necessary to define the region which, at a given time, will produce reverberation.

4.5.1 Train length and ping length

If the sound from an acoustic source is turned on and off in a time which is short compared to the time necessary to receive echoes from distant targets, the system is said to be in a pulse-echo mode of operation. The interval of time that the source is transmitting is called the pulse duration τ, whereas the time between consecutive pulses is called the pulse repetition rate PRR. It is usually convenient to establish as "zero time" that time when the pulse is on for one-half its total duration; for example, if the pulse is to be on for 100 msec, "elapsed" time would begin after 50 msec of transmitting.

Consider the x, ct diagram shown in Figure 4.7. This is a diagram that shows the location of the start and end of the pulse at any time. The pulse is traveling with velocity c and is of duration τ. It can be seen that at time $t/2$ the center of the pulse is at a distance $ct/2$ from the source, whereas the total range ensonified is given by $c\tau$. The distance $c\tau$ is known as the train length.

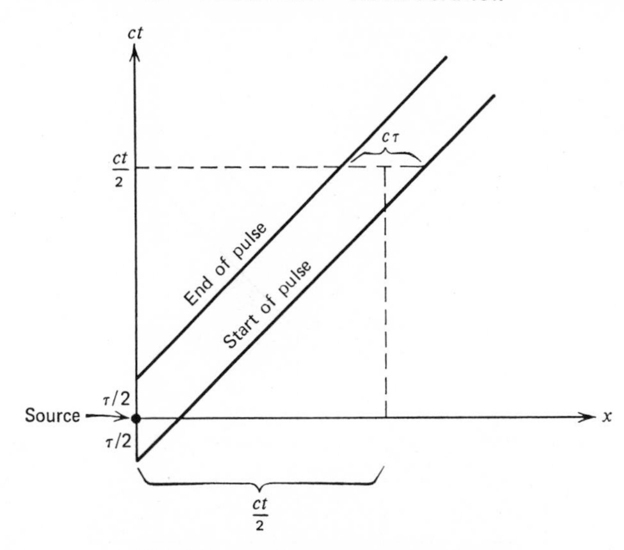

FIG. 4.7 x-ct Diagram of sound pulse.

Although at time $t/2$ the range ensonified is given by the train length, not all particles in this region are capable of causing reverberation at time t when the sound is scattered back to the source. Consider the x, ct diagram in Figure 4.8. Only those particles within the range $c\tau/2$ can scatter sound that will *simultaneously* reach the source at time t, thus causing reverberation. The distance $c\tau/2$ is known as the ping length and is the key factor in determining the amount of reverberation.

4.5.2 Simplified theory for volume reverberation

A simplified theory for volume reverberation can be obtained from (4.12) under the following conditions: (*a*) all the scatterers have equal target area σ, (*b*) there are N scatterers/unit volume, and (*c*) each scatterer is independent of every other scatterer; that is, each acts as if there were no other scatterers present. Consider the source and the receiver to be at the same location, then the intensity due to scattering from one scatterer is given by

$$I_s = \frac{I_0 \sigma}{4\pi r^4}$$

Under the above conditions the intensity of the volume reverberation must be

$$I_V = \frac{I_0 \sigma}{4\pi r^4} NV$$

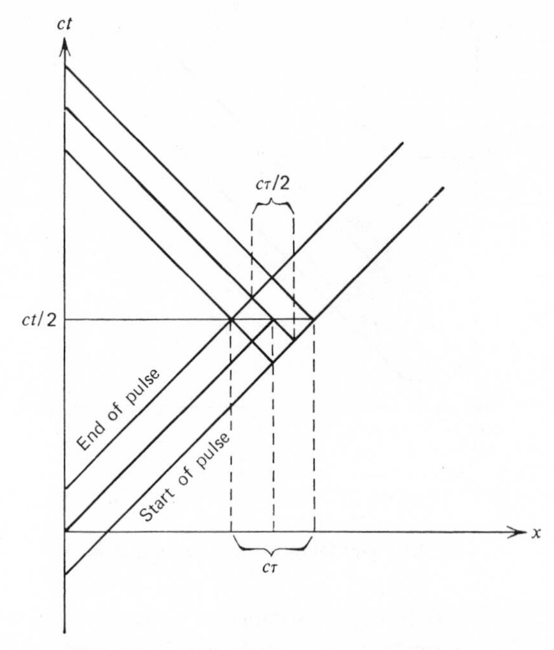

FIG. 4.8 ***x-ct*** **Diagram of reflected pulse.**

where V is the volume capable of producing reverberation at time t. If the directivity of the source has a half angle α, then $V = 2\pi r^2 r_0(1 - \cos \alpha)$, where r_0 is the ping length $c\tau/2$ as shown in Figure 4.9. Since $r = ct/2$, the intensity of the volume reverberation is given by

$$I_V = \frac{I_0 \sigma N \tau_0}{ct^2}(1 - \cos \alpha) \tag{4.14}$$

It is apparent that increased output, ping length, or a broad projector beam tends to increase the volume reverberation. In addition, the intensity decreases as the square of the time compared with the fourth power of

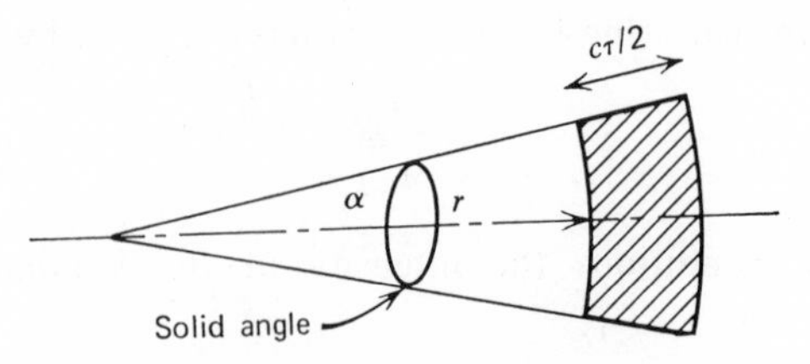

FIG. 4.9 Condition for volume reverberation.

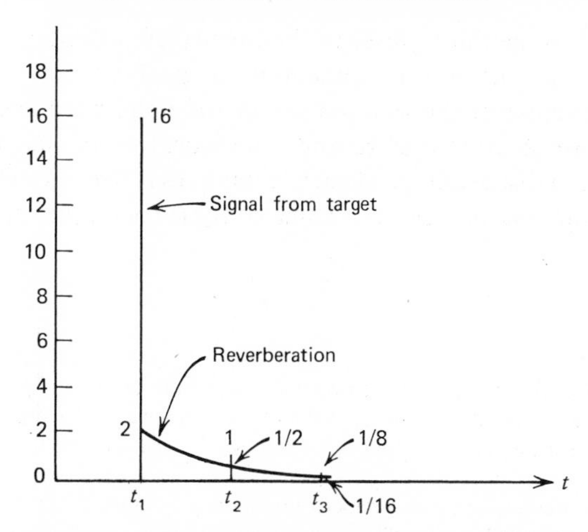

FIG. 4.10 Loss of target signal due to reverberation.

the travel time as is the case for a single target. Figure 4.10 illustrates the importance of this fact. Assume that at some distance r the intensity of an echo from a target is eight times larger than the background volume reverberation. If the target moves away to a distance of $2r$, then $3r$, the intensity of the target, is quickly lost in the background reverberation.

4.5.3 Simplified theory for surface reverberation

In a manner similar to that of Section 4.5.2 it is possible to obtain a simplified theory for surface reverberation. It is once again assumed that all scatterers have equal target area σ and each scatterer is independent of all other scatterers but in this case N is the number of scatterers per unit area. Thus

$$I_{su} = \frac{I_0 \sigma}{4\pi r^4} NA$$

where I_{su} is the intensity of the surface reverberation and A is the surface area ensonified. The surface area can be shown to be $A = 2\alpha r r_0$, where again α is the projector beam half angle and r_0 is the ping length. Thus the surface reverberation is given by

$$I_{su} = \frac{2I_0 \sigma N \alpha \tau}{\pi c^2 t^3} \tag{4.15}$$

Again, to increase the output the beam width or the pulse length would only increase the reverberation. In the case of surface reverberation the intensity is proportional to the third power of the time (therefore range). It is easy to see that sonar ranging is very susceptible to problems involving signal to noise ratio. Because of the randonmess of the noise produced by reverberation, it is however quite easy to apply statistical techniques to recover the signal. The main disadvantage of such techniques is that they require time for gathering in data and processing. Some of these techniques are covered in a later chapter.

PROBLEMS

4.1 Obtain the velocity of sound in seawater for a salinity of 30 parts/1000 at a depth of 1000 m and a temperature of 6°C. What is the velocity gradient at this depth if the temperature has stabilized? What is the radius of curvature of a horizontal ray at this depth?

4.2 If a layer of water has a negative velocity gradient, g, obtain an expression for the range, R, at which a ray which starts off horizontally at the surface reaches a depth, d. Compare this with the actual path length of the ray.

4.3 Plot the acoustic attenuation (in decibels per yard) of a 20-, 40-, 60-, 80-, 100-, 120-kHz signal. Separate the attenuation due to relaxation processes from that due to viscous losses.

4.4 Obtain the ratio of the received scattered intensity to the transmitted intensity from a rigid sphere of diameter 10 m (assume the scattering factor $k = 1$) at a distance of 1000 m. What is the target strength?

4.5 If the echo from the sphere in problem 4.4 competes with volume reverberation ($\sigma N = 10^{-5}\ \mathrm{m}^{-1}$) what must be the directivity of the source such that the echo from the sphere equals the noise from the reverberation? Assume a frequency of 60 kHz and a ping length of 100 periods.

CHAPTER 5

Electroacoustic Transduction

5.1 THE ELECTROACOUSTIC TRANSDUCER AS A TWO-PORT NETWORK

The technology of electroacoustics, like that of many other branches of applied science, has its beginning in the discovery of a number of devices and methods whereby mechanical power in terms of force and velocity could be converted into electrical power in terms of current and electromotive force. Many of these devices are capable of reversing the direction of conversion. From each of the available methods we may design an electroacoustic transducer which differs from an ordinary transducer in that the energy interchange involves two steps—a scheme for arranging the electromechanical interaction, and one for the mechanoacoustical interaction. Sometimes networks are drawn to include both steps,[1] but we indicate the mechanoacoustic interchange by means of force and impedance components.

Almost universally our observation of the performance of an electroacoustic transducer is made at electrical terminals. Although the pressure developed at the mechanical terminals of a sound projector is the quantity of interest, this quantity will most likely be determined at the electric terminals of a pressure-sensitive instrument. Naturally this dependence on electrical measurements has encouraged the development and use of electrical equivalents in the analysis of electromechanical devices. Further, the extensive background of electric circuit theory has been a great aid. Therefore we begin with an all-electric two-port network as an easy approach to circuits equivalent to electroacoustic systems.

The network of Figure 5.1 consists of passive linear elements. The elements Z_1 and Z_2 are the impedances at the two ports when the opposite port is open, whereas Z_{12} and Z_{21} relate voltages in one circuit to currents in another. There are no restrictions on the number or kind of elements in the box

[1] L. L. Beranek, *Acoustics*, McGraw-Hill, New York, 1954, p. 81.

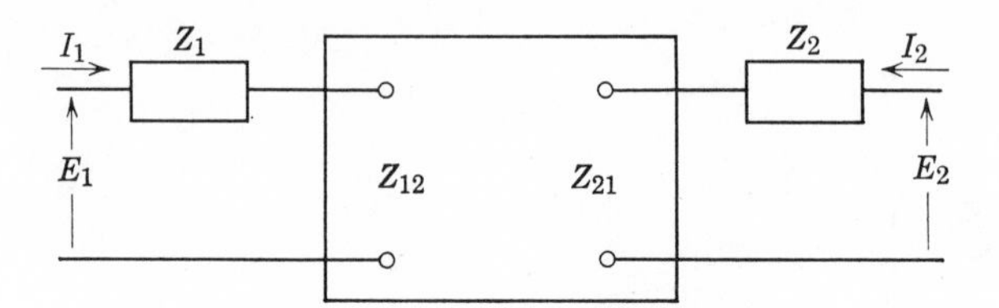

FIG. 5.1 A two-port net.

except that they must be bilateral, linear, and passive. Since there are no external connections between the terminal pairs, the current into one terminal of a pair must equal that out of the other terminal of the same pair. We may express the relations between potentials and currents by the equations

$$\begin{aligned} E_1 &= Z_1 I_1 + Z_{12} I_2 \\ E_2 &= Z_{21} I_1 + Z_2 I_2 \end{aligned} \tag{5.1}$$

so long as the elements in the box meet the restrictions listed above, the reciprocity theorem states that $Z_{12} = Z_{21}$. We may complete the circuit of Figure 5.1 by connecting a load Z_l across the right-hand port, thus replacing E_2 by $-Z_l I_2$ and solving for the driving point impedance Z_i

$$Z_i = Z_1 - \frac{Z_{12}^{\ 2}}{Z_2 + Z_l} \tag{5.2}$$

No attempt is made here or in later chapters to distinguish complex quantities by bold type. In general, whether a quantity is complex may be determined from the context and no distinction is necessary unless the absolute value is involved.

The box of Figure 5.2 is gifted with a property not possessed by networks of electrical components only. It relates causes and effects on the two sides though the left-hand port is purely electric while the right hand is mechanical. Following the procedure used in writing (5.1),

$$E = Z_e I + T_{em} V \tag{5.3a}$$

$$F = T_{me} I + Z_m V \tag{5.3b}$$

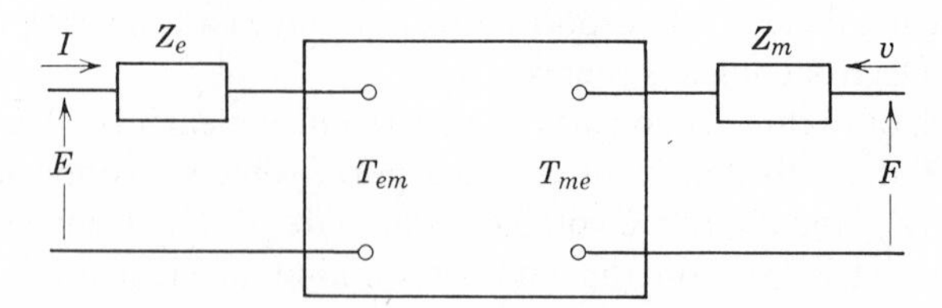

FIG. 5.2 Electromechanical two-port net.

The coefficients[2] of current and velocity in (5.3) have the following meaning:

$Z_e = (E/I)_{v=0}$ an open-circuit driving point impedance,

$T_{em} = (E/v)_{I=0}$ a transduction coefficient which relates the electromotive force developed at the electrical port to the velocity in the mechanical net,

$T_{me} = (F/I)_{v=0}$ a transduction coefficient which relates the force developed at the mechanical port to the current in the electrical mesh, and

$Z_m = (F/v)_{I=0}$ an open-circuit driving point impedance.

In the network of Figure 5.1, in which all elements are electrical impedances, Z_{12} and its equal Z_{21} are termed open-circuit transfer impedances. They perform no function other than to relate one electrical quantity to another. Transduction coefficients which are also open-circuit transfer impedances perform the additional function of converting mechanical units to electrical units or the reverse. Another distinction is that in systems involving magnetic rather than electric fields, the transduction coefficients are equal in magnitude but opposite in sign.

If a load Z_l is placed across the terminals of the mechanical mesh, the driving point impedance becomes

$$Z_i = Z_e - \frac{T_{em}T_{me}}{Z_m + Z_l} \tag{5.4}$$

5.2 THE MOVING COIL TRANSDUCER

To examine the meaning of (5.4) it will be an economical procedure to apply it to a transducer that has found wide use as a passive acoustic instrument in water. Looking at Figure 5.3, we immediately recognize the basic structure, a coil moving in a magnetic field, so common in loudspeakers. The essential difference is in the type of radiating surface required for the two media, air and water. The moving-coil principle is chosen as a first approach to electromechanical coupling because it presents the fewest technical problems. A current-carrying conductor in a magnetic field is acted on by a force equal to the product of the magnitudes of the field, the current, and the length of conductor normal to the direction of the field. The converse phenomena is that an electromotive force is developed in a conductor moving across a magnetic field equal to the product of the magnitudes of the field, the velocity of the conductor, and its length normal to the field.

The two foregoing statements define the magnitude of the transduction

[2] See F. V. Hunt, *Electroacoustics*, Harvard Press and John Wiley and Sons, 1954, for a more complete and enlightening discussion of electromechanical coupling.

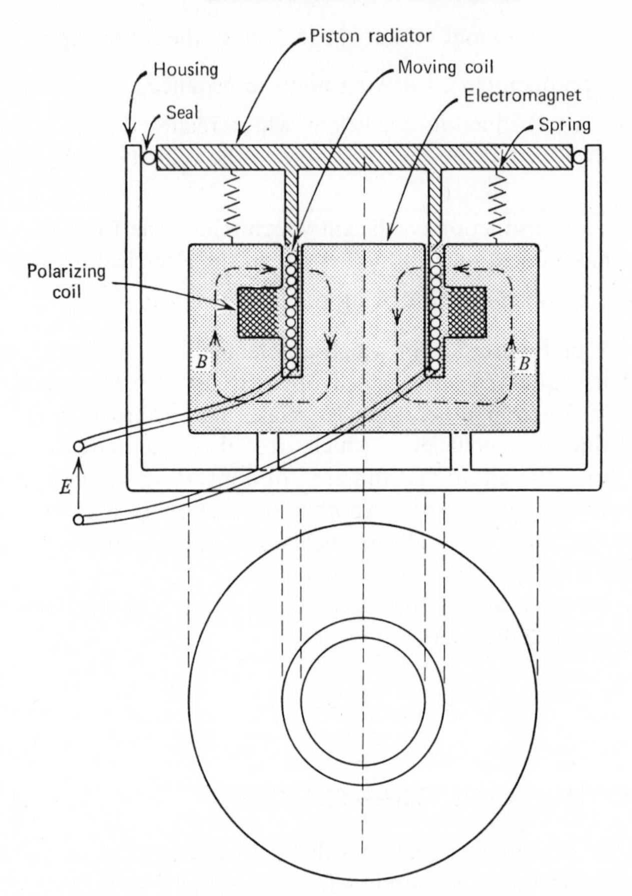

FIG. 5.3 Schematic of an electromagnetic transducer.

coefficients without establishing a sign convention. Figure 5.4 shows the sets of orthogonal quantities involved. If a positive sign is given to T_{me}, the proper direction in space require a negative sign for T_{em}. Equations for the equivalent electromechanical two-port net are

$$\begin{aligned} E &= Z_e I - Blv \\ F &= BlI + Z_m v \end{aligned} \tag{5.5}$$

The asymmetry of the transduction coefficients leads to a difficulty in drafting suitable equivalent circuits; for example, the electrical network shown in Figure 5.5 has the common impedance Z_B coupling the two meshes. Unlike the coupling coefficients of (5.5) Z_B will show up with the same sign

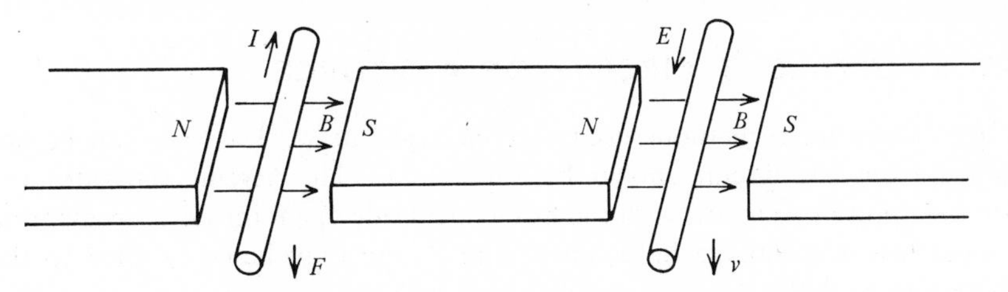

FIG. 5.4 Space orientation of quantities involved in electromagnetic transduction.

in the two network equations. Among the various arrangements for avoiding this awkward dilemma, that proposed by Professor Hunt,[3] which includes a space operator k, corrects a symbolic failure to express the orthogonality of the force reactions in electromagnetism. Referring again to Figure 5.4, we write the transduction coefficient as Blk where k rotates the physical quantity immediately following it counterclockwise in space 90° about the field direction as an axis. The positive direction of that quantity after rotation will then point in the positive direction of the product. In Figure 5.4 the counterclockwise rotation of I about B results in its pointing downward as shown by F, and the counterclockwise rotation of v about B results in its pointing in the direction shown by E. With this interpretation of the function of k and the additional understanding that $k^2 = -1$, we may rewrite (5.5) as

$$\begin{aligned} E &= Z_e I + Blkv \\ F &= BlkI + Z_m v \end{aligned} \tag{5.6}$$

and the driving point impedance with F replaced by a load $-Z_L v$ is

$$Z_i = Z_e - \frac{(Blk)^2}{Z_m + Z_L} \tag{5.7a}$$

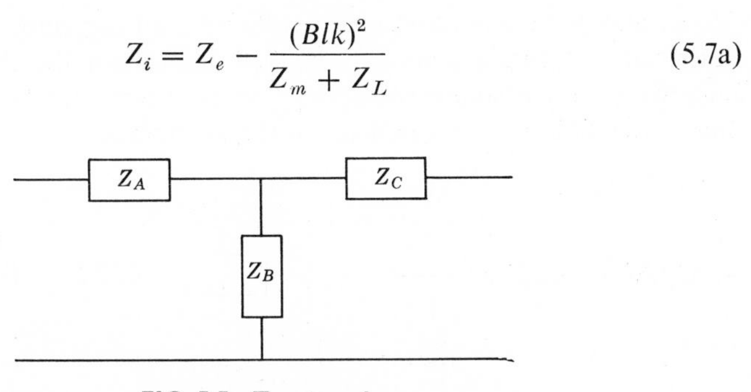

FIG. 5.5 T network.

[3] F. V. Hunt, "Symmetry in the Equations for Electromechanical Coupling," in *J. Acoust. Soc. Am.*, **22**, 672(A) (September 1950).

5.3 EQUIVALENT CIRCUIT OF A SOURCE

By conventional methods used with electrical networks, (5.6) can be obtained from the circuit shown by Figure 5.6. This circuit separates the driving point impedance, with the mechanical side open, into Z_e, representing copper loss and leakage inductance, and Z_c the impedance coupled to the induction field B.

Writing (5.7a) in the form

$$Z_i = R_e + j\omega L_e + \left(\frac{R_m}{(Bl)^2} + \frac{j\omega M}{(Bl)^2} + \frac{K}{j\omega(Bl)^2} + \frac{Z_L}{(Bl)^2}\right)^{-1} \tag{5.7b}$$

we recognize the last term as being the impedance of a number of elements in parallel; that is, if $R_1 = (Bl)^2/R_m$, $C_1 = M/(Bl)^2$, $L_1 = (Bl)^2/K$, and $Z_1 = (Bl)^2/Z_l$, (5.7b) may be rewritten as

$$Z_i = R_e + j\omega L_e + \left(\frac{1}{R_1} + j\omega C_1 + \frac{1}{j\omega L_1} + \frac{1}{Z_1}\right)^{-1}$$

The equivalent circuit for this form of (5.7) is readily seen to be represented by Figure 5.7.

The theory and also the practice of electrodynamic systems as applied to loudspeakers is well developed. The same may be said of another important application, the electrodynamic shaker used for environmental testing. The ease of controlling the displacement, velocity, and acceleration amplitudes adapts it particularly well to the execution of complex test programs. As might be expected, the design requirements for each of the three important applications, as loudspeakers, as shakers, or as underwater acoustic instruments, are quite different. In the case of the loudspeaker the parameters M, K, and R, representing mass, stiffness, and damping, are not precisely described. These elements may be well defined for the shaker and for the underwater sound source, due partly to less severe bandwidth requirements and to the difference in character of the medium into which they work.

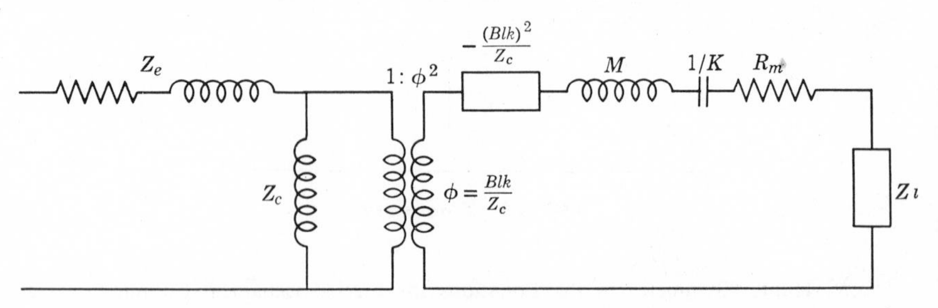

FIG. 5.6 Equivalent circuit of a moving coil transducer.

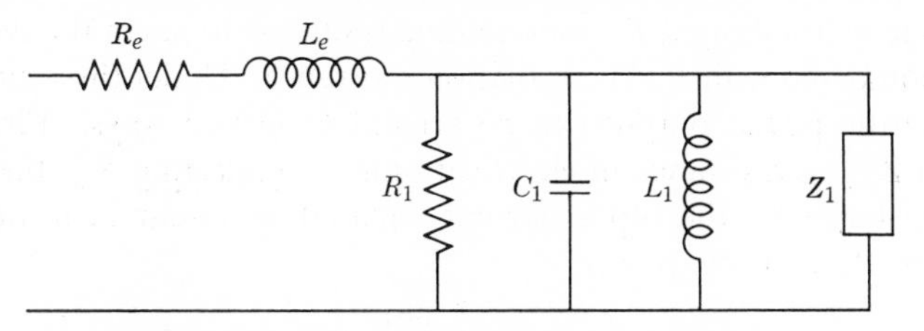

FIG. 5.7 Equivalent circuit for moving coil.

The following somewhat casual approach to the design of an electrodynamic projector may give a better feel for the character of this instrument. Suppose that Figure 5.3 represents a transducer to operate at a resonant frequency of 1500 Hz. Other arbitrary decisions to be made in the beginning involve size, weight, depth of operation, source of the induction field B, and the desired driving-point impedance. Figure 5.3 shows B as being supplied by sending a dc current through a coil—in other words, an electromagnet. Stronger fields are obtainable this way, but supplying the dc current is rather awkward. Permanent magnets will develop a field of about 1 Wb/m². In addition to the design of the magnetic circuit, other problems are the design of a piston radiating face, the moving coil support, and a spring system that will support static forces as well as resonate the mass of piston and coil at the desired frequency. Assuming these problems solved with the following parameters,

$B = 1$ Wb/m²

$M = 30$ kg (weight of piston and coil)

$K = 3.36 \cdot 10^9$ N/m

$l = 200$ m #10 gage

Piston diameter $= 0.3$ m

$L_e = 0.02$ H, $R_e = 0.65$ ohms

$Z_l = (4.55 + j8.32)10^4$ mechanical ohms

$Z_e = 188$ ohms

notice that K, the spring stiffness, must be adjusted to the sum of the mass of the radiator, the moving coil, and the reactive component of the radiation impedance.

We wish to determine the maximum acoustic power output, the bandwidth at constant force, power response, and its highest possible efficiency. At mechanical resonance, the driving point impedance is

$$Z_i = R_e + j\omega L_e + \frac{(Bl)^2}{R_m + R_l}$$

For a successful design, R_m representing frictional losses in the system will be small compared with R_l, the radiation resistance. Under these circumstances the acoustic power output can be arrived at in two ways. First: the last term in Z_i is a resistance in electrical ohms. Neglecting R_m, the product of this resistance by the input current squared represent acoustic output in watts, or P_a, acoustic power

$$P_a = \frac{(Bl)^2}{R_l} I^2 \tag{5.8a}$$

Second, at the mechanical terminals the acoustic power out is given by the product of the radiation resistance in mechanical ohms by the velocity squared. Since

$$V = \frac{.F}{Z_m + Z_l} = \frac{BlI}{Z_m + Z_l}$$

or at resonance and neglecting R_m

$$V = \frac{BlI}{R_L}$$

$$P_a = R_L \frac{(BlI)^2}{R_l^2} = \frac{(Bl)^2}{R_l} I^2 \tag{5.8b}$$

In the preceding discussion, rms values were used throughout and a slight phase difference between the current and driving force was ignored. The maximum safe current for the moving coil is about 15 A. Therefore

$$P_a(\max) = \frac{(200 \times 15)^2}{4.55 \times 10^4} \simeq 200 \text{ W}$$

There will also be a loss in the coil of $0.65 \times (15)^2 = 146$ W. Under these ideal conditions the efficiency will be

$$\text{eff} = \tfrac{200}{346} = 58\%$$

which is considerably in excess of a likely achievement. The bandwidth may be determined from the two definitions:

$$Q_m = \frac{M\omega_0}{R_l} \qquad Q_m = \frac{f_0}{f_2 - f_1}$$

therefore the frequency spread between the half-power points is

$$f_2 - f_1 = \frac{f_0}{Q_m} = \frac{1500}{7.85} = 191 \; Hz$$

It must be remembered that M in the above represents the sum of the mass of the moving parts and that equivalent to the reactive component of the radiation impedance.

Vibrating systems resonant at low frequencies tend to be rather massive affairs so the ease with which a moving coil vibrator can be adapted to any size is an attractive feature. Because it has been so successful as a loud-speaker it is appropriate at this point to assess the difficulty of adjusting this instrument to underwater use. Although the system just discussed is a highly idealized version, it will serve to illustrate the effects introduced by the difference in the two media. The series impedance looking into the driving terminals of the transducer in the two cases are

$$Z_i = Z_e + \frac{(Bl)^2}{Z_m + Z_l}$$

In water

$$Z_i = 0.65 + j188 + \frac{4 \times 10^4}{Z_m + (4.55 + j8.32)10^4}$$

In air

$$Z_i = 0.65 + j188 + \frac{4 \times 10^4}{Z_m + 23}$$

At the resonant frequency the motional impedance in water is about one ohm, while in air the comparable value is 1740 ohms. One immediately recognizes the power problem with the underwater application. As a hydrophone, the situation is not so serious.

5.4 THE MOVING COIL TRANSDUCER AS A HYDROPHONE

The transducer discussed in the last section was designed to produce a maximum practical power level with good efficiency. As a hydrophone, the criteria of good performance are a high open-circuit voltage per unit pressure in the sound field relative to a given input impedance and an open circuit voltage independent of the sound frequency. We shall take a look at the behavior of the preceding design in this respect.

From (5.6) the open-circuit voltage is

$$E = BlkV$$

and from (1.44)

$$V = \frac{F}{Z_m}$$

where F is the product of the sound pressure p by twice the effective piston area S'.[3] Therefore

$$E = \frac{BlkS'p}{Z_m} \tag{5.9}$$

and the open-circuit voltage response

$$\frac{E}{p} = \frac{BlkS'}{R + j(\omega M - K/\omega)} \tag{5.10}$$

Examination of (5.10) shows the response would be independent of frequency only with R very large compared with the reactive component of the impedance. Since a very large R reduces the sensitivity, the two requirements must be met by some other expedient. By the application of electrical network techniques to the mechanical circuit of a moving coil microphone, Winte and Thuras[4] obtained a response, constant from 100 to 10,000 Hz. This was done at the expense of a considerable decrease in sensitivity, but there was also a corresponding increase in the mechanical impedance, two compensating changes. While instruments working in the ocean environment are not so flexible as those used in the air, the application of suitable techniques can improve the characteristics of a hydrophone needed in the low frequency range.

5.5 MOTIONAL IMPEDANCE AND ADMITTANCE

So far, in the analysis of electroacoustic systems, we have made available information derived from the differential equation of the mechanical system, and from electromechanical networks which define the interaction between electrical and mechanical quantities. Equation 5.7 will be used to introduce another tool that has been extensively developed.[5,6]

$$Z_i = Z_e - \frac{(Blk)^2}{Z_m + Z_l} \tag{5.7a}$$

which in general may be written as

$$Z_i - Z_e = \frac{T^2}{Z_m + Z_l} \tag{5.8}$$

where T is the proper transduction coefficient for the system under study. The right-hand term of (5.8) is reduced to zero when mechanical motion is blocked. As the mechanical impedance is reduced, this term and correspondingly, the velocity amplitude produced by a driving force on the

[3] Summary Technical Report of Division 6, NDRC, pp. 59–60.

[4] Wints and Thuras, *J. Acoust. Soc. Am.*, **3**, 44 (1931).

[5] See Summary Technical Report of Division 6, NDRC, pp. 49–61.

[6] Cady, *Piezoelectricity*, McGraw-Hill, New York, 1946, ch. 14.

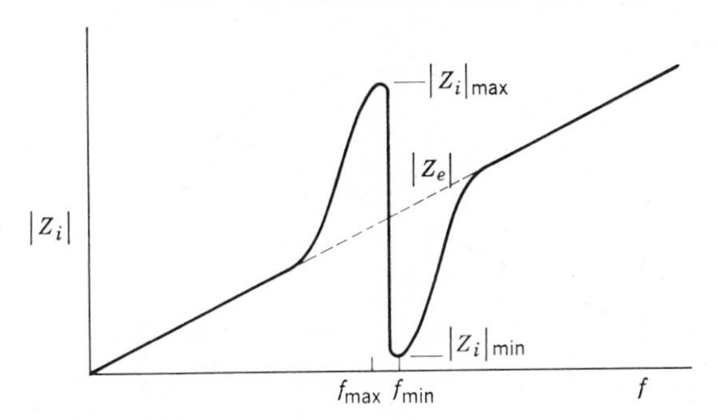

FIG. 5.8 Absolute value of Z_i as a function of frequency.

mechanical system increases. As a component of the input impedance due to motion it is known as motional impedance, designated Z_{mot};

$$Z_{\text{mot}} = Z_i - Z_e = \frac{T^2}{Z_m + Z_l} \tag{5.9}$$

where T^2 in some cases is a complex number.

The ease and clarity of one's grasp of information presented in graphical form is the basis for the importance of developing and using these forms. There are a number of ways in common use, each of which adds something to our understanding, of graphing the variation of the input impedance of an electroacoustic transducer as a function of frequency. The first and least informative is illustrated by Figure 5.8. The data is easily taken from a voltmeter across the transducer terminals while a constant current with varying frequency passes through. This is a handy bench test, for the ratio $|Z_i|_{\text{max}}/|Z_i|_{\text{min}}$ and the frequency difference, $f_{\text{min}} - f_{\text{max}}$, are indications of quality, and the interpolation of $|Z_e|$ gives the value of clamped impedance at the resonance frequency.

To show independently the three variables, frequency and the two impedance components, it is the practice to use the abscissa as the real axis and the ordinate as the reactive axis with frequency written in along the curve as a parameter. Figure 5.9 illustrates this method. In the picture, the vertical line represents the electrical impedance with the mechanical side blocked, and the circle represents the electrical impedance due to motion of the coils for a moving coil system with input impedance

$$Z_i = R_e + j\omega L_e + \frac{T^2}{R_m + j(\omega M - K/\omega)}$$

At the frequency marked f_0, $\omega M = K/\omega$, and the motional impedance has its maximum value, T^2/R_m, the diameter of the circle. To show that the

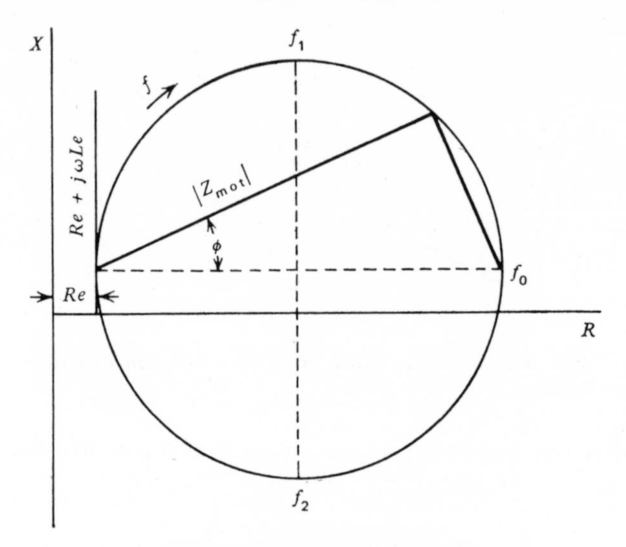

FIG. 5.9 Impedance diagram.

locus of the motional impedance is a circle, consider the right triangle in Figure 5.9 having $|Z_{\mathrm{mot}}|$ as the base and T^2/R_m as the hypotenuse:

$$|Z_{\mathrm{mot}}| = \frac{T^2}{R_m} \cos \varphi$$

which is the equation of a circle, in this set of coordinates, described as φ varies from $\pi/2$ to $-\pi/2$. The angle φ is also the phase angle

$$\varphi = \tan^{-1} \frac{X_{\mathrm{mot}}}{R_{\mathrm{mot}}}$$

which has the value 45° at f_1 and $-45°$ at f_2. We recognize these as the half-power point frequencies discussed in Chapter 1. Mention has been made of the fact that the transduction coefficient may be a complex number such as $Te^{-j\theta}$. In such cases, the diameter of the motional impedance circle makes an angle -2θ with the real axis.

Graphical displays of admittance are equally useful in the analysis of vibrating systems; for example, Figure 5.10 is the equivalent circuit of one type of piezoelectric vibrator. For this circuit the blocked driving point admittance represented by a leakage R_e across a capacitor is

$$Y_e = \frac{1}{R_e} + j\omega C_0 \tag{5.10}$$

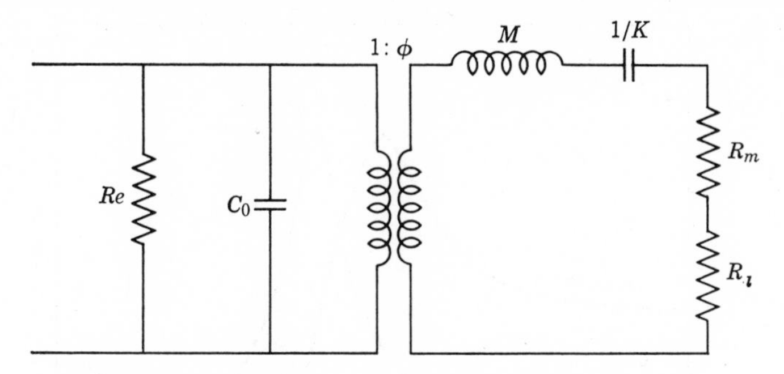

FIG. 5.10 Equivalent circuit of a piezoelectric vibrator.

and the motional admittance is

$$Y_{\text{mot}} = \frac{\varphi^2}{R_m + R_l + j(\omega M - K/\omega)} \tag{5.11}$$

Equation 5.11 is identical in form with (5.9), and its locus when plotted in a plane having conductance G and susceptance B as coordinates will be a circle, as shown in Figure 5.11.

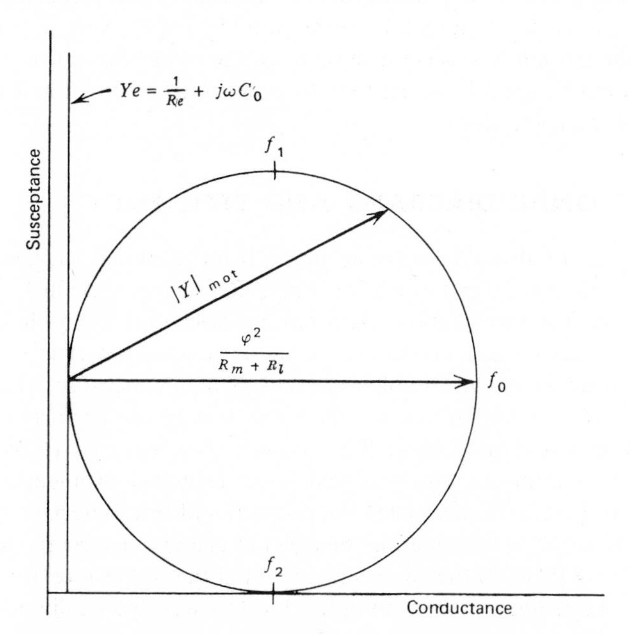

FIG. 5.11 Motional admittance in water for a piezoelectric vibrator.

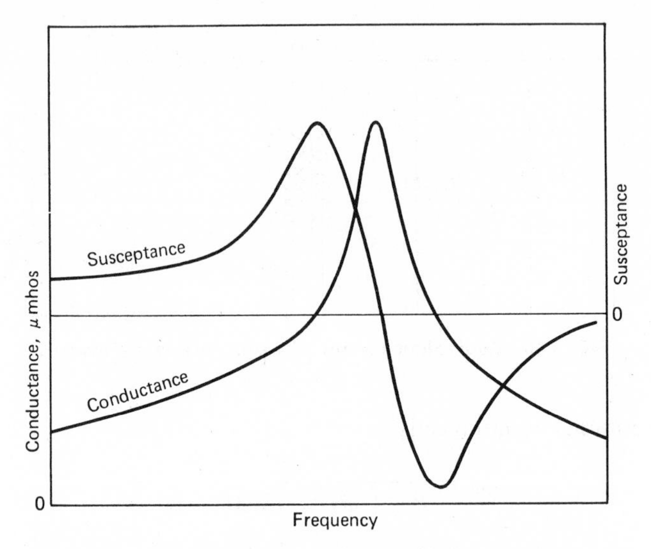

FIG. 5.12 Admittance of piezoelectric transducer in water.

A favorite method of plotting data, particularly for piezoelectric transducers is shown by Figure 5.12 using similog paper with a conductance scale on the left and a susceptance scale on the right. Since this is the same data as that in Figure 5.11 we are thereby enabled to compare the two methods for ease of interpretation.

5.6 MOTIONAL DIAGRAMS AND EFFICIENCY

The theory of deductions to be made from motional diagrams is well developed, interesting, and worthy of study. References previously mentioned cover the subject thoroughly. Appropriate selections from this literature will be introduced where needed. In years past, when power levels were low and little heed given to complex element structure because structure was simple, the temptation to use the theory in testing the performance of the final product was great because the necessary data was easy to come by. It required less equipment and involved fewer technical problems. The diagrams are still useful in measuring the properties of raw materials, the quality of single elements, and evaluating the effect of changes in circuit components. When data are taken in the environment and at the power level for which the transducer is designed, it is meaningful, but this was practically never true in the past and is not so easy to do.

One familiar use of motional diagrams is for the determination of efficiency of energy conversion. A proper definition of efficiency of energy conversion is the ratio of acoustic power developed to driving power in. Taking a look at the mechanical mesh of Figure 5.6, one sees that power consumption here involves R_m which represents unavoidable damping effects in the mechanical structure, and the real part of Z_l which is the impedance due to acoustic radiation. There is also power loss on the electrical side, but we hope that the real part of Z_l consumes the major part of the power. This question reveals a defect of the circle diagram. It does not separate R_m and R_l.

To continue with our determination of efficiency, the definition is that

$$\text{efficiency} = \eta = \frac{R_l}{R_i}\left|\frac{v}{I}\right|^2 \tag{5.12}$$

where R_i is the total input resistance and R_L is a resistance resulting from the radiation of acoustic energy. Repeating (5.5) in the form

$$\begin{aligned} E &= Z_e I + Tv \\ -Z_l v &= TI + Z_m v \end{aligned} \tag{5.5}$$

at resonance

$$\left|\frac{v}{I}\right|^2 = \frac{|T|^2}{|R_m + R_l|^2}$$

and

$$\eta = \frac{R_l}{R_i}\frac{|T|^2}{|R_m + R_l|^2} \tag{5.13}$$

Two sets of impedance measurements, one with the transducer in air when the only dissipative load is supposedly R_m and one with it suspended in a volume of water so large that reflections from bounding surfaces do not interfere, will supply the data for (5.13).

On the motional impedance circle formed by the data taken with the instrument in air the diameter D_a strikes the locus at the frequency of resonance of the mechanical mesh, and has the value

$$D_a = \frac{|T|^2}{R_m} \tag{5.14}$$

Similarly, from the measurements with it immersed, the diameter D_W is

$$D_W = \frac{|T|^2}{R_m + R_l} \tag{5.15}$$

using (5.14) and (5.15) in the following combination:

$$D_W\left(\frac{D_a - D_W}{D_a}\right) = R_L \frac{|T|^2}{(R_m + R_l)^2}$$

Therefore

$$\eta = \frac{D_W}{R_i}\left(\frac{D_a - D_W}{D_a}\right) \tag{5.16}$$

A serious fault with (5.16) and conclusions drawn from it, is that it is based on the assumption that moving from air to water results only in a change in the mechanical arm and that that is properly represented by the change from Z_m to $Z_m + Z_l$. As a matter of fact there may be changes in the electrical side due to environmental effects and in most instances the impedance Z_m is quite different in air from its value in water.

5.7 MAXIMUM EFFICIENCY AND POWER LIMITS

As pointed out in the preceding section, deductions from impedance and admittance data are frequently invalidated by the failure of a reasonably explicit theory for performance at high levels, and further by the fact that mechanical properties do not remain constant with changes in loading conditions. Theory and related data however, are pertinent to the discussion by Professor Hunt of conditions for maximum efficiency.[7] A brief presentation of his development begins with the generalized equivalent circuit shown in Figure 5.13 and a restatement for convenience of the associated terminology.

Loop equations for the network of Figure 5.13 are

$$E = (Z_e'' + Z_e')I + Tv \tag{5.17a}$$

$$F = TI + \left(Z_m'' + \frac{T^2}{Z_e'}\right)v \tag{5.17b}$$

Comparing these with the general two-port network (5.3a) and (5.3b)

$$E = Z_e I + T_{em} v \tag{5.3a}$$

$$F = T_{me} I + Z_m v \tag{5.3b}$$

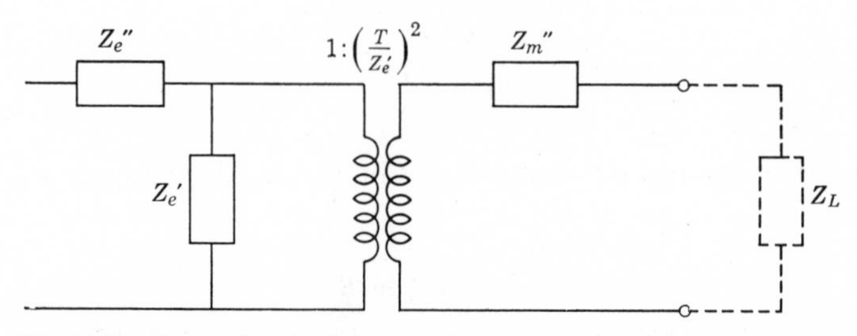

FIG. 5.13 Equivalent circuit with an electromechanical element in the self-impedance of each mesh.

[7] F. V. Hunt, *Electroacoustics*, Wiley, New York, 1954, pp. 126–142.

We see that the open-circuit, driving-point impedance Z_e consists of a component Z''_e not associated with the mechanical mesh and a component Z'_e which is so associated. The first, Z''_e, as shown by (5.17b) contributes to the self-impedance of the mechanical mesh. Again, following Professor Hunt's terminology,

$$T = T_{em} = T_{me} = (j \text{ or } k)Z_{em} = (j \text{ or } k)(R_{em} + jX_{em}) \qquad T^2 = -Z_{em}^{\ 2} \tag{5.18}$$

Also the open-circuit driving point impedance of the mechanical arm consists of two parts: Z''_m and $-Z_{em}^{\ 2}/Z'_e$. The first part Z''_m is the mechanical impedance in the absence of transduction, while the second which will be termed Z'_m is an impedance transduced from the electrical side. From its definition $Z'_m Z'_e = -Z_{em}^{\ 2}$, and the mechanical impedance from (5.3b), $Z_m = Z'_m + Z''_m$.

If a load Z_l, representing a radiation impedance, is placed across the mechanical terminals, the driving point impedance becomes

$$Z_i = Z''_e + Z'_e + \frac{Z_{em}^{\ 2}}{Z''_m + Z_l - Z_{em}^{\ 2}/Z'_e} \tag{5.19}$$

and the corresponding admittance is

$$y_i = \frac{1}{Z''_e + Z'_e} - \frac{Z_{em}^{\ 2}/Z_e^{\ 2}}{Z''_m + Z_l} \tag{5.20}$$

Equation 5.19 may be rewritten as

$$Z_i - Z_e = Z_{\text{mot}} = \frac{Z_{em}^{\ 2}}{Z''_m + Z_l - Z_{em}^{\ 2}/Z'_e} \tag{5.21}$$

Similarly (5.20) becomes

$$Y_i - Y_e = Y_{\text{mot}} = \frac{Z_{em}^{\ 2}/Z_e^{\ 2}}{Z''_m + Z_l} \tag{5.22}$$

The frequencies at which the reactive components of the denominator vanish will be represented by f_R and f_y in the following discussion of efficiency. At the frequency f_R the motional impedance is a maximum and

$$X''_m + X_l = \text{imaginary part of } Z_{em}^{\ 2}/Z'_e$$

or by the definition

$$Z'_e Z'_m = -Z_{em}^{\ 2}, \qquad X''_m + X_l + X'_m = 0 \tag{5.23}$$

At the frequency f_y the motional admittance is a maximum and

$$X''_m + X_l = 0 \tag{5.24}$$

The effective electroacoustic efficiency at the frequency f_R is given by (5.13) rewritten as

$$\eta_R = \frac{R_l}{R_i} \frac{|Z_{em}|^2}{|R_m + R_l|^2} \tag{5.13}$$

To determine the relation of this expression to that for optimum efficiency, let us examine the conditions for a maximum value of the general expression for efficiency given by

$$\eta = \frac{R_e R_l (R_{em}^2 + X_{em}^2)}{[R_e(R_m + R_l) + R_{em}^2][R_e(R_m + R_l) - X_{em}^2] + [R_e(X_m + X_l) + R_{em}X_{em}]^2} \tag{5.25}$$

It is a fair assumption to be made in the case of reasonably efficient electroacoustic sources that only the squared term in the denominator will change appreciably with frequency over a major portion of the motional circle. Therefore a maximum of η occurs at the frequency f_m, where

$$X_m + X_l = -\frac{R_{em}X_{em}}{R_e} \tag{5.26}$$

Recalling that at the resonant frequency f_R

$$X''_m + X'_l + X_m = 0 \tag{5.23}$$

or

$$X_m + X_l = 0$$

we see that the frequency of maximum efficiency may not be the resonant frequency, or it may not be the frequency of maximum admittance. In addition to efficiency, a critical property of transducers, namely power-handling capability, is involved here. The effective magnitude of current into a magnetic field type of transducer is limited by magnetic saturation, so the limit to the power acceptance capability of a given source is set by the real component of the motional impedance. It is possible, however, that the efficiency at the frequency of maximum resistance is so poor that no increase in power output is gained by the greater power input at that frequency. Specifically, for a magnetostrictive transducer, the term on the right of (5.26) is negative, therefore requiring a frequency f_m greater than f_R to satisfy the stated condition. And it is to be observed from the motional impedance locus that its real component is decreasing significantly as the frequency parameter increases from f_R toward f_m.

Using (5.25), (5.26), and (5.23) we may compare the efficiency of f_m to that at f_R:

$$\frac{\eta_m}{\eta_R} = \frac{[R_e(R_m + R_l) + R_{em}^2][R_e(R_m + R_l) - X_{em}^2] + R_{em}^2X_{em}^2}{[R_e(R_m + R_l) + R_{em}^2][R_e(R_m + R_l) - X_{em}^2]} \tag{5.27}$$

This ratio may be rewritten in terms of the resistances involved in computing input power handling capacity as:

$$\frac{\eta_m}{\eta_R} = \frac{(\text{total resistance at resonance}) \cos^2 \theta_m}{\text{total resistance at maximum efficiency}} \tag{5.28}$$

where θ_m is the phase angle of $Z_m + Z_l$ at f_m.

Let I_m represent the maximum effective current for a given magnetostrictive transducer. Then its maximum acoustic output power will be given by the product of I_m^2 by the total resistance and efficiency. Therefore letting P_m and P_R represent the maximum acoustic power output capabilities at the frequencies f_m and f_R, respectively,

$$P_m = \eta_m I_m^2 \times \text{total resistance at maximum efficiency}$$

$$P_R = \eta_R I_m^2 \times \text{total resistance at resonance}$$

or

$$\frac{P_m}{P_R} = \cos^2 \theta_m \tag{5.29}$$

According to (5.29), a magnetostrictive transducer is capable of producing more acoustic power at the resonant frequency irrespective of efficiency. To examine a typical value of this ratio, we recall some relevant expressions:

$$Z_{\text{mot}} = \frac{|Z_{em}|^2}{Z_m + Z_l} = \frac{|Z_{em}|^2}{(R_m + R_l)[1 + j(X_m + X_l)/(R_m + R_L)]}$$

Defining

$$\frac{X_m + X_l}{R_m + R_l} = 2Q_m p$$

where $2p = (\omega/\omega_R - \omega_R/\omega)$, we may express the phase angle θ of $Z_m + Z_l$ as

$$\theta = \tan^{-1} 2Q_m p$$

Accordingly

$$\cos^2 \theta = \frac{1}{1 + 4Q_m^{\ 2} p^2} \tag{5.30}$$

and at the frequency f_m

$$\cos^2 \theta_m = \left[1 + Q_m^{\ 2}\left(\frac{f_m}{f_R} - \frac{f_R}{f_m}\right)^2\right]^{-1} \tag{5.31a}$$

It may be shown[8] for a magnetostrictive ring style of transducer, that the frequency of maximum admittance f_y coincides with the frequency f_m of

[8] F. V. Hunt, *Electroacoustics*, Wiley, New York, 1954, p. 140.

optimum efficiency. For such (5.31a) may be written as

$$\cos^2 \theta_m = \left[1 + Q_m{}^2\left(\frac{f_y}{f_R} - \frac{f_R}{f_y}\right)^2\right]^{-1} \tag{5.31b}$$

and

$$2p = \frac{f_y{}^2 - f_R{}^2}{f_R f_y} \simeq \frac{f_y{}^2 - f_R{}^2}{f_y{}^2} = k_{\text{eff}}^2 \tag{5.32}$$

The last term of (5.32) is an indicator of the closeness of electromagnetic coupling between the electrical and mechanical meshes and is known as the effective electromechanical coupling coefficient. Its value has a range of 0.001 for a tube and plate transducer to 0.5 for a cobalt–nickel ring. As a typical value to be used in (5.31) and (5.29), we shall assume that $k_{\text{eff}}^2 = 0.16$ and $Q_m = 4$.

$$\frac{P_m}{P_R} = \cos^2 \theta_m = \frac{1}{1 + Q^2 k_{\text{eff}}^4} = \frac{1}{1.41}$$

Thus one may see that a source of the parameters used here is capable of delivering about 40% more acoustic power at resonance over that at maximum efficiency. The preceding discussion, however, does not define a frequency of optimum acoustic power capability. For magnetostrictive units it will lie between f_R and f_m. Two important factors involved are magnetic hysteresis and eddy current losses. These are subject to the level of the unidirectional magnetic bias imposed on the active material.

The problem of optimum frequency for power is made quite simple for the better piezoelectric materials, because transduction losses are low, and electromechanical efficiency is hardly dependent on frequency. Voltage breakdown is the relevant limiting factor. For a limiting voltage maximum power will be accepted at the frequency f_y.

The preceding discussion of transduction is generally applicable to electromechanical devices. In later chapters we shall see how the theory can be used to predict the character of the more widely employed types.

CHAPTER 6

Magnetostrictive and Piezoelectric Systems

6.1 INTRODUCTION

In earlier chapters we discussed the general principles of vibrating systems as related to electroacoustic transducers, and also their interaction with the medium through radiation. We now proceed to the study in greater detail of the application and performance of the two most successful materials. Specific examples will be limited to the piezoceramics and the magnetostrictive metals but this is not an inference that in a certain instance another material could not be superior. The piezoelectric crystals have been in use since World War I and the student should consult the vast literature on the subject.[1,2] There are also some applications suitable to the properties of the piezomagnetic ceramic ferrites.[3]

The fundamentals of ferromagnetism are accorded more space than given the electric field. This is no measure of the relative importance of the two systems but of the freedom of the design engineer in setting up operating conditions. If specifications indicate a ceramic, he has a choice of several materials, but the characteristics of the material chosen will have been built in by its manufacturer and not subject to change. On the other hand, if a magnetostrictive metal is used, there is beyond the selector of a material a choice of lamination thickness, manner of polarization if any, degree of polarization, kind of consolidation, etc., which significantly affect material constants.

In the review of magnetic and electric fields that follows, we shall be restricted to just such definitions and illustrations of the basic concepts that

[1] W. G. Cady, *Piezoelectricity*, McGraw-Hill, New York, 1946.

[2] W. P. Mason, *Piezoelectric Crystals and Their Applications to Ultrasonics*, Van Nostrand, Princeton, New Jersey, 1950.

[3] C. M. van der Burgt, "Matronics," Philips Technical Information Bulletin No. 15.

permit a coherent development of electromechanical phenomena. The only magnetostrictive transducer to be discussed is the scroll or ring. The magnetostrictive longitudinal vibrator, predominately used in other years, is rapidly being supplanted by the ceramic. When reliability and endurance considerations are paramount to bandwidth and power capability, the scroll is still a likely choice for a sound source. However, the use of ceramics as rings, cylinders, disks, and laminar arrangements is described because all of these forms are in current use.

6.2 THE MAGNETIC FIELD

The motion of the individual electrons about the atomic nucleus that give rise to the properties of ferromagnetic materials is well explained elsewhere.[4] We begin with the definition of magnetic parameters as they exist in structures that will be of interest to us. Figure 6.1 shows a simple laboratory arrangement for investigating the static magnetic properties of the ring enclosed by the primary winding. The primary circuit contains a current source, a variable resistor for controlling the magnitude of current, and an ammeter. The secondary circuit contains a fluxmeter calibrated to measure changes in flux density. A current I in the primary yields a magnetizing field H given by

$$H = \frac{NI}{2\pi r} \quad \text{ampere turns/meter} \tag{6.1}$$

This magnetizing field produces a magnetic flux density β of

$$\beta = \mu H \text{ Webers/meter}^2 \tag{6.2}$$

where $\mu = \mu_r\mu_0$, $\mu_0 = 4\pi \times 10^{-7}$ henries/meter, and μ_r = relative permeability of ring core (dimensionless). The permeability μ is a constant μ_0 when the core material is absent, but for a ferromagnetic core it is a function of the flux density β. Unfortunately this function can be described satisfactorily only by a chart such as that shown in Figure 6.2. A number of terms can be defined and identified by the use of this β-H chart. First, beginning at the

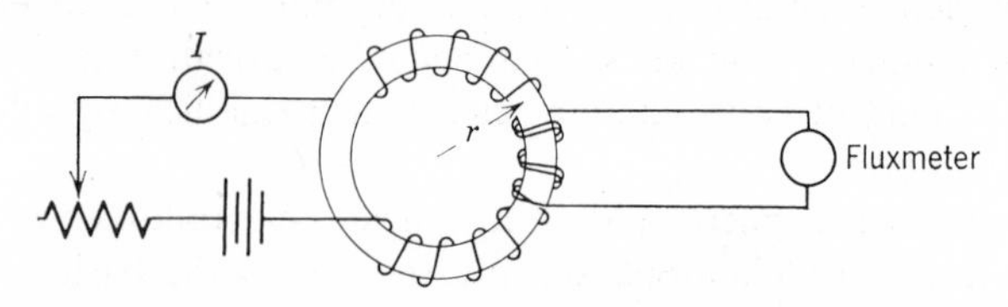

FIG. 6.1 Apparatus for obtaining a magnetization curve.

[4] R. M. Bozorth, *Ferromagnetics*, Van Nostrand, Princeton, New Jersey, 1951.

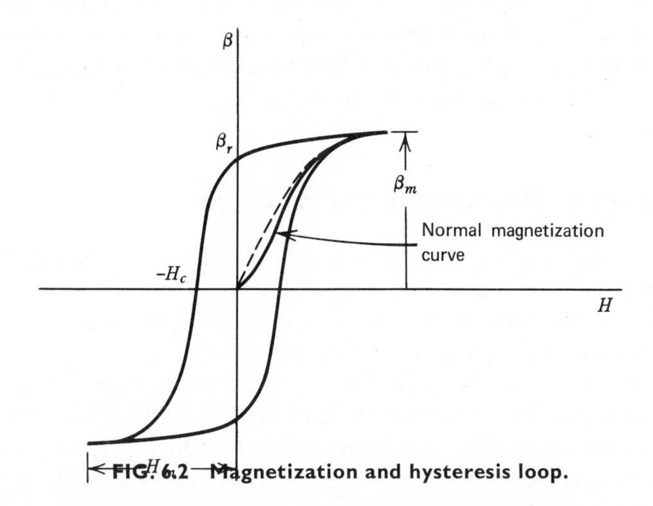

FIG. 6.2 Magnetization and hysteresis loop.

zero of coordinates, we describe the initial magnetization curve for a sample with no memory of a previous magnetic history by increasing the magnetic field to a value $H_{\max}$ where the curve is flat, a condition called *magnetic saturation.* At the origin the slope of the curve is the *initial permeability.* As the current is decreased, the β-H locus is the upper curve intersecting the ordinate at the point $(0, \beta_r)$. This *residual flux density of remanence* B_r is called the *retentivity* when it represents a decay from full saturation. A reversal of the magnetizing field of sufficient magnitude to reduce β to zero reaches the point $(-H_c, 0)$. This value of H is called the coercive force. As shown by the chart, the material can be brought to magnetic saturation in the reverse direction, reduced to $(0, \beta_r)$, to $(H_c, 0)$ and to (H_m, β_m) to close what is called a *major hysteresis loop.* Minor hysteresis loops may be traced by cycling the H field over values smaller than $H_{\max}$. All points on such loops lie within the major hysteresis loop. The normal magnetization curve shown by Figure 6.2 is the locus of tips of successive hysteresis loops obtained by cycling H over decreasing values. The permeability as defined by the tangent line from the origin has a significant meaning because it is repeatable. One can see that other constants characteristic of the material must be defined by the major hysteresis loop for the same reason.

Two other properties of the ferromagnetic core that may be determined from the hysteresis loop are the energy stored in the magnetic field and that lost through hysteresis. These are both of vital concern to the design engineer. The energy per unit volume stored in the core as the magnetization curve is being traced is

$$W_m = \int_0^{\beta} H \, d\beta \text{ joules/meter}^3 \tag{6.3}$$

which is the area between the curve and the β axis. A complete hysteresis loop encloses an area which represents energy lost. Since a magnetostrictive transducer is driven by a magnetizing field which traces this loop, each cycle thereby generating heat in the core, hysteresis is a major contributor to its energy losses.

6.3 MAGNETIC CIRCUIT WITH AIR GAP

While a current is flowing in the windings of Figure 6.3, the magnetizing field H_i induces in the ring a flux of density β. A fundamental property of magnetic flux is its continuity, for example, flux lines form closed loops. In a homogeneous ring of high permeability these loops are circles contained entirely within the core. For materials of lower permeability there will be some leakage with a nonuniform winding as in Figure 6.3. There would also be some fringing at the gap. But for a uniformly wound core and a narrow gap, β will be continuous across the gap. Therefore we can write

$$H_g = \frac{\beta}{\mu_0} \tag{6.4}$$

$$H_i = \frac{\beta}{\mu_r \mu_0} \tag{6.5}$$

or

$$\frac{H_g}{H_i} = \mu_r \tag{6.6}$$

where μ_r is the relative permeability of the core.

As demonstrated later, the performance of a magnetostrictive ring as an

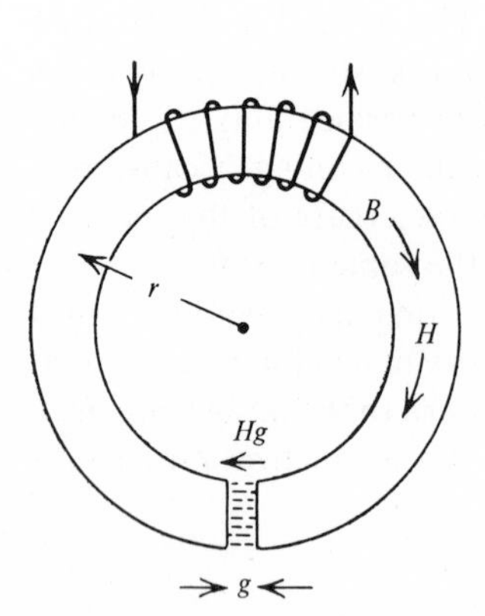

FIG. 6.3 Ferromagnetic ring with gap.

electroacoustic vibrator may be subject to an imposed unidirectional magnetic bias. So we have the simple problem of establishing this bias both for a ring with no gap and for a ring with gap. From a *B-H* curve of the ring material we may determine the value of H required to produce the designed flux density. With no gap the magnetizing field is given by

$$\frac{NI}{2\pi r} = H_i \tag{6.7}$$

where r is the mean radius of the core, and N the total number of turns about it. With the air gap of length g as shown in Figure 6.3, we resort to the fact that the line integral of H around the magnetic circuit must equal the number of ampere turns or

$$\oint H\,dl = NI \tag{6.8}$$

Since $\oint H\,dl = H_i(2\pi r - g) + H_g g$ and $H_g = \mu_r H_i$, we have

$$NI = [2\pi r + (\mu_r - 1)g]H_i \tag{6.9}$$

The introduction of a gap into a core tends to decrease the variation in the ratio of β to NI. This may be an advantage in some cases.

Figures 6.4 and 6.5 will assist in illustrating the properties and applications of permanent magnets in electroacoustics. The core may be a permanent magnet establishing a flux in the gap. The loudspeaker is such an application. On the other hand, a permanent magnet placed in the gap will supply a magnetizing field for a core of permeable magnetic material. This is the method of supplying the unidirectional bias for the magnetostrictive transducers used with shipboard sonar. Given a continuous core of *hard* magnetic material, an expression for a material of high coercivity $-H_c$ suitable for a

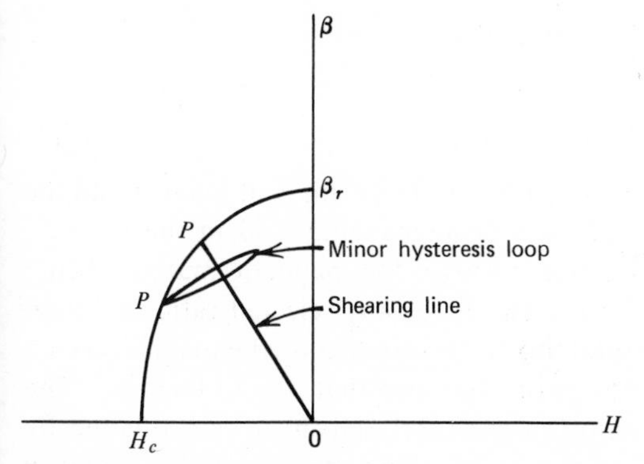

FIG. 6.4 Demagnetization curve for permanent magnet.

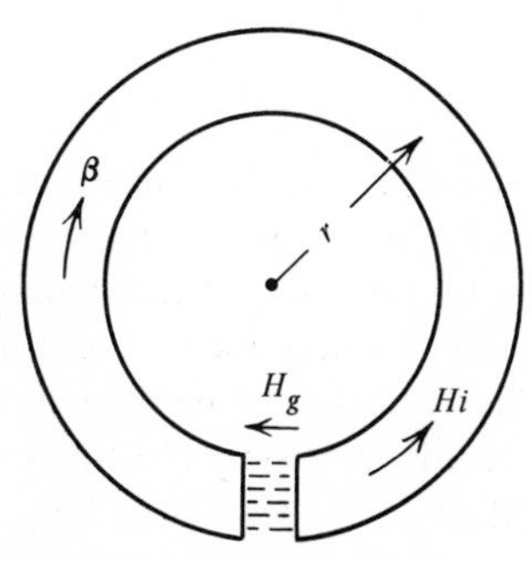

FIG. 6.5 Permanent magnet with gap.

permanent magnet: when a magnetizing field H_m applied to this core is reduced to zero, the induction flux decays to its remanence value β_r. Now, if a gap is introduced without otherwise affecting the material, the induction flux will decay further, as shown in Figure 6.4 to the point P on the β-H curve. Since there is no externally applied magnetizing field the line integral of H around the circuit of Figure 6.5 is

$$\oint H\,dl = H_i(2\pi r - g) + H_g g = 0 \tag{6.10}$$

or

$$H_i(2\pi r - g) = -H_g g \tag{6.11}$$

Since β is continuous across the gap $\mu_0 H_g = \beta$ and by multiplying (6.11) through by μ_0, we get

$$\mu_0 H_i(2\pi r - g) = -\beta g$$

or

$$\left(\frac{\beta}{H_i}\right) = \frac{-(2\pi r - g)\mu_0}{g} \tag{6.12}$$

With the introduction of the gap, H_i becomes the demagnetizing field in the core and β is the resulting induction flux density. In other words, the β and H_i of (6.12) are the coordinates of the point P and their ratio is the slope of the shearing line.

To illustrate the use of Figure 6.4 and (6.12), let us consider the problem of supplying polarizing flux to a magnetostrictive longitudinal vibrator consisting of a stack of laminations similar to that shown in Figure 6.6. From experimental information we determine the desired flux density β_m to be induced in the stack and the required magnetizing field. Calling the total flux path l and the magnet thickness g, we take the integral of H around the magnetic circuit

$$\oint H \cdot dl = H_p g + H_m l = 0$$

or

$$H_p g = -H_m l \tag{6.13}$$

where H_p and H_m are the magnetizing fields in the permanent magnet and the metal stack, respectively. Now H_p is a demagnetizing field in the magnet, so by referring to the demagnetization curve of the magnetic material being used, we may locate the point P from (6.13) for any selected ratio of l to g. The other factor involved is the total flux to be supplied by the magnet, which includes not only the flux over the path l but also that due to leakage. The latter is likely to be a large fraction of the total and must be estimated by use of the relative reluctances of the leakage and path l. If the point P falls low on the demagnetization curve, flux available from the magnet may

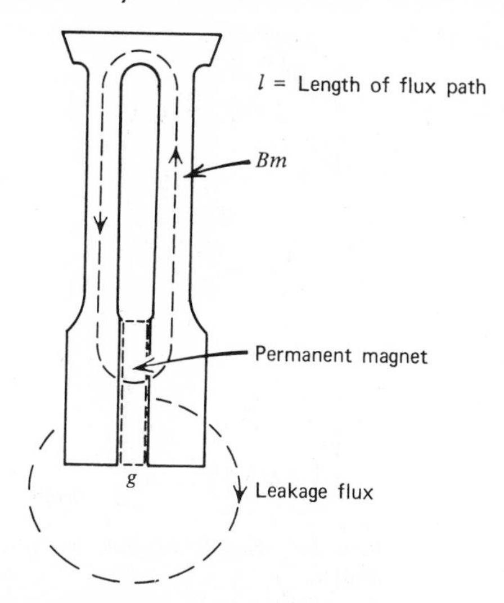

FIG. 6.6 P — M polarization of magnetostrictive stacks.

be insufficient. Choice of the cross-sectional area of the magnet controls the total flux that it supplies. It is rather unusual to utilize a magnet polarized through the thin dimension. To do so requires material of very high coercivity; also, because it is preferable that the ac flux follow a path through the magnet, the material should be a nonconductor. Some of the barium ferrites[5] are quite good in this respect though lacking in mechanical strength.

6.4 EDDY CURRENTS

In a previous section the problem of magnetic hysteresis losses was mentioned. Another energy loss which could be quite serious is due to eddy currents developed in a conducting metal by time changing flux. To reduce this loss, magnetostrictive metals are laminated. Also, increasing the resistivity by changes in composition may help. Eddy currents not only introduce heat losses but also block the penetration of an imposed magnetic field. How this comes about is explained simply by use of Figure 6.7 which shows a cross section of a metal rod. Increasing flux into the plane of the paper develops a magnetic flux opposing the imposed flux. The opposing flux increases toward the center, thereby decreasing the resultant of the two. There is a

[5] *American Institute of Physics Handbook*, McGraw-Hill, New York, 1957, Table 5, h-13, pp. 5–129.

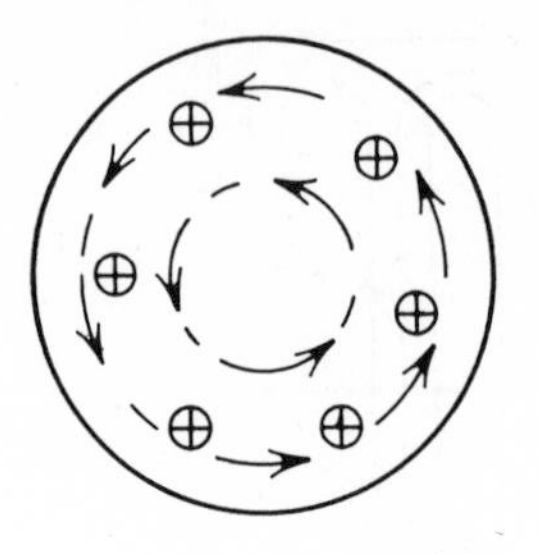

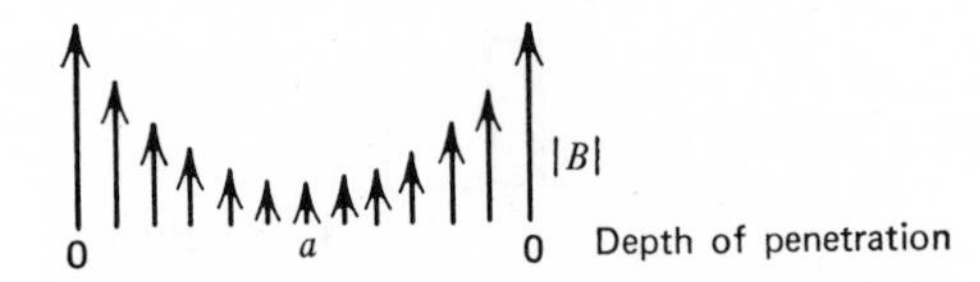

FIG. 6.7 Resultant flux as a function of depth.

characteristic depth expressed as a function of material constants

$$d = \left(\frac{4\pi\rho_e}{\mu f}\right)^{1/2} \tag{6.14}$$

such that a magnetic field applied tangentially to a sheet is reduced to $e^{-2\pi}$ times its surface value at the depth d. The material constants in (6.14) are

ρ_e = resistivity in ohm-meters

$\mu = \mu_r\mu_0$ = magnetic permeability in henries/meter

f = frequency of applied field

In addition to the effects already described, eddy currents introduce a phase lag into the dependence of β on time. As a result the magnetic permeability becomes a complex number dependent on frequency as well as the material constants. Whenever permeability is involved, it is convenient to take care of this dependency through use of an eddy current factor χ, where

$$\chi = \chi_0 e^{-j\zeta} = \chi_R - j\chi_I \tag{6.15}$$

The eddy current factor is used in the expression for the clamped core impedance of the magnetostrictive ring. If this impedance without eddy currents is $j\omega L_0$, the factor χ corrects for the eddy currents replacing $j\omega L_0$ with Z_e', where

$$Z_e' = R_e' + jX_e' = j\omega L_0\chi \tag{6.16}$$

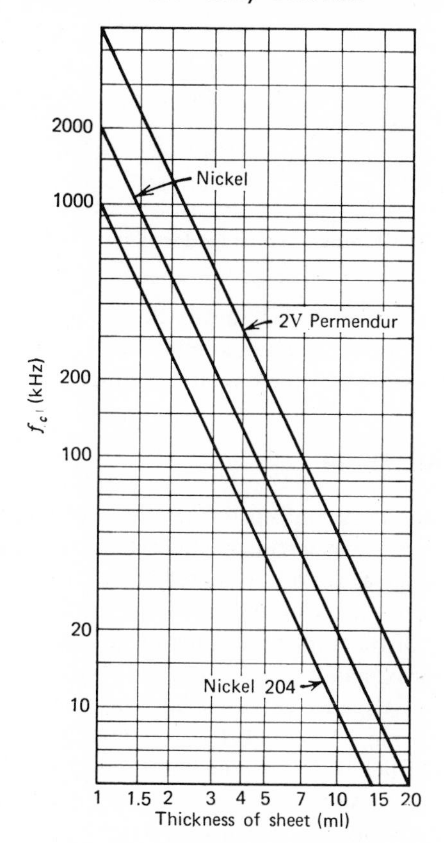

FIG. 6.8 Eddy current parameter, f_c.

For a flat sheet of thickness t there is a characteristic frequency defined as

$$f_c = \frac{2\rho_e}{\mu\pi t^2} h_z \tag{6.17}$$

Figure 6.8 is a chart[6] for magnetostrictive materials showing f_c as a function of thickness t. This chart is based on arbitrary choices of material constants substituted into (6.17). Changes in composition, heat treatment, or polarization of material may alter these constants significantly.

The clamped core impedance Z_e', as given by (6.16), deserves special attention because of its role in the determination of physical constants. Normally the locus of an equation of this form which uses R_e' and X_e' as coordinates and frequency as a parameter would be a straight vertical line displaced a distance

[6] Summary Technical Report, Div. 6, NDRC, Vol. 13, p. 37.

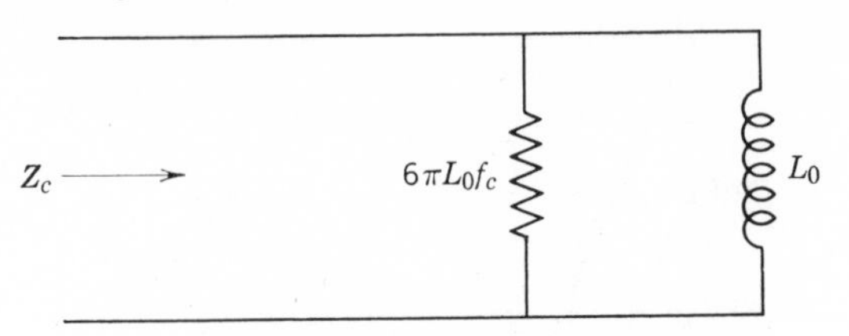

FIG. 6.9 Clamped core impedance for $f < fc$.

R'_e from the ordinate. Since R'_e and X'_e are both functions of frequency, the locus at frequencies lower than f_c follows a semicircle. In this region the core impedance may be represented by the circuit of Figure 6.9.

6.5 MAGNETOSTRICTION

Magnetostriction is the term applied to the effects of magnetic induction on the dimensions of ferro- and ferrimagnetic materials. The utilization of these effects in underwater acoustics is limited to the Joule and the Villari effect. The Joule effect is the name given to the change in length of the material along the axis of the applied magnetic field. The change in length per unit length is approximately proportional to the square of the flux density. For some materials such as the cobalt–iron and nickel–iron alloys, the change is an expansion. For nickel the change is a contraction. A reversal of the induction field brings about the same change; for example, during each half cycle of an alternating field, a nickel rod would undergo a complete cycle of contraction and return to normal length, a performance called frequency doubling. Figure 6.10 contains a dotted line showing strain as a function of magnetic induction β and solid lines illustrating the lag introduced by hysteresis.

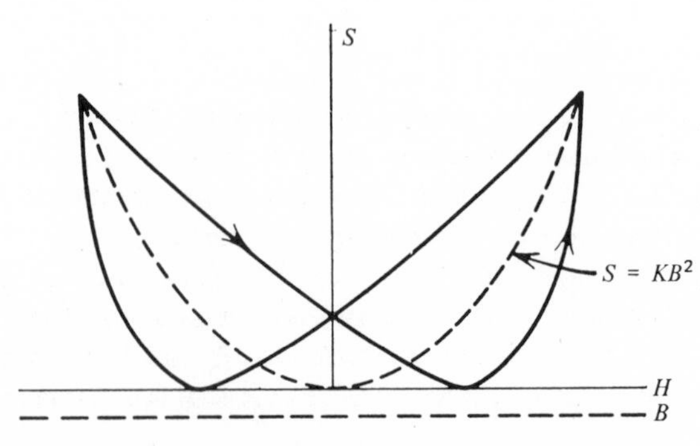

FIG. 6.10 Hysteresis in magnetostrictive strain.

The Villari effect is the change in the induction field resulting from an externally applied stress. The effect occurs only in the presence of an existing field. In addition to changes in the direction of applied stresses and fields, there are also transverse and volume effects of no significance in underwater transducers. A quantitative description of magnetostriction for our purpose is greatly simplified by the importance of only a few of the possible physical and electrical coefficients.

Referring again to Figure 6.10, we recognize the dotted line as the locus of the equation:

$$S = K\beta^2 \tag{6.18}$$

with K an empirical constant, this is an idealization of the Joule effect. The ferromagnetic material, soft-annealed nickel which is one of the best, will serve as an introduction to the magnitudes of numbers involved. As the normal magnetization curve of Figure 6.2 approaches β_m, which for nickel is just over 0.5 Webers/meter², the strain for a static field approaches -40×10^{-6}, and (6.18) no longer applies. The elastic coefficient corresponding to Young's modulus for nickel is about 2.1×10^{11} N/m² so the stress required to develop a strain equivalent to that provided by a saturation field is slightly under 1000 lb/in.² Under dynamic conditions at the resonant frequency, the strain developed could exceed the yield point, resulting in work-hardening of the metal.

As the preceding discussion indicates, dimensional changes are practically infinitesimal. Taking the differential of (6.18),

$$S_i = 2K\beta_0\beta_i \tag{6.19}$$

where β_i is a small change in the induction field from β_0 and S_i a small change in the strain. The only variables involved in our applications are the differentials T, S, H, and β so the equations of state of a magnetostrictive core may be written in the matrix form as

$$T = c^\beta S - h\beta \tag{6.20a}$$

$$H = -hS + \left(\frac{1}{\mu^s}\right)\beta \tag{6.20b}$$

where c^β is the elastic modulus under constant induction; μ^S is the permeability of the clamped core and h, the magnetostrictive coefficient, is a function of the existing induction field as given by

$$h = 2c^\beta K\beta_0 \tag{6.21}$$

6.6 ELECTROMECHANICAL COUPLING COEFFICIENT

Other properties of the material may be defined by manipulation of (6.20a) and (6.20b). The elimination of S gives

$$c^\beta H + hT = \frac{1}{\mu^S}\left(1 - \frac{\mu^S h^2}{c^\beta}\right)c^\beta \beta \tag{6.22}$$

The stress-free permeability may be obtained from (6.22) by setting $T = 0$

$$H = \frac{1}{\mu^S}\left(1 - \frac{\mu^S h^2}{c^\beta}\right)\beta \tag{6.23}$$

The coefficient of β in (6.23) is the inverse of the stress-free permeability or

$$\frac{1}{\mu^S}(1 - k^2) = \frac{1}{\mu^T} \tag{6.24a}$$

$$\mu^S = \mu^T(1 - k^2) \tag{6.24b}$$

where

$$k^2 = \frac{\mu^S h^2}{c^\beta} \tag{6.25}$$

The elastic modulus c^H may be defined by a similar procedure. The elimination of β from (6.20a) and (6.20b) gives

$$T + \mu^S hH = c^\beta\left(1 - \frac{\mu^S h^2}{c^\beta}\right)S \tag{6.26}$$

and setting $H = 0$ yields

$$T = c^\beta(1 - k^2)S \tag{6.27}$$

where the coefficient of strain must be the modulus with constant magnetizing field c^H. Expressed explicity

$$c^H = c^\beta(1 - k^2) \tag{6.28}$$

The quantity k, as defined by (6.25), is called the *electromechanical coupling coefficient*. It is a measure of the degree of coupling that can be obtained between the electrical and mechanical sides of a transducer. It is to be observed that this is a property of the material under conditions where no losses are sustained by motion. It can be shown that for an increase in β, the ratio of energy stored in the form of increased strain to the input stored energy is k^2 or

$$k^2 = \frac{\frac{1}{2}c^H S^2}{\beta^2/2\mu^T} \tag{6.29}$$

For a stress-free rod (6.20a) gives

$$S = \frac{h\beta}{c^\beta} \tag{6.30}$$

and inserting this into (6.29) we obtain

$$k^2 = \frac{c^H \mu^T h^2}{(c^\beta)^2} \tag{6.31}$$

By the use of (6.24b) and (6.28) we find that

$$\mu^T c^H = \mu^S c^\beta$$

which inserted in (6.31) yields the desired expression.

6.7 POLARIZATION

Although not universally true, in most cases an optimum value of the coupling coefficient is desirable. Unlike piezoelectric materials within which this property is subject in practice to little alteration, k^2 for magnetostrictive materials is determined by the factors $\mu^S h^2$, each of which is a nonlinear function of the magnetic field. If the imposed ac field has a small amplitude limited to the region about the origin of Figure 6.10, the magnetostriction constant will have almost a zero value. A large amplitude about the origin results in frequency doubling. Three disadvantages of the frequency-doubling excitation are excessive harmonic distortion, greater hysteresis losses, and the fact that efficient power conversion is obtained only at high power levels. These handicaps may be avoided by maintaining a fixed magnetizing field in the material on which the ac exciting field is imposed. This unidirectional field may be positioned at any desired point on the magnetization curve either by a direct current or by a permanent magnet. A *best point* depends on the manner of operation, though it is likely to be at the point of maximum coupling. To get a clear notion of the excitation obtained with a dc magnetic bias, we may examine Figure 6.11, where H_0, representing the unidirectional magnetizing field, is positioned at the knee of the normal magnetization curve. An exciting field of very small amplitude about this point will develop a minor hysteresis loops with negligible loss. The product μh^2, though not necessarily a maximum, is constant and the strain developed will have the frequency of the exciting field. Under these conditions, the performance conforms to the linear equations (6.20a) and (6.20b). The exciting field may have an amplitude as great as H_0 without frequency doubling but, of course, none of the associated parameters will be constant during a cycle; the minor hysteresis loop will produce losses, and the exciting

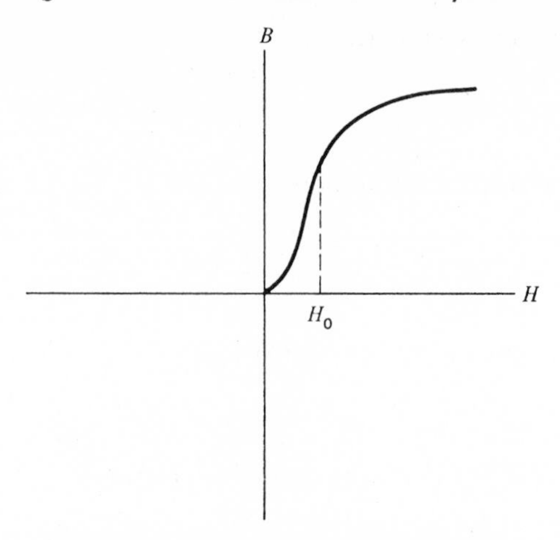

FIG. 6.11 dc Bias of the magnetic field.

force will be harmonically distorted.[7] There is also a limited application of materials operating at remanent polarization. Where severe reliability specifications are coupled with low power requirements over a narrow frequency range, metals such as nickel and 2V-Permendur annealed to a semihardness answer the purpose.[8] Figure 6.12[9] shows the upper half of a major hysteresis loop for the alloy 2-V Permendur and a minor loop representing an exciting field. Greater exciting fields than the one shown would

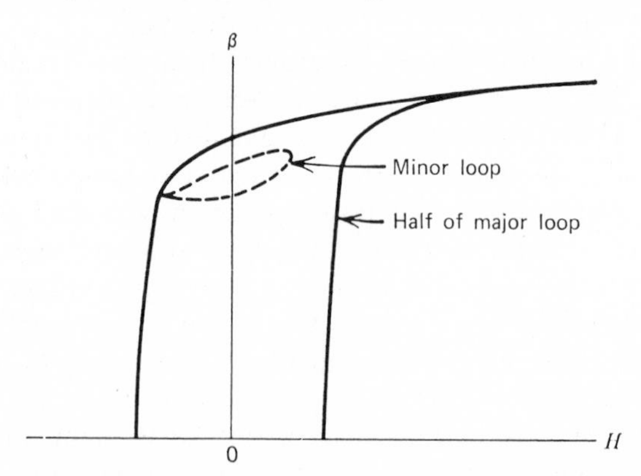

FIG. 6.12 Hysteresis of semihard magnetic material.

[7] R. W. Whymark, "Utilization of Magnetostrictive Materials in Generating Intense Sound," *J. Acoust. Soc. Am.*, **33,** 725–732 (1961).

[8] Summary Technical Report, Vol. 13, pp. 96–101.

[9] Summary Technical Report, vol. 13, Figure 27.

result in operations at lower average-flux densities. Loss of the remanent flux will result from over driving. The desirable remanence may be restored, however, with a dc surge to 100 Oe for the sample of Figure 6.12.

6.8 THE MAGNETOSTRICTIVE SCROLL

The engineer has a number of choices in designing a cylindrical magnetostrictive transducer to vibrate in the radial mode, but in most cases an assembly of scrolls is the best answer. The term *scroll* is applied to what is basically a ring with a thin wall of rectangular cross section. It is made by winding a metal strip about a mandrel and using a bonding agent to hold it permanently. Figure 6.13 locates the scroll dimensions. The metal strip is rolled to a thickness determined from the chart of Figure 6.8 or (6.17). The width of the strip fixes the wall height l and, of course, the number of strip layers fixes t. The technique of assembly requires some learning experience. The metal strip must be given the proper heat treatment before consolidation, and subsequent handling must be such that the metal does not suffer work-hardening. Additional caution must be observed to ensure that the material in its permanent form is not under a residual strain.

It was shown in Chapter 1 that a ring in its fundamental mode has a resonant frequency given by $f_r = c_m/2\pi a$, where c_m is the speed of sound in the metal structure. In Chapter 4 a derivation of the radiation impedance of cylindrical transducers yields a reactive term equivalent to a mass that lowers the resonant frequency of an immersed scroll. Since these effects apply to all sorts of rings, regardless of material, they are discussed later.

6.9 THE EQUIVALENT CIRCUIT

With the aid of the general discussion of two-portnet works of Chapter 5, we shall develop specifically the equivalent electric circuit for the scroll just

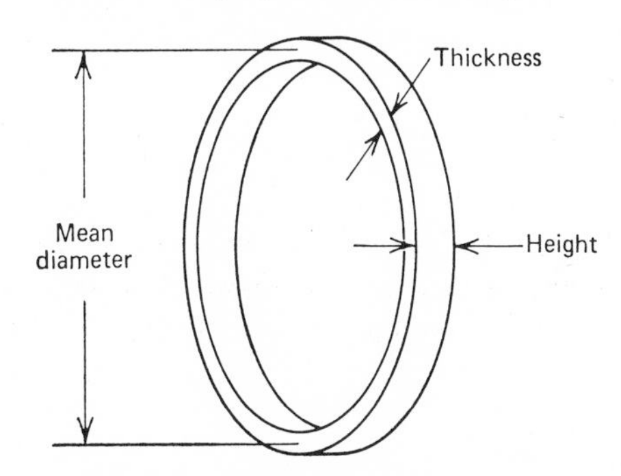

FIG. 6.13 Magnetostrictive scroll.

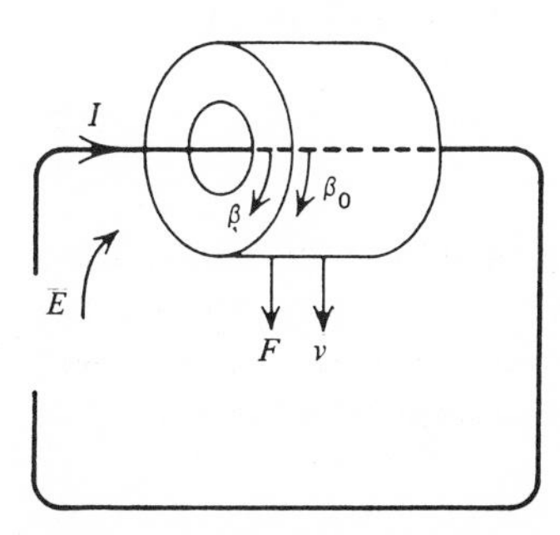

FIG. 6.14 Electromechanical sign conventions.

described. Repeating for convenience the canonical equations

$$E = Z_e I + Z_{em} v \tag{5.3a}$$

$$F = Z_{me} I + Z_m v \tag{5.3b}$$

with the help of Figure 6.14 and (6.20a) and (6.20b) the transduction coefficients

$$Z_{em} = \left(\frac{E}{v}\right)_{I=0} \tag{6.32}$$

$$Z_{me} = \left(\frac{F}{I}\right)_{v=0} \tag{6.33}$$

may be determined. It is necessary first to establish sign conventions in accordance with Figure 6.14. Positive current is in a direction to increase β in the direction of the existing polarizing flux β_0 and positive E is in the direction of I when the core is blocked. Positive F is acting outward in the direction of positive radial velocity.

With a current change I there is an induction change β and in (6.20a) setting, $S = 0$, $T = -h\beta$. If N is the number of turns conducting I and the dimensions of the core are thickness b, height l, radius a,

$$T = -\frac{N\mu^S \chi h I}{2\pi a} \tag{6.34}$$

The total radial force $F = 2\pi b l T$ is

$$F = -\frac{b l N \mu^S \chi h I}{a} \tag{6.35}$$

Therefore from (6.33)

$$Z_{me} = -\frac{b l N \mu^S \chi h}{a} \tag{6.36}$$

To find the other transduction coefficient Z_{em} consider the core vibrating with simple harmonic motion with the electric circuit open. The strain S is given by $S = v/j\omega a$ and, with open circuit, applied $H = 0$. So from (6.20b) the change in induction

$$\beta = \frac{\mu^S \chi h v}{j\omega a} \tag{6.37}$$

The open-circuit voltage developed by this flux change is $E = j\omega Nbl\beta$, or

$$E = \frac{blN\mu^S\chi h}{a} v \tag{6.38}$$

From (6.32)

$$Z_{em} = \frac{blN\mu^S\chi h}{a} \tag{6.39}$$

To discuss the equivalent circuit in detail we must first look at the mechanical impedance of the ring as affected by magnetostriction. Without magnetostriction

$$Z''_m = R''_m + jX''_m = R''_m + j\left(\omega m - \frac{s_0}{\omega}\right)$$

where m is the mass of the ring and s_0 its stiffness. We can show that R_m and s_0 are affected by magnetostriction, so we shall rewrite the mechanical impedance Z_m as consisting of two parts:

$$Z_m = Z'_m + Z''_m \tag{6.40}$$

where Z''_m is of purely mechanical origin and Z'_m results from the electromechanical coupling of the core material. It can be shown[10] that

$$Z'_m = -\frac{k^2 s_0}{j\omega}\chi \tag{6.41}$$

In explanation of (6.41) it is readily understood that a core in motion is generating a flux in the core which sets up eddy currents resulting in heat losses. This same flux, of course, affects the elastic modulus, and since Z'_m is an addition to the mechanical impedance due to the core impedance Z_e we can express it in terms of the core impedance and the transduction coefficient. Using (6.41), the core impedance $Z'_e = j\omega N^2\mu\chi bl/2\pi a$ and Z_{em} from (6.39), we have

$$Z'_m = -\frac{Z_{em}^2}{Z'_e} \tag{6.42}$$

In summary, the impedance coefficients of the two-port net of the magnetostrictive scroll are

$$Z_e = Z''_e + Z'_e$$

[10] Summary Technical Report, Vol. 13, p. 40.

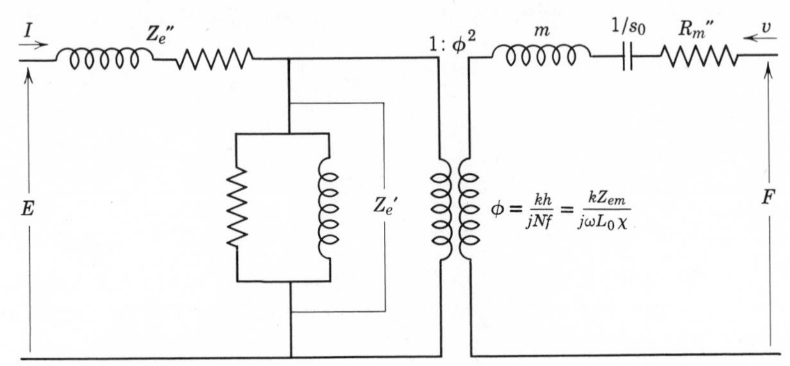

FIG. 6.15 Equivalent circuit for a magnetostrictive scroll.

Where Z_e'' is the leakage and copper loss and Z_e' is

$$Z_e' = j\omega L_0\chi = j\omega \frac{N^2 bl\mu\chi}{2\pi a}$$

$$Z_{em} = \frac{blN\mu\chi h}{a}$$

$$Z_{me} = -Z_{em}$$

$$Z_m = Z_m'' + Z_m' = R_m'' + j(\omega m - s_0/\omega) - \frac{Z_{em}^{\;2}}{Z_e'}$$

The corresponding equivalent circuit, using F. V. Hunt's operator k, is shown by Figure 6.15. The network equations are

$$E = Z_e I + kZ_{em}$$

$$F = kZ_{em} + Z_m v$$

where

$$kZ_{em} = kZ_{me}$$

6.10 DESIGN AND PERFORMANCE OF A FREE-FLOODING SCROLL SOUND SOURCE

The building block of a scroll sound source is illustrated schematically by the sample shown in Figure 6.13. Elements such as this are assembled coaxially with variable parameters consisting of the height of each, the number, and the distance between them. The effectiveness of the assembly depends on the successful manipulation of these variables in solving the radiation problems. Such problems are common both to magnetostrictive scrolls and to ceramic rings.

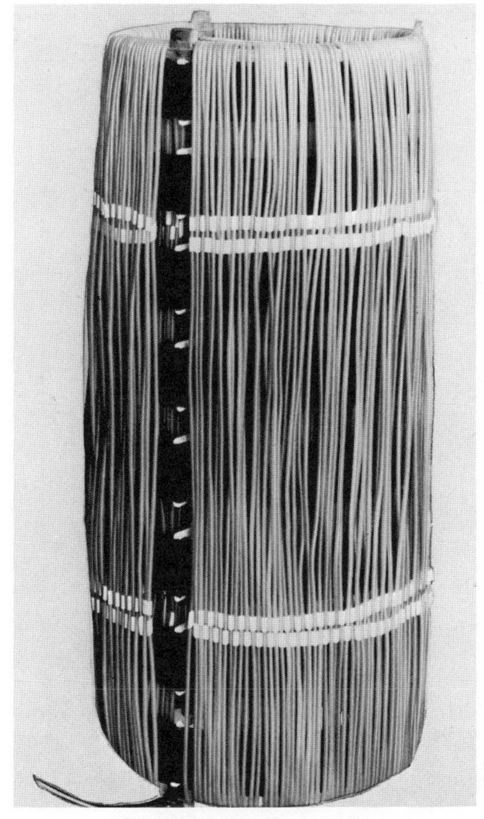

FIG. 6.16 Scroll assembly.

An assembly of 16 scrolls is shown by Figure 6.16. The active material is International Nickel's nickel 204, an alloy of commercially pure nickel with approximately 4% cobalt. The properties of this material have been investigated in detail by C. A. Clark.[11] The metal strip 0.005 in. in thickness, and 0.9 in. wide, had been annealed for one hour at 750°C in a reducing atmosphere, and then exposed briefly at that temperature to air. The latter operation develops a very thin oxide coat on the strip surface which possesses excellent insulating properties. The strip was then coiled and consolidated into a ring of the dimensions defined by Figure 6.15:

$$\text{radius } a = 0.146 \text{ m}$$

$$\text{thickness } b = 0.00762 \text{ m}$$

$$\text{height } l = 0.039 \text{ m}$$

[11] C. A. Clark, "The Dynamic Magnetostriction of Nickel–Cobalt Alloys," *Brit. J. Appl. Phys.*, **7**, 355–360 (October 1956).

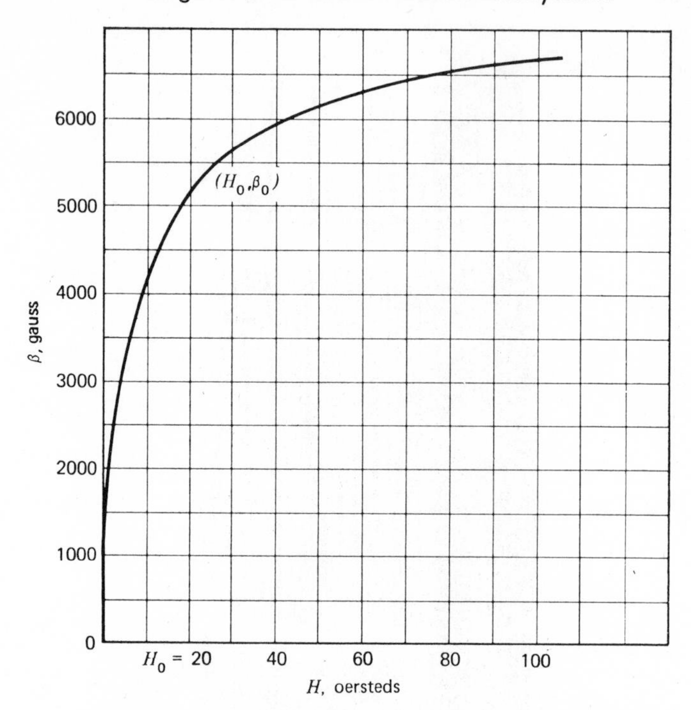

FIG. 6.17 Normal magnetization curve for nickel 204.

Although consolidation is not difficult, observance of certain details[12] adds to its ease and success.

As mentioned previously, the performance characteristics of magnetostrictive material is quite sensitive to controls such as power level and dc polarization. A normal magnetization curve and the dc polarization used are shown in Figure 6.17 as a part of the data collected in assessing the transducer of Figure 6.16. The original measurements were made in units of øersteds and gauss and are shown for convenience. The driving point impedance of Figure 6.18 represents measurements made at a low power level and indicate, therefore, small changes of β and H about the point (H_0, β_0) involving constant values of c^{β} and μ^S. We may readily see that at another point of polarization, say $H_0 = 2\text{Oe}$, all coefficients including h would have a quite different value.

The data from Figures 6.18, 6.19, and 6.20 outline the behavior of the given assembly under one set of conditions possessing no claim to best performance. From these figures we may compute constants of the material

[12] Bulmer, Dickson, Maples and Smith, Audio Engineering Society Preprint No. 465; Audio Engineering Society, New York, New York.

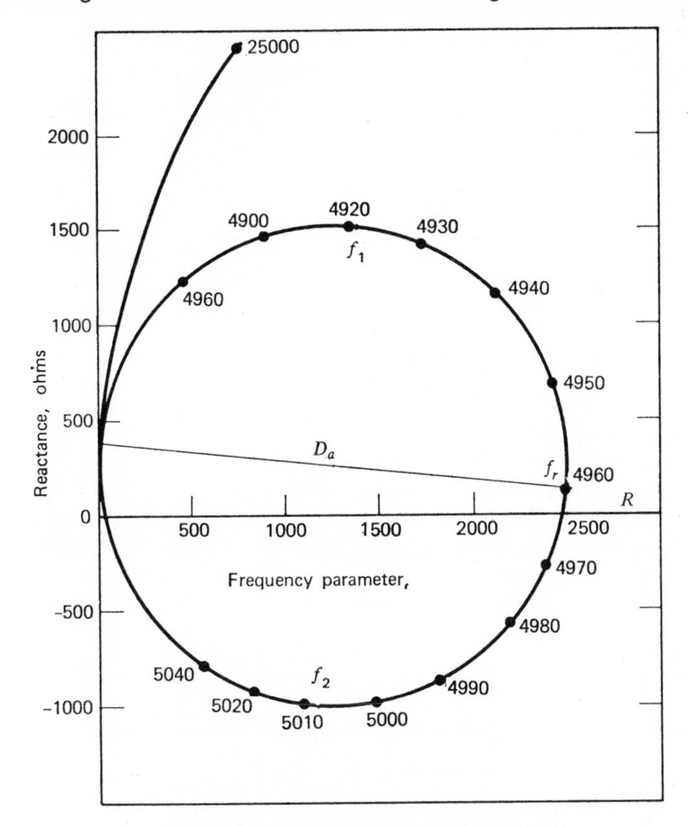

FIG. 6.18 Transducer impedance in air.

and the assembly; for example, Figure 6.18 is the driving point impedance

$$Z_i = Z_e + \frac{Z_{em}^2}{Z_m}$$

of the assembly in air, and Figure 6.19 is the locus of the driving point impedance of the same assembly in water:

$$Z_i = Z_e + \frac{Z_{em}^2}{Z_m + Z_l}$$

We are interested in using the data from these loci in evaluating the constants h, μ^S, c^β, k of the material and ϕ, Z_e, R_m, Q_m for the assembly.

The permeability of the core and the low ohmic resistance of the winding permits us to ignore Z''_m as one of the components of Z_e. Therefore the clamped core impedance is the locus of the parallel components given by

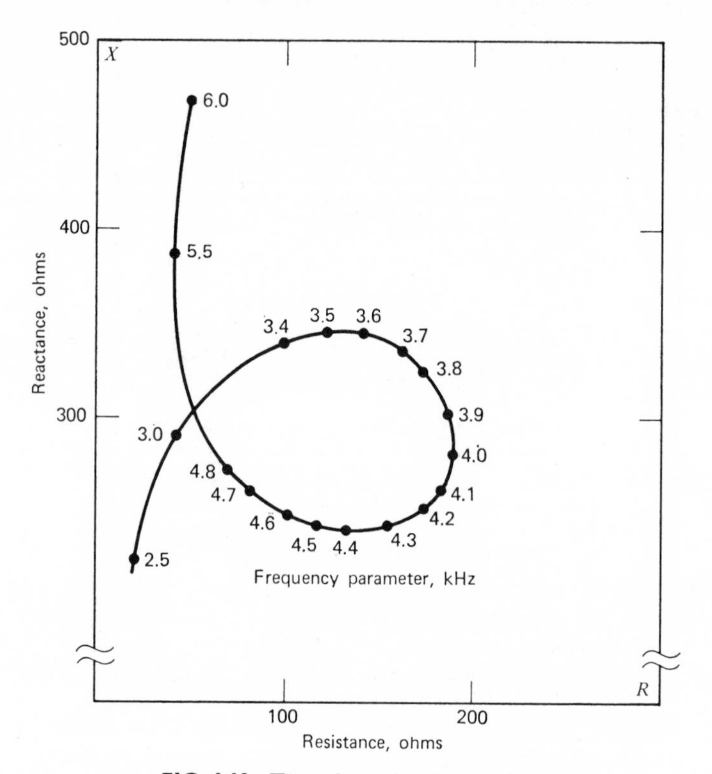

FIG. 6.19 **Transducer impedance in water.**

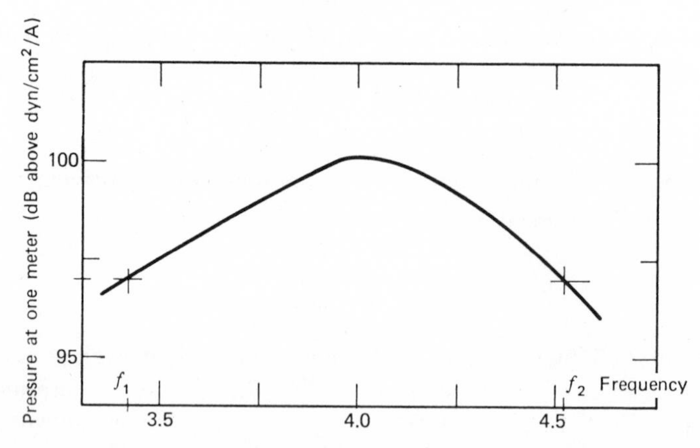

FIG. 6.20 **Constant current transmitting response.**

Figure 6.9. This locus is the circular arc described by the frequency range from zero to 25 kHz. The circle is the locus of the motional impedance. The line marked D_a passes through the center of the circle, locates the series resonant frequency f_r of the mechanical mesh, and extended locates the center of curvature of the clamped core impedance locus. The diameter perpendicular to D_a locates the *quadrantal* frequencies f_1 and f_2. These facts may be confirmed by an analysis of the equivalent circuit impedance. The data from Figures 6.18 and 6.19 are inserted into the following equalities which have been renumbered for convenience:

$$m = \rho 2\pi abl \qquad (l \text{ is total for 16 scrolls}) \tag{6.43}$$

$$Q_m = \frac{f_r}{f_2 - f_1} = \frac{m\omega_r}{R_m} \tag{6.44}$$

$$|Z_{em}|^2 = D_a R_m = \frac{D_a m\omega}{Q_m} \tag{6.45}$$

$$jL_0\chi\omega = Z_e' \simeq jX_e' = j\frac{N^2 bl\mu^S}{2\pi a}\omega \tag{6.46}$$

Figure 6.19 we get $\mu^S/\mu_0 = 46$:

$$c^H = \rho(2\pi a f_r)^2$$

Though the electromechanical coupling coefficient k_{eff}^2 may be computed from Figure 6.18, a more reliable value may be obtained by a simple measurement of f_y, the frequency of maximum admittance and using:

$$1 - \frac{f_r^2}{f_y^2} = k_{\text{eff}}^2 \tag{5.32}$$

This measurement gave $f_y = 5.4$ kHz, or $k_{\text{eff}}^2 = 0.15$ as a value for the electromechanical coupling coefficient for nickel 204 under these operating conditions.

We may also use the approximations:

$$k_{\text{eff}}^2 \simeq \frac{\mu^S h^2}{c^\beta} \tag{6.47}$$

and $c^\beta = c^H/(1 - k^2)$

$$(c^\beta = 2.1 \cdot 10^{11} \text{ N/m}^2)$$

to find h from

$$h^2 = \frac{k^2 c^\beta}{\mu^S} \tag{6.48}$$

$$(h = 2.35 \cdot 10^7 \text{ newtons/weber})$$

[13] Summary Technical Report, Div. 6, NDRC, Vol. 13, p. 56.

The data in Figures 6.19 and 6.20 may be used to answer empirically the question discussed in Chapter 5 of frequency versus efficiency and acoustic output capability. Although the general relation of power to response is reserved for a later chapter, we may readily understand that with constant current input as used in the response of Figure 6.20 that the electrical power into the transducer at any given frequency is given by the real component of the impedance shown by Figure 6.19. The acoustic power developed is proportional to the transmit response. The two curves show a maximum power in and power out at the resonant frequency of 4 kHz. The frequency of maximum efficiency $= f_E = f_y$ for this example, may be found from (5.32). The change in efficiency over this frequency range is slight, so with a Q_m of the order of 4, there would be a considerable disadvantage in driving at a frequency of maximum admittance.

6.10.1 Bandwidth and frequency of free-flooding scrolls

The downward shift in the resonant frequency indicated by Figure 6.19 means that the load Z_l added by placing the transducer in water has a fairly large reactive component equivalent to an additional mass. Under the assumption that the resonant frequency in air is given by

$$f_a = \frac{1}{2\pi a}\left(\frac{s}{m}\right)^{1/2}$$

and the resonant frequency in water by

$$f_w = \frac{1}{2\pi a}\left(\frac{s}{m + \Delta m}\right)^{1/2}$$

the data from Figures 6.18 and 6.19 show

$$\Delta m = 0.54m$$

as the equivalent mass.

Another result of free flooding is the effect on bandwidth. The increase in effective mass and introduction of a reactive component into the radiation load both reduce the bandwidth. Because of the difficulty of establishing boundary conditions for assemblies of this sort, there is little in addition to empirical data for predicting quantitative effects. Since there will be standing wave systems inside the cylinder, these may be used to control bandwidth and frequency.[14]

[14] Theodore J. Meyers, USN Underwater Sound Laboratory Reports 723–754, 1966.

6.10.2 Performance under high power

In choosing a sound source for a particular application, the engineer has the difficult task of matching his requirements to the performance characteristics of available materials. The difficulty is increasing because of the increasing number of materials available. Fundamental among the performance characteristics are reliability and power handling capability. Since the cost of installing a sound source may exceed the cost of the source, assurance of long performance life is of paramount importance. On the other hand, the cost of installation escalates with weight, so power capability per unit weight is of equal importance. Unfortunately these two desirable properties are not found in the same material. The free-flooding magnetostrictive scroll is quite satisfactory in uniformity and reliability. We may use the experimental data from Figures 6.19 and 6.20 for estimating its acoustic power output per unit weight at a level of polarization requiring a dc current of 6.1 A. A driving current of 6.1 rms amperes may be used without reversing the magnetic induction. At 4 kHz, from Figure 6.19, the driving point resistance is 188 ohms so the power input as a function of driving current I is $188\,I^2$, under the assumption that the driving point resistance is not affected by power level. With a driving current of 6.1 rms A, the input power will be 7 kW and with a conversion efficiency of 60%, the acoustic power output will be 4.2 kW, or about 82 W/lb. Additional comments necessary for the proper use of the above results are:

1. The resistance increases with power level, resulting in a higher input power than that given; however, the acoustic output in watts per pound is experimentally correct, which means that the discrepancy is adjusted by the efficiency figure.
2. Acoustic power output per unit weight varies with polarization, and the optimum for this assembly is approximately doubled by using a dc polarizing current of 10 A.
3. As proved in a later section, the acoustic power output per unit weight varies directly as the frequency and inversely as the bandwidth.

6.11 THE ELECTROMECHANICAL EQUATIONS OF STATE

In the preceding discussion of magnetostriction, applicable only to the structures of laminated metal strip, it was possible and convenient to treat the coefficients of (6.20a) and (6.20b) as scalar quantities. When strong coupling exists between strains in orthogonal directions as with the piezoelectric crystals and polycrystalline ceramics, the use of tensor, or matrix,

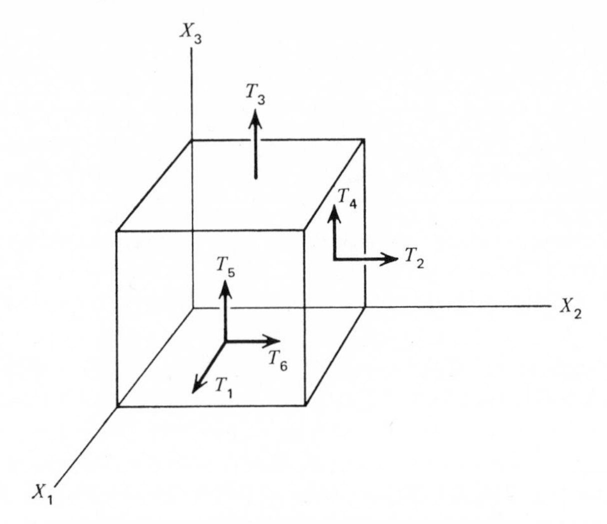

FIG. 6.21 Notation for stress components.

notation permits a systematic treatment of what otherwise could be a confusing procedure. The electromechanical equations of state give the relation between electrical and mechanical effects under constant temperature conditions. The literature on the subject is extensive,[15–17] and we shall repeat only that required to clarify the choice of coefficients in specific cases.

The elementary volume in the rectangular coordinate system of Figure 6.21 illustrates the nine possible stress components. Of these, three are normal to the faces of the cube and six are shear stresses parallel to the cube faces. Since the body is in rotational equilibrium, there are only three independent shear stresses. It is conventional to arrange the six stress components in a column matrix as

$$[T] = \begin{bmatrix} T_1 \\ T_2 \\ T_3 \\ T_4 \\ T_5 \\ T_6 \end{bmatrix} \tag{6.49}$$

where $T_4 = T_{23}$, $T_5 = T_{13}$, and $T_6 = T_{12}$.

[15] "Standards on Piezoelectric Crystals," in *Proc. IRE*, **37,** 1378 (Dec. 1949).

[16] H. W. Katz, *Solid State Magnetic and Dielectric Devices*, Wiley, New York, 1959.

[17] W. G. Cady, "Piezoelectric Equations of State and Their Application to Thickness-Vibration Transducers," *J. Acoust. Soc. Am.* **22,** 579 (Nov. 1950).

There may also exist in the elementary cube six independent deformations or strains; three in shear, and three of elongation which may be arranged in a column matrix as

$$[S] = \begin{bmatrix} S_1 \\ S_2 \\ S_3 \\ S_4 \\ S_5 \\ S_6 \end{bmatrix} \tag{6.50}$$

The elements of the strain matrix are defined as follows:

In conformity with the notation of Chapter 1, ξ_1 represents a displacement in the x_1 direction, and $\partial\xi_1/\partial x_1$ represents a strain that direction. Accordingly, $\partial\xi_1/\partial x_1$, $\partial\xi_2/\xi x_2$, and $\partial\xi_3/\partial x_3$ represent displacement strains S_1, S_2, and S_3. Likewise, the shear strains are:

$$S_4 = 2S_{23} = \frac{\partial\xi}{\partial x_3} + \frac{\partial\xi_3}{\partial x_2} \tag{6.51}$$

$$S_5 = 2S_{13} = \frac{\partial\xi_1}{\partial x_3} + \frac{\partial\xi_3}{\partial x_1} \tag{6.52}$$

$$S_6 = 2S_{12} = \frac{\partial\xi}{\partial x_2} + \frac{\partial\xi_2}{\partial x_1} \tag{6.53}$$

At this point for those not conversant with matrix algebra, we will write (6.20a) and (6.20b) in matrix form and show where these components belong. A fairly well-adopted convention is that a symbol enclosed in brackets represents a matrix.

$$[T] = [c^D][S] - [h_t][D] \tag{6.54a}$$

$$[E] = -[h][S] + [\beta^s][D] \tag{6.54b}$$

are the equations of state with strain S and electric displacement D as independent variables. To write (6.54a) in full we have the following:

$$\begin{bmatrix} T_1 \\ T_2 \\ T_3 \\ T_4 \\ T_5 \\ T_6 \end{bmatrix} = \begin{bmatrix} c_{11}^D & c_{12}^D & & \cdots & c_{16}^D \\ c_{21}^D & & & \cdots & \\ c_{31} & c_{32} & c_{33} & \cdots & \\ \cdot & \cdot & \cdot & \cdots & \cdot \\ \cdot & \cdot & \cdot & \cdots & \cdot \\ c_{61}^D & \cdot & \cdot & \cdots & c_{66}^D \end{bmatrix} \begin{bmatrix} S_1 \\ S_2 \\ S_3 \\ S_4 \\ S_5 \\ S_6 \end{bmatrix} - \begin{bmatrix} h_{11} & h_{21} & h_{31} \\ {}_{12} & \cdot & h_{13} \\ \cdot & \cdot & \cdot \\ \cdot & \cdot & \cdot \\ \cdot & \cdot & \cdot \\ h_{16} & \cdot & h_{66} \end{bmatrix} \begin{bmatrix} D_1 \\ D_2 \\ D_3 \end{bmatrix} \tag{6.55}$$

Equation 6.55 indicates the possibility of 36 elastic coefficients and 18 piezoelectric constants. The expansion of (6.54b) involves a column matrix E of three elements, a rectangular matrix h of 18 elements, a square matrix β^S of 9 elements and a column matrix D of three elements.

The complexity indicated by the discussion of equations of state is quite real for two reasons:

1. Each element of the matrix is a quantity to be determined experimentally.
2. At power levels in common use secondary effects become significant.

Fortunately, in important applications, a few of the elements of these matrices exist; for example, T_1 may be the only stress involved, S_1 the only important strain, and D_3 the only field component. So the expansion of (6.55) reduces to

$$T_1 = c_{11}{}^D S_1 - h_{31} D_3 \qquad (6.56)$$

Omitting the brackets for convenience, other independent variables may be used in writing the equations of state in matrix form:

$$S = s^D T + g_t D \qquad (6.57a)$$

$$E = -gT + \beta^T D \qquad (6.57b)$$

$$S = s^E T + d_t E \qquad (6.58a)$$

$$D = dT + e^T E \qquad (6.58b)$$

$$T = c^E S - e_t E \qquad (6.59a)$$

$$D = eS + t^S E \qquad (6.59b)$$

In these equations

T = stress in N/m²

S = strain

E = electric field in V/m

D = electric displacement in coulombs/m²

and the coefficients are defined by their relation to the other quantities.

Use may be made of the choice of variables to eliminate the largest number of coefficients in writing the equations of state. It is the custom to establish coordinates in a structure as subscripts 1, 2, 3, where 3 is the direction of the applied field and 1, the direction of stress.

6.12 THE THIN-WALLED CERAMIC RING

The thin-walled ceramic ring is an almost perfect example of the vibrating system with one degree of freedom and our discussion is an emphasis of that property; for example, the equation of motion of a vibrating system with

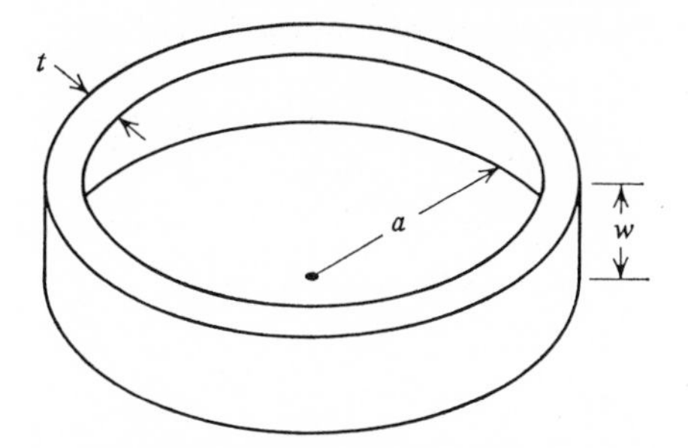

FIG. 6.22 Thin-walled ceramic ring.

one degree of freedom subjected to a damping and driving force is

$$m \frac{d^2x}{dt^2} + R \frac{dx}{dt} + sx = F$$

Referring to the developments of Chapter 1 and Figure 6.22, we may find the following definitions for the constants in the preceding equation:

$$m = 2\pi awt\rho \tag{6.60a}$$

$$R = \gamma_r 2\pi aw \tag{6.60b}$$

$$s = \frac{2\pi wt}{as_{11}{}^E} \tag{6.60c}$$

The ceramic ring has been provided with electrodes on the outside and inside surfaces and has been polarized by the application of a saturation voltage to these electrodes. The wall dimensions t and w are very small compared with the radius a. In the frequency range of interest motion and forces in these directions are negligible. We may set up a system of coordinates in which 1 is in the circumferential direction and 3, in the radial direction. So the boundary conditions are that all stress components vanish except T_1 and likewise S_1 is the sole strain component; also, $E_2 = E_1 = 0$ and $D_1 = D_2 = 0$. Therefore a convenient set of piezoelectric equations for the thin-walled ring are

$$D_3 = d_{31}T_1 + \epsilon_{33}{}^T E_3 \tag{6.61a}$$

$$S_1 = s_{11}{}^E T_1 + d_{31}E_3 \tag{6.61b}$$

If a harmonic electric field $E_3 = E_0 e^{j\omega t}$ is applied across the electrodes, the ring will respond harmonically with a motion $\xi = \xi_0 e^{j\omega t}$. Let us examine an instantaneous situation in which the ring has undergone a radial extension ξ and is moving with a radial velocity $d\xi/dt$. A stress T_1 in the wall is exerting a total radial force of $-2\pi wtT_1$, which is in equilibrium with the damping and accelerating force. The equation of motion for the ring is

$$m\frac{d^2\xi}{dt^2} + R_m\frac{d\xi}{dt} = -2\pi wtT_1 \tag{6.62}$$

From (6.61b)

$$T_1 = \frac{\xi}{s_{11}{}^E a} - \frac{d_{31}E_3}{s_{11}{}^E}$$

and (6.62)

$$m\frac{d^2\xi}{dt^2} + R_m\frac{d\xi}{dt} + \frac{2\pi wt}{s_{11}{}^E a}\xi = \frac{2\pi wtd_{31}}{s_{11}{}^E}E_3 \tag{6.63}$$

$$(-m\omega^2 + j\omega R_m + s)\xi = \frac{2\pi wtd_{31}}{s_{11}{}^E}E_3 \tag{6.64}$$

Eliminating T_1 from (6.61a) and (6.61b), we have

$$D_3 = \frac{d_{31}}{s_{11}{}^E a}\xi + \epsilon_{33}{}^T E_3 - \frac{d_{31}{}^2}{s_{11}{}^E}E_3$$

$$D_3 = \frac{d_{31}}{s_{11}{}^E a}\xi + \epsilon_{33}{}^T(1 - k_{31}^2)E_3 \tag{6.65}$$

where

$$k_{31}{}^2 = \frac{d_{31}{}^2}{s_{11}{}^E\epsilon_{33}{}^T}$$

The total charge on the ring electrodes is $Q = 2\pi awD$ and the instantaneous current into the electrodes is the time derivative of Q, or $I = j\omega 2\pi awD$. Therefore from (6.64) and (6.65)

$$I = j\omega 2\pi aw\left[\frac{2\pi wtd_{31}{}^2}{(s_{11}{}^E)^2 a(j\omega R_m - m\omega^2 + s\omega^2)} + \epsilon_{33}{}^T(1 - k_{31}{}^2)\right]E_3 \tag{6.66}$$

$$I = j\omega C_0 V + \left(\frac{2\pi wd_{31}}{s_{11}{}^E}\right)^2\frac{V}{R_m + j(\omega m - s/\omega)} \tag{6.67}$$

where

$$C_0 = \frac{2\pi aw}{T}\epsilon_{33}{}^T(1 - k_{31}{}^2)$$

The second term of (6.67) may be written as ϕ^2V/Z_m, where ϕ is the equivalent turns ratio for an electromechanically coupled circuit and V is the voltage across the input terminals. The equivalent circuit is shown in Figure 6.23. Also in Figure 6.24 we have a diagram of the admittance of an unloaded ring with R_m of (6.67) quite small. The two frequencies on this diagram, easily determined and useful as a criterion of the quality of the piezoelectric material, are f_m and f_n. These two frequencies are only a few cycles removed from f_r and f_a, the resonant and antiresonant frequencies. The resonant

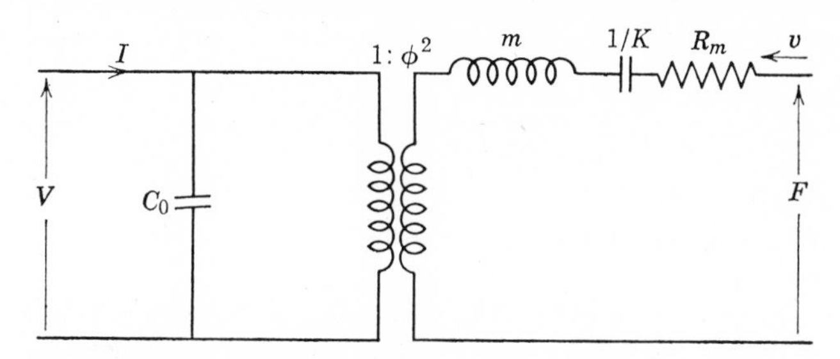

FIG. 6.23 Equivalent circuit for a piezoceramic ring. $\phi = 2\pi w d_{31}/s_{11}^E$; $m = 2\pi awtp$; $R_m = \gamma_r 2\pi aw$; $C_0 = (2\pi aw/t)\epsilon_{33}^T \times (1 - k_{31}^2)$; $k_{31}^2 = d_{31}^2/s_{11}^E\epsilon_{33}^T$.

frequency f_r is defined as the frequency at which the susceptance term vanishes. Antiresonance occurs at that fequency f_a, where the susceptance again vanishes. From (6.67), and Fig. 6.23 omitting R_m as negligibly small,

$$Y = j\omega C_0 + \frac{\phi^2}{j(\omega m - s/\omega)} = j\omega C_0 + \left[j\left(\omega L - \frac{1}{\omega C}\right)\right]^{-1}$$
$$Y = \frac{-\omega^2 LC_0C + C_0 + C}{jC(\omega L - 1/\omega C)} \tag{6.68}$$

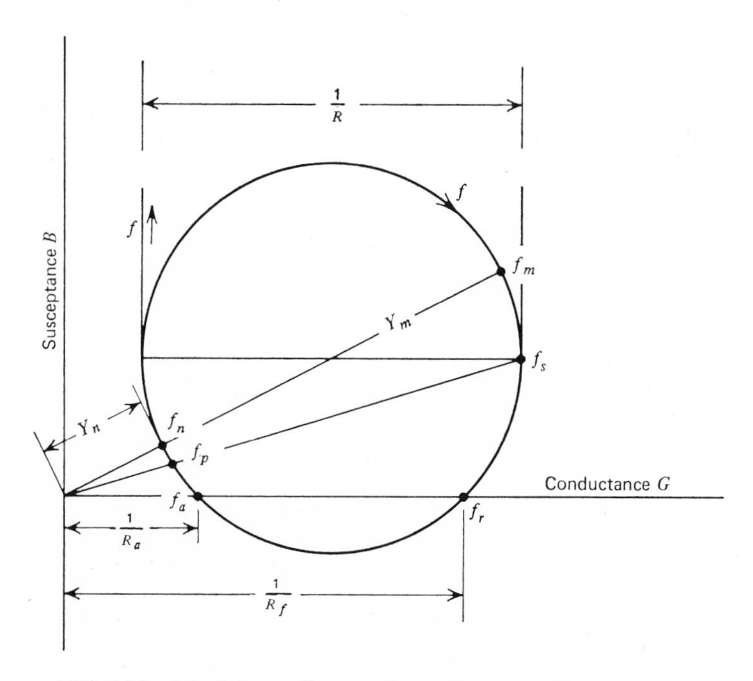

FIG. 6.24 Admittance diagram for a piezoceramic transducer.

for minimum admittance Y_a the numerator of (6.68) is set to zero, or

$$\frac{C}{C_0 + C} = k_{31}^2 = \frac{f_a^2 - f_r^2}{f_a^2} \tag{6.69}$$

6.12.1 The segmented ceramic ring

Figure 6.25 shows a ceramic ring consisting of equally dimensioned segments. Electrodes between segments are properly connected to external circuitry so that the exciting electric field has a polarity in agreement with the established polarity of the segments. The two advantages to be gained by this type of construction are an improved electromechanical coupling coefficient and it avoids the size limitation due to the problems of firing large ceramic pieces. To compensate for the weaker wall structure the ring is held in compression by some means such as an outside layer of glass fiber in tension. This is not a perfect solution to the problem because it is not only difficult to apply but it reduces the electromechanical coupling and introduces a discontinuity in the wall structure.

The electromechanical equations of state for the fundamental mode are quite simple for a thin-walled, segmented ring because there is no important cross coupling and the fields are uniform about the ring circumference. In the coordinate system shown by Figure 6.25 we have

$$S_3 = s_{33}{}^E T_3 + d_{33} E_3 \tag{6.70a}$$

$$D_3 = d_{33} T_3 + \epsilon_{33}{}^T E_3 \tag{6.70b}$$

The one troublesome coefficient in these equations is the elastic modulus $s_{33}{}^E$ because of the segmented wall. An empirical value may be established

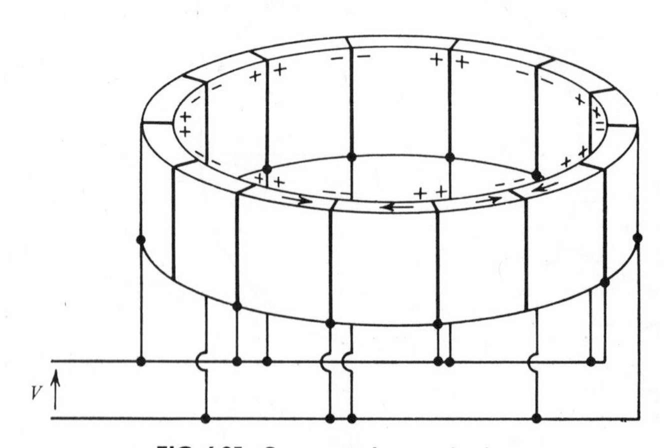

FIG. 6.25 Segmented ceramic ring.

by experiment. Without going into detail we shall rewrite the equations of the previous section for a ring of N segments. For a radial displacement ξ the stress developed in the ring wall is

$$T_3 = \frac{\xi}{a s_{33}^E} - \frac{d_{33}E_3}{s_{33}^E} \tag{6.71}$$

and the equation of motion for the ring is

$$m \frac{d^2\xi}{dt^2} + R_m \frac{d\xi}{dt} + K\xi = -2\pi w t T_3 \tag{6.72}$$

where the coefficients are, as before

$$m = 2\pi a w t \rho$$

$$R_m = \gamma_r 2\pi a w$$

$$K = \frac{2\pi w t}{a s_{33}^E}$$

From (6.71) and (6.72) we have

$$m \frac{d^2\xi}{dt^2} + R_m \frac{d\xi}{dt} + \frac{2\pi w t}{a s_{33}^E}\xi = \frac{2\pi w t d_{33}}{s_{33}^E} E_3 \tag{6.73}$$

If the applied field is a simple harmonic function of time, the steady rate solution to (6.73) is

$$\xi = \frac{2\pi w t d_{33}}{s_{33}^E j\omega Z_m} E_3 \tag{6.74}$$

and from (6.70a) and (6.70b)

$$D_3 = \frac{d_{33}}{a s_{33}^E}\xi + \epsilon_{33}^T(1 - k_{33}^2)E_3 \tag{6.75}$$

With this ring construction, the electromechanical coupling coefficient has the value

$$k_{33}^2 = \frac{d_{33}^2}{s_{33}^E \epsilon_{33}^T} \tag{6.76}$$

which is usually quite superior to k_{31}^2. The current I into the ring is the time derivative of the product of D_3 by the electrode area Nwt or

$$I = j\omega N w t D_3 \tag{6.77}$$

or

$$I = j\omega C_0 v + \left(\frac{N w t d_{33}}{a s_{33}^E}\right)^2 \frac{v}{Z_m} \tag{6.78}$$

The equivalent circuit for the segmented ring is that shown by Figure 6.23

with the difference that

$$\phi = \frac{Nwtd_{33}}{as_{33}{}^E} \tag{6.79}$$

$$C_0 = \frac{N^2wt}{2\pi a}\epsilon_{33}{}^T(1 - k_{33}{}^2) \tag{6.80}$$

$$k_{33} = \frac{d_{33}{}^2}{s_{33}{}^E\epsilon_{33}{}^T} \tag{6.76}$$

6.13 THE PIEZOCERAMIC LONGITUDINAL VIBRATOR

Recent developments in sound sources of a directional type have been almost exclusively involved with piezoceramic materials. The physical structure of these devices may be quite simple, consisting of a center section of active material which works between an inertial mass and a radiating diaphragm. Figure 6.26 shows a diagrammatic arrangement of the components. In addition to the parts shown, there is usually a rod under tension through the center attached to front and back components for the purpose of holding the system under compression at all vibration levels. The relative dimensions and masses of the components determine the bandwidth and power capability of the set. A large number of these units may be used to form an array, planar, cylindrical, spherical, or conformal so that the radiating face will have a geometry suited to the desired mosaic into which it must fit.

The active center section may be one continuous tube with electrodes on inner and outer surfaces with polarization and electric field applied normal to these surfaces. This utilizes the 3–1 coupling. The active section may consist of an assembly of rings with electrodes on the ends, one of which is shown in Figure 6.27. A number of these rings bonded together and held under compression by a tension rod make a tube of whatever length desired

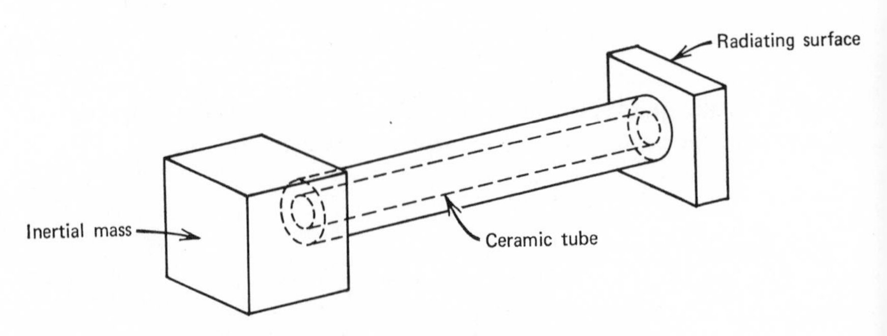

FIG. 6.26 Piezoceramic longitudinal vibrator.

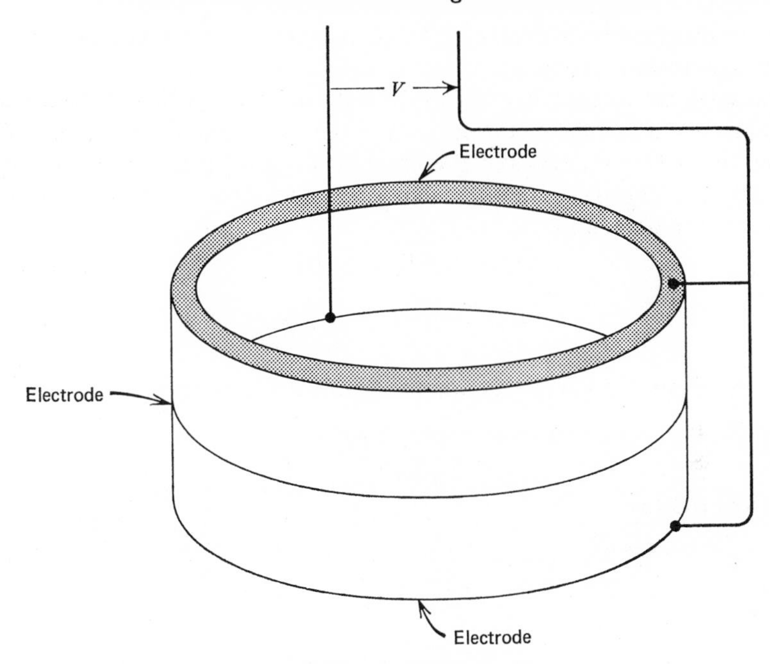

FIG. 6.27 End-plated rings.

utilizing the 3–3 coupling. In designing these structures we may approach them as systems with either lumped or distributed constants. The former involves less work and usually is as accurate as our knowledge of the physical constants of the materials. We may observe at this point that the electric field conditions for the segmented rod with electrodes connected in parallel are not those of a continuous end-plated rod.

As an introduction to these systems, let us examine the fundamental mode of vibration of the thin uniform bar driven by an electric field parallel to its length shown in Figure 6.28. One is not likely to find an exact duplicate of this bar in use because its length would require an extremely high voltage for polarization and the impedance at the electrical terminals would also be too

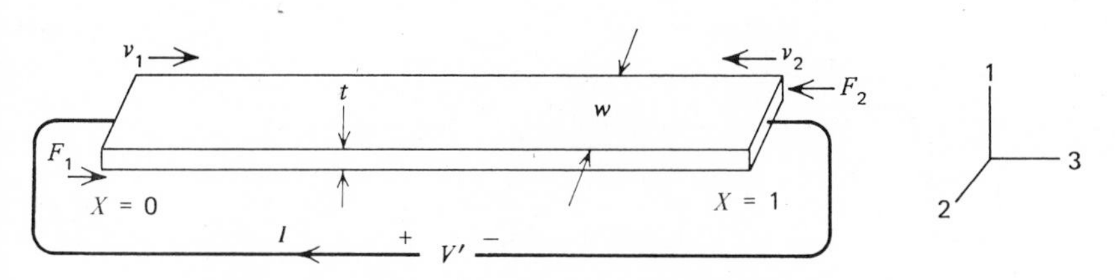

FIG. 6.28 Bar with electric field parallel to motion.

high. A modified form consisting of an assembly of short, end-plated segments avoids these problems.

The small coordinate diagram to the right in Figure 6.28 and the arrows adjacent to F and v define directions and sign conventions. Since the sides of the bar are free of stress and D is constant along the length of the bar, it is convenient to choose T and D as independent variables so the piezoelectric equations are limited to

$$S_3 = s_{33}{}^D T_3 + g_{33} D_3 \tag{6.81a}$$

$$E_3 = -g_{33} T_3 + \frac{1}{\epsilon_{33}{}^T D_3} \tag{6.81b}$$

From the equation of motion for an incremental section of the uniform bar

$$\rho \frac{\partial^2 \xi}{\partial t^2} = \frac{\partial T_3}{\partial x}$$

and from (6.81a)

$$\frac{\partial T^3}{\partial x} = \frac{1}{s_{33}{}^D} \frac{\partial}{\partial x}\left(\frac{\partial \xi}{\partial x} - g_{33} D_3\right) = \frac{1}{s_{33}{}^D} \frac{\partial^2 \xi}{\partial x^2}$$

yielding

$$\rho \frac{\partial^2 \xi}{\partial t^2} = \frac{1}{s_{33}{}^D} \frac{\partial^2 \xi}{\partial x^2} \tag{6.82}$$

or

$$\frac{\partial^2 \xi}{\partial t^2} = c^2 \frac{\partial^2 \xi}{\partial x^2}$$

It was also shown in Chapter 2 that for a sinusoidal excitation of the bar, the solution to the above equation of wave motion is

$$\xi = (A \sin kx + B \cos kx) e^{j\omega t} \tag{2.21}$$

where A and B from the boundary conditions established by Figure 6.28 have the values

$$A = -\frac{1}{j\omega}\left(\frac{v_2}{\sin kl} + \frac{v_1}{\tan kl}\right) \tag{6.83}$$

$$B = \frac{1}{j\omega} v_1 \tag{6.84}$$

at $x = 0$, $F_1 = -wtT_3$ and using (6.81a), (6.83), and (2.21)

$$F_1 = Z_0\left(\frac{v_1}{j \tan kl} + \frac{v_2}{j \sin kl}\right) + \frac{g_{33}}{j\omega s_{33}{}^D} I \tag{6.85}$$

where Z_0 is the characteristic mechanical impedance of the bar.

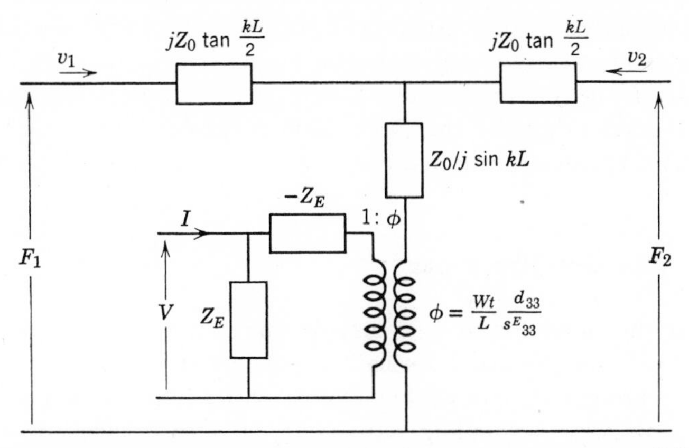

FIG. 6.29 Equivalent circuit end-plated bar. $k = \omega/c = \omega(\rho s_{33}^D)^{1/2}$; $Z_0 = wt(\rho/s_{33}^D)^{1/2}$; $Z_\epsilon = 1/j\omega C_0$; $C_0 = (wt/L)\epsilon_{33}^T(1 - k_{33}^2)$.

Also at $x = l$, $F_2 = -wtT_3$, or

$$F_2 = Z_0\left(\frac{v_1}{j \sin kl} + \frac{v_2}{j \tan kl}\right) + \frac{g_{33}}{j\omega s_{33}{}^D} I \tag{6.86}$$

The voltage across the bar is obtained from (6.81b)

$$V = \int_0^l E_3\, dx = \int_0^l \left(-g_{33}\frac{\partial \xi}{\partial x} + \frac{I}{\epsilon_{33}{}^T wtj\omega D_3}\right) dx \tag{6.87}$$

$$V = \frac{g_{33}}{j\omega s_{33}{}^D}(v_1 + v_2) + \frac{I}{j\omega C_0}$$

where

$$C_0 = \epsilon_{33}{}^T(1 - k_{33}{}^2)\frac{wt}{l} \tag{6.88}$$

and

$$k_{33}{}^2 = \frac{g_{33}{}^2}{s_{33}{}^D \epsilon_{33}{}^T} \tag{6.89}$$

The equivalent six-terminal network is shown in Figure 6.29.

6.14 VARIATIONS OF THE PIEZOCERAMIC LONGITUDINAL VIBRATOR

When the materials available for piezoelectric transducers were limited to the naturally occurring crystals such as quartz or the artificially grown crystals such as ammonium dihydrogen phosphate, practical variations were limited by size, strength, and dielectric properties; for example, the end-plated bar of Figure 6.28 would be found only in a quartz-metal sandwich. With the development of the piezoceramics, possibilities in design have exploited the engineer's imagination.

6.14.1 The side-plated bar

The bar shown by Figure 6.30 may be one of a number of piezoelectric crystals,[18] with silver electrodes usually applied by an evaporation technique, or it may represent a thin-walled cylindrical shell of one of the piezoceramic materials. Because of coupling between the possible resonant modes of the ceramic cylinder, the analysis presented for the long, thin bar may require modification.[19]

We shall limit the discussion to the more popular ceramics. The bar of Figure 6.30 has been polarized in the three-direction by the application of an electric field between the electrodes. This is also the direction of the exciting field. Since the electric field E and the stress T have components in only one direction, it is most convenient to set up the electromechanical equations as

$$S_1 = s_{11}{}^E T_1 + d_{31} E_3 \tag{6.90a}$$

$$D_3 = d_{31} T_1 + \epsilon_{33}{}^T E_3 \tag{6.90b}$$

Proceeding as with the end-plated bar, we arrive at the equivalent three-port net shown by Figure 6.31.

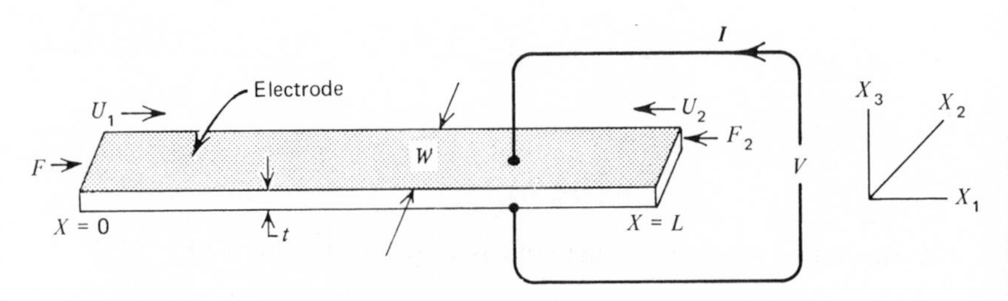

FIG. 6.30 Side-plated length expander.

[18] W. P. Mason, *Piezoelectric Crystals and their Applications to Ultrasonics*. Van Nostrand, Princeton, New Jersey, 1950.

[19] J. F. Haskins, and J. L. Walsh, "Vibrations of Ferroelectric Cylindrical Shells with Transverse Isotrophy: I Radial Polarized Case," in *J. Acoust. Soc. Am.*, **29**, 729 (1957).

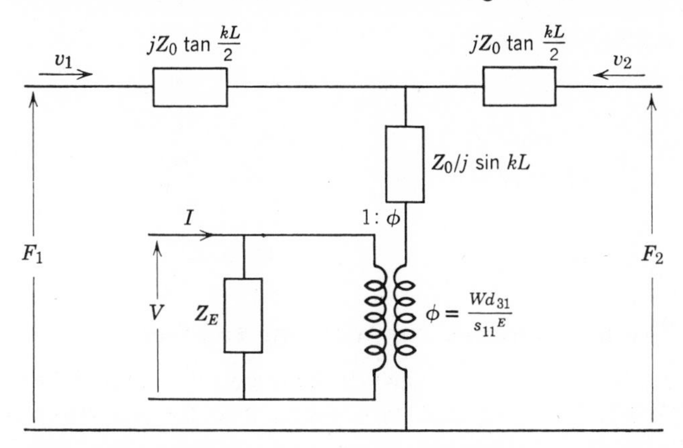

FIG. 6.31 Equivalent circuit side-plated bar. $k = \omega/c = \omega(Ps_{11}^{E})^{1/2}$; $Z_0 = wt(p/s_{11}^{E})^{1/2}$; $C_0 = (wt/t)\epsilon_{33}^{T}(1 - k_{31}^{2})$; $Z_E = 1/j\omega C_0$.

6.14.2 The piezoelectric bar with symmetrical load

If the piezoelectric bar is used as a sound source, the forces F_1 and F_2 are due to radiation impedance loading. Just as in the case of the ring transducer, this impedance may be strictly resistive in nature or it may contain a positive reactive component which acts as an additional mass. If the ends of the bar are equally loaded with a radiation impedance $\underline{Z_L}$, F_1, and F_2 may be replaced by $-\underline{Z_L}v$. Additional changes that convert the network of Figure 6.31 into that of Figure 6.32 are made through substitution of lumped constants for the mechanical impedances, as demonstrated in Chapter 2, Section 2.3.3. Of course, this substitution is valid only in the neighborhood of the

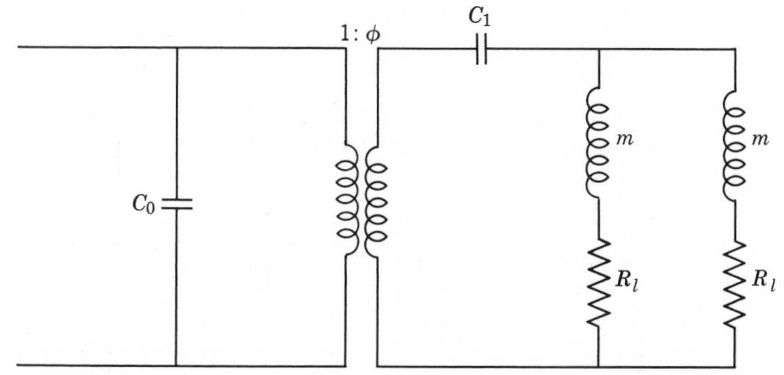

FIG. 6.32 Symmetrically loaded bar. $C_0 = wl\epsilon_{33}^{T}(1 - k_{31}^{2})$; $k_{31}^{2} = d_{31}^{2}/s_{11}^{E}\epsilon_{33}^{T}$; $Z_0 = Pwtc^{E}$; $\phi = wd_{31}/s_{11}^{E}$; $m = (pw + 1)/4$; $C_1 = 8ls_{11}^{E}/\pi^{2}wt$.

half-wave resonance of the bar.

$$C_0 = \frac{lw\epsilon_{33}^T}{t}(1 - k_{31}^2) \qquad k_{31}^2 = \frac{d_{31}^2}{s_{11}^E \epsilon_{33}^T}$$

$$Z_0 = \rho wtc^E \qquad \phi = \frac{wd_{31}}{s_{11}^E}$$

$$m = \frac{wlt\rho}{2}, \qquad C_1 = \frac{4ls_{11}^E}{\pi^2 wt}$$

6.14.3 The piezoceramic bar with one end free

Setting the force F_1 to zero converts the circuit of Figure 6.29 to that of Figure 6.33, which is simplified by conversion to the equivalent circuit of

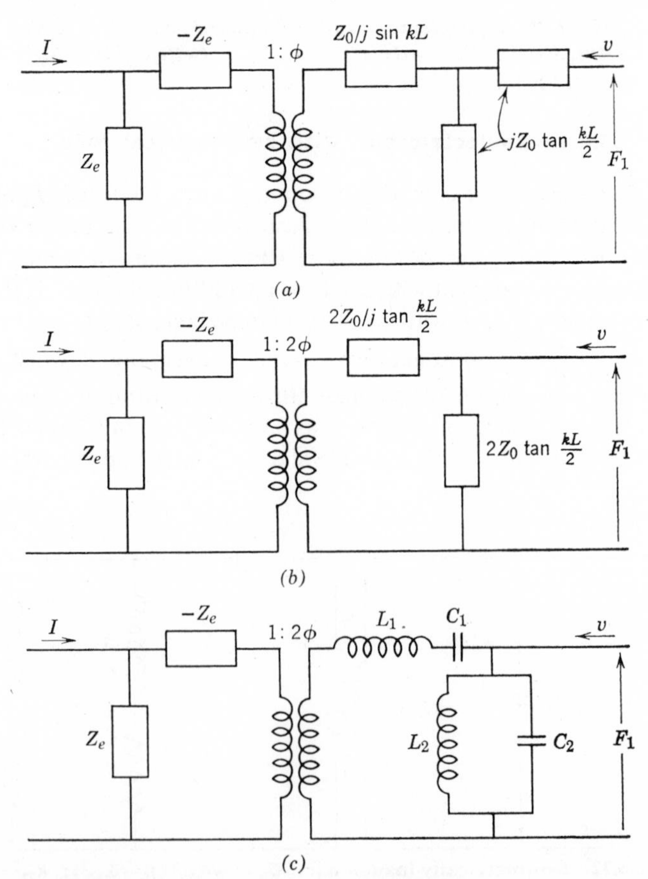

FIG. 6.33 Equivalent circuits; end-plated piezoceramic bar with one end free. $C_1 = 2ls_{33}^E/\pi^2 wt$; $C_2 = ls_{33}^E/8wt$; $\phi = wtd_{33}/ls_{33}^E$; $L_1 = wtl\rho/2$; $L_2 = (8/\pi^2)wtl\rho$.

Figure 6.33*b*. Again in the vicinity of resonance the mechanical impedances may be represented by the equivalent lumped constants shown in Figure 6.33*c*, where

$$C_1 = \frac{2ls_{33}^E}{\pi^2 wt} \qquad L_1 = \frac{wtl\rho}{2}$$

$$C_2 = \frac{ls_{33}^E}{8wt} \qquad L_2 = \frac{8}{\pi^2} wtl\rho$$

$$\phi = \frac{wtd_{33}}{ls_{33}^E}$$

6.14.4 The composite piezoelectric resonator

The structure shown diagrammatically by Figure 6.34 has several advantages:

1. Because of its great compressive strength, the ceramic tube may be prestressed by the tension rod which adds to its stability under changing ambient pressures, but assures compression under all driving conditions.
2. The active section consists of an assembly of rings which being end-electroded and polarized axially employ the highest available coupling coefficient.
3. The choice of the number and manner of connecting the rings offers a wider range of impedance and frequency.
4. As outlined in Chapter 2, the use of head and tail masses permits control by the designer of transducer bandwidth.

A general definition of the equivalent circuit for the composite resonator is not practical because each application is a specific case. If the electrodes are connected in parallel and are numerous, we have the condition that E_3 is approximately constant along the bar while D_3 is a function of the electrode position. An interesting but rather involved representation would be that of

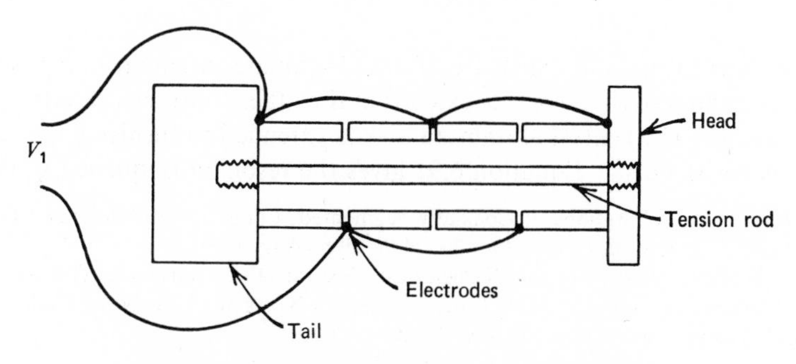

FIG. 6.34 Composite piezoceramic resonator.

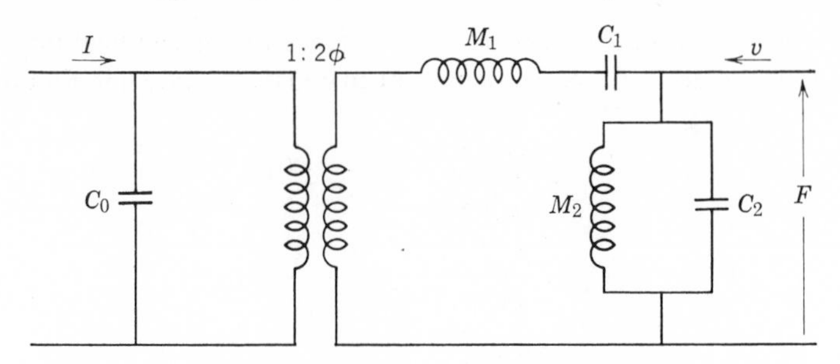

FIG. 6.35 Equivalent circuit for a composite resonator.

a multiport net.[20] A less formidable but an approximate approach is shown by Figure 6.35, where $M_1C_1 = M_2C_2$ is the one necessary condition. Since the composite bar is not a symmetrical structure, the impedances in the mechanical meshes are unequal.

6.15 THE PIEZOELECTRIC FLEXURAL DISK TRANSDUCER

One of the most difficult problems confronting the engineer in underwater acoustics is that of developing a low-frequency source of reasonable size. The smallest piezoceramic ring, of lead-zirconate-titanate, at 350 Hz has a diameter of about 10 ft. A longitudinal resonator at this frequency would have a length approaching 15 ft. In looking for a resonant system of a smaller size, one must not ignore the area requirement for an effective radiation impedance. The piezoelectric flexural disk is an unusually tractable system with permissible parameters covering a wide frequency range and easily predictable radiation-impedance characteristics. The critical design problems of bandwidth, reliability, directivity, and power capability are discussed in unpublished reports.[21] For an introduction to the possibilities of this design we may consider the example of Figure 6.36 from a USL report by Johnson and Woollett.

$$f_{ra} = \frac{813}{a}\left(\frac{h}{a}\right) \quad \text{using aluminum with PZT and} \quad \frac{m}{h} = 3 \qquad (6.91)$$

The two outer layers of the disk are of a lead–titanate–zirconate composition poled and excited so that one side extends as the other contracts, resulting in a flexing action. The central aluminum disk is extended to furnish a free edge mounting for the disk. Equation 6.91 gives the resonant frequency f_{ra} in air

[20] G. E. Martin, "Vibrations of Coaxially Segmented, Longitudinally Polarized Ferroelectric Tubes," in *J. Acoust. Soc. Am.*, **36,** 1496–1506 (Aug. 1964).

[21] R. S. Wollett "Theory of the Piezoelectric Flexural Disk Transducer with Applications to Underwater Sound." USL Research Report No. 490. U.S. Navy Underwater Sound Laboratory, Fort Trumbull, New London, Conn.

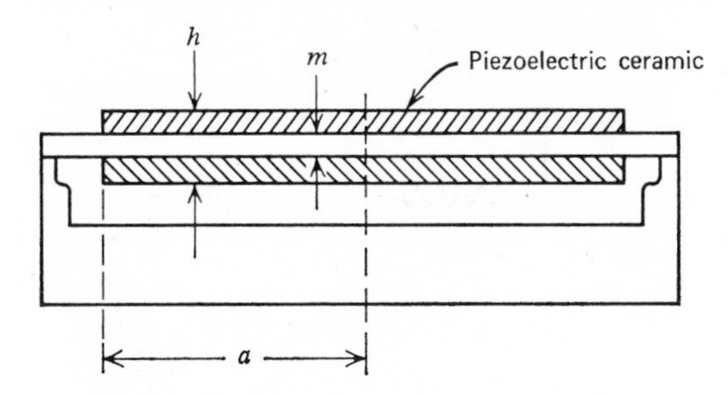

FIG. 6.36 Supported edge trilaminar disc.

for a family of disks. The dimensions are in meters. Methods[21] for mounting to eliminate back radiation and to support high ambient pressure have been developed. Unfortunately the effect of these measures on bandwidth and frequency are not accurately predictable.

6.16 IRE STANDARDS ON PIEZOELECTRIC CRYSTALS

Even a superficial coverage of the properties of the piezoelectric materials used for underwater sound sources and receivers would require an extended text. Instead, we may take advantage of the quite generally available, carefully planned and edited discussions published in the Proceedings of the IRE. Briefly, these begin with the page number listed below with the volume number: Vol. 37, p. 1378; Vol. 45, p. 354; Vol. 46, p. 765; Vol. 49, p. 1162. The later issues are involved principally with piezoceramics, so the first named is suggested as an essential introduction to the general topic.

In order to pass judgment on the quality of a material before it becomes a transducer component, a limited number of its physical constants must be determined and compared with accepted standards. For the materials manufacturer, disks or blocks may serve as suitable samples. Particularly for the ceramics where uniformity is poor, tests should be made when possible on parts as used. Both static and resonance methods of testing are outlined in the standard procedures.[22] The latter are to be preferred. Figure 6.24 will serve to clarify useful parameters defined by the IRE Standards.

In order of appearance on the chart

f_m = frequency at maximum admittance
f_s = motional series resonance frequency
f_r = resonance frequency
f_a = antiresonance frequency
f_p = parallel resonance frequency
f_n = frequency at minimum admittance

[22] *Ibid.*

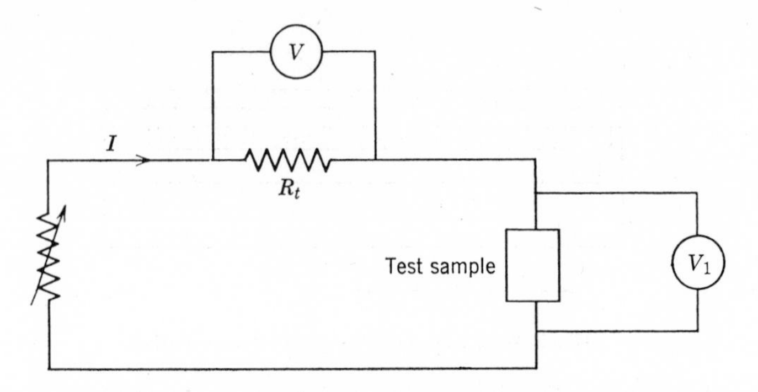

FIG. 6.37 Admittometer.

The data for Figure 6.24 acquired by the use of an admittance bridge involves an amount of effort that may be justified only in the evaluation of a finished product. Using Figures 6.24 and 6.37, we can show how the critical parameter may be more easily determined. The effective electromechanical coupling factor k is given by

$$k^2 = \frac{f_p^{\ 2} - f_s^{\ 2}}{f_p^{\ 2}} \tag{6.92}$$

An examination of Figure 6.24 shows that

$$\frac{f_n^{\ 2} - f_m^{\ 2}}{f_n^{\ 2}} \simeq \frac{f_p^{\ 2} - f_s^{\ 2}}{f_p^{\ 2}} \tag{6.93}$$

This condition holds for unloaded parts of sufficiently good quality to be usable. We may see that if the signal generator of Figure 6.37 applies a constant voltage V_1 and the series resistor $\underline{R_T}$ is smaller than the minimum impedance of the test sample, the current through R_T and the voltage across it will vary with frequency as the absolute value of the admittance of the sample, thereby describing the admittance locus shown by Figure 6.38. The information from this locus and an evaluation of the capacity C_0 of the specimen at a frequency well below resonance is usually all the data needed for proper inspection of the part. The locus over this limited frequency range resembles that of a resonator having one degree of freedom, and departures from this typical shape may be due to minor resonance effects from faulty structures. The difference between the absolute values of admittance at f_m and f_n is the diameter of the admittance circle $1/\underline{R}$. This value allows a computation of Q_m, the mechanical quality factor, from

$$Q_m = \frac{1}{2\pi f_s CR} \tag{6.94}$$

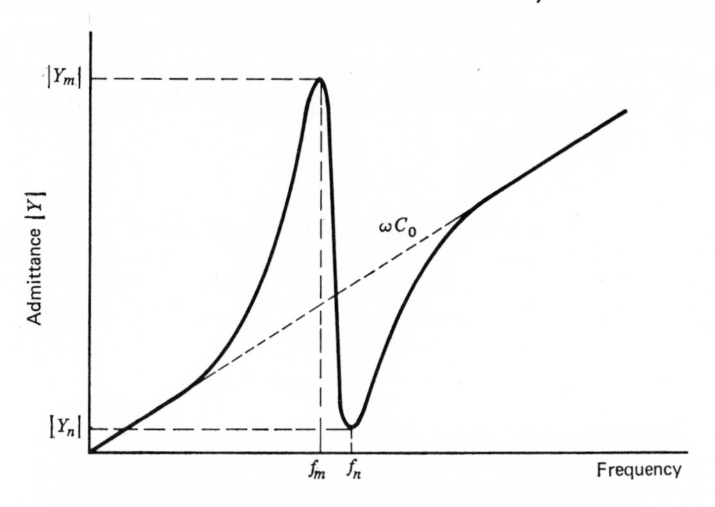

FIG. 6.38 Admittance locus indicated by admittometer.

where

$$C = \frac{C_0(f_p^2 - f_s^2)}{f_s^2} \tag{6.95}$$

From measurements made on side-electroded thin ceramic cylinders or thin bars

$$\frac{k_{31}^2}{1 - k_{31}^2} = \frac{\pi}{2}\frac{f_p}{f_s}\tan\frac{\pi}{2}\frac{\Delta f}{f_s} \tag{6.96}$$

where $\Delta f = f_p - f_s$. Also

$$k_{31}^2 = \frac{d_{31}^2}{s_{11}^E \epsilon_{33}^T} \tag{6.97}$$

and

$$g_{31} = \frac{d_{31}}{\epsilon_{33}^T} \tag{6.98}$$

From end-electroded thin ceramic bars or cylinders

$$k_{33}^2 = \frac{\pi}{2}\frac{f_s}{f_p}\tan\frac{\pi}{2}\frac{\Delta f}{f_p} \tag{6.99}$$

$$d_{33} = k_{33}(\epsilon_{33}^T s_{33}^E)^{1/2} \tag{6.100}$$

and

$$g_{33} = \frac{d_{33}}{\epsilon_{33}^T}$$

Table 6.1 lists a few of the properties of the most commonly used piezoelectric materials. A great deal of additional information may be obtained from the manufacturers, each of whom, in the case of the ceramics, has his own proprietary composition.

TABLE 6.1

Characteristics of Piezoelectric Material

Property symbol		Ba^{2+}, Ca^{2+}, TiO_3^{2-}	Pb^{2+}, Zr^{2+}, TiO_3^{2-}
Density,	ρ kg/m^3	5500	7600
Curie point		120°C	300°C
	Q_m	425	500
Dielectric Constants	$\epsilon_{33}{}^T/\epsilon_0$	1250	1300
	$\epsilon_{33}{}^S/\epsilon_0$	950	675
Piezoelectric	$d_{31}(10^{-12}$ m/v)	−58	−125
Constants	$d_{33}(10^{-12}$ m/v)	150	270
	$g_{31}(10^{-3}$ vm/n)	−5.5	−11
	$g_{33}(10^{-3}$ vm/n)	15	25.5
Compliance	$s_{11}{}^E(10^{-12}$ m^2/n)	8.6	12.3
Compliance	$s_{33}{}^E(10^{-12}$ m^2/n)	9.1	15.5
Coupling	k_{31}	−0.19	−0.31
Coefficients	k_{33}	0.46	0.65

		Quartz X-Cut	ADP 45°Z Cut	Rochelle Salt 45°X Cut
Density	ρ kg/m^3	2650	1800	1700
Dielectric Constant	ϵ^T/ϵ_0	4.5	15.3	350
Coupling Coefficient	k	0.1	0.29	0.67
Compliance	$s(10^{-12}$ m^2/N)	12.7	52	56

PROBLEMS

6.1 Under static conditions the strain $\Delta l/l$ in nickel is given approximately by $S = KB^2$ where $K = -1.0 \times 10^{-4}$ $(m^2/\text{Wb})^2$. According to Figure 6.17, what is the elongation in a 0.2-m bar resulting from an increase of H to 20 oersteds? The yield strength of annealed nickel is 7000 lb/in.^{2}, and its elastic modulus is 20×10^{10} N/m^2. Show that forces induced by static magnetic flux cannot exceed the yield strength. Also give conditions under which dynamic forces produced by an alternating flux may exceed the yield strength.

6.2 (*a*) Using the data from Figure 6.17, calculate the number of ampere turns needed to establish a flux density of 5600 gauss in a laminated ring core having a mean diameter of 11.5 in. (*b*) After introducing a $\frac{1}{4}$-in. gap into the ring of part (*a*) plot the locus of flux density in the ring versus total ampere turns, using the data of Figure 6.17.

6.3 Given the following data for nickel 204, $\rho_m = 8700$ kg/m^3, $c_m = 4600$ m/sec, $u_r = 78$, (*a*) find the resonant frequency in air of the ring described in problem 1.2. This ring has a wall 0.3-in. thick and 0.96-in. high. Sixteen of these are stacked to form a cylinder and wound with 172 turns. (*b*) Plot the clamped core impedance from 0 to 10 kHz using the equivalent circuit of Figure 6.9. The lamination thickness is 0.005 in.

6.4 Figure 6.18 is the motional impedance of the assembly described in problem 6.2. (*a*) Compare the value of $|Z_{me}|$ calculated from the circle with that given by (6.39). (*b*) Given $f_y = 5.4$ kHz and $f_r = 5.0$ kHz, compute the effective electromechanical coupling coefficient. (*c*) Find the magnetostrictive stress constant, h, for $u^S = 78 \times 4\pi \times 10^{-7}$ H/m and $c^B = 2 \times 10^{11}$ N/m^2.

6.5 The quality of piezoceramic materials as measured by voltage or stress breakdown varies a great deal from one material to another and from sample to sample of the same material. One procedure for polarizing barium titanate is to impose upon it a dc electric field of the order of 40,000 V/in. Find the resulting strain, and the compressive force required to prevent this strain. Suppose that an alternating field of 7 V/ml were imposed on a half-wave bar at its resonant frequency, and the bar is so loaded resistively that it has a mechanical Q of 4. Find the maximum strain in the bar.

6.6 The frequency of mechanical resonance of a half-wave longitudinal resonater is affected by its electrical boundary conditions. Find the lengths for 15 kHz half-wave bars of lead titanate zirconate: (*a*) electric field (E constant) perpendicular to length, (*b*) electric field (D constant) parallel to length.

6.7 A segmented piezoceramic ring consists of 18 barium titanate bars with proper dimensions to give the ring a mean diameter of 0.125 m, a wall thickness of 0.01 m, and a height of 0.03 m. The electrodes are in parallel as in Figure 6.25. Given the dielectric loss tangent as 0.02 and the mechanical quality factor $Q_m = 200$, find all components of the equivalent circuit.

6.8 A composite piezoceramic resonator similar to that shown by Figure 6.34 has an active section of PZT tubes of outside diameter 0.075 m and wall thickness 0.0125 m. A forward mass of 0.8 kg acts as the radiating face. The rear inertial mass of 24 kg is free. The mass and stiffness of the tension rod may be ignored. (*a*) Find the length of tube needed to maintain the resonant frequency at 1500 kg. (*b*) A three-eighths inch rod with an elastic modulus of 2×10^{12} N/m^2 imposes a static pressure of 2000 psi on the tube. Find the maximum tensile stress on the rod for a displacement of the radiating face of 10^{-5} m.

CHAPTER 7

Radiation Patterns

7.1 INTRODUCTION

The American Standard Acoustical Terminology defines the directional response pattern of a transducer used for sound emission or reception as the "description, often presented graphically, of the response of the transducer as a function of the direction of the transmitted or incident sound waves in a specified plane and at a specified frequency." Simple sources, as described in Chapter 3, and uniformly vibrating spherical shells radiate energy equally in all directions, and therefore have no directional characteristics, but distributed arrays of sources or receivers, or extensive surfaces serving as such, have unique radiation patterns. Pattern control serves a number of useful purposes and is a major design problem. Most obvious is that the concentration of energy in a searchlight type of beam not only provides a higher intensity level but also gives directional information. This is particularly useful in target location and for acoustically guided missiles. By the use of phasing techniques beams may be shifted, split, rotated, or otherwise manipulated to get information or guidance advantage. A directional receiver also discriminates against isotropic noise in the sea, thereby improving the signal-to-noise ratio.

In this chapter we shall outline methods of approach to pattern design. Since antennas for the radiation of electromagnetic energy involve similar problems, we may profit by contributions from that area.

7.2 RECORDING A DIRECTIVITY PATTERN

The directivity pattern of a transducer not only contains the information essential for computing its efficiency in energy conversion, but also its performance in meeting its unique design specifications. To assure a universal acceptance of data-measuring performance, American Standards has described in detail the procedure for calibration of electroacoustic transducers in Z24.24-1957.

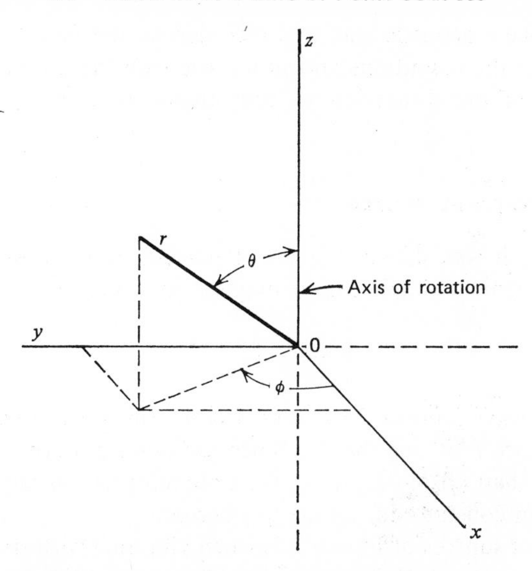

FIG. 7.1 Coordinate system for transducer calibration.

Theoretically the transducer under test is suspended in an unlimited three-dimensional space with no reflecting surfaces. Calibration stations have employed elaborate arrangements to achieve this condition. The coordinate system, shown in Figure 7.1, is a result of practical means of handling equipment. A transmitting transducer stationed at the origin of this system develops a sound field at a fixed distance r that is a function only of the angles ϕ and θ when r is quite large compared with any dimension of the transducer. Instead of moving a measuring instrument about in space at a fixed distance from the origin, it is more practical to rotate the transmitter about an axis while holding the measuring instrument fixed. This yields a pattern in only the one plane normal to the axis of rotation so patterns in several other planes may be made by shifting the transducers about this axis to give a more complete picture of the radiation field. The pattern of a receiving transducer or hydrophone is obtained by replacing the measuring instrument by a sound source.

Most electroacoustic transducers are reciprocal in the sense that their receiving and transmitting patterns are identical.

7.3 PATTERNS OF A LINE OF POINT SOURCES

Effective acoustic pressure is the quantity usually recorded as the radiation pattern of a transmitter, and the open circuit voltage developed by a receiver in a field of plane acoustic waves is recorded as the pattern of the latter. First we outline the conditions and procedure with the use of point sources and then extend the discussion to continuous, or arrays of continuous, surfaces.

7.3.1 A two-point source

In Chapter 3 it was shown that the effective acoustic pressure developed at a distance r from a simple source may be expressed by

$$p = \frac{A}{r} e^{j(\omega t - kr)} \tag{7.1}$$

where k is the wave number $2\pi/\lambda$, and A is the rms pressure amplitude at a reference distance r of one meter. Since radiation patterns are records of relative rather than absolute values, the only information required about A is that it remain constant during the test period.

The two point sources of Figure 7.2, when vibrating with simple harmonic motion in phase and with equal amplitudes, each develop at P, a distance r_0 from the coordinate center, chosen midway between the sources, an effective acoustic pressure defined by (7.1).

The total pressure at P is

$$p = \frac{A}{r_1} e^{j(\omega t - kr_1)} + \frac{A}{r_2} e^{j(\omega t - kr_2)} \tag{7.2}$$

As shown by Figure 7.2, lines from S_1 and S_2 to the point P would not be parallel. This is the condition existing in the Fresnel radiation zone, frequently referred to as the near field. However, if the distance to P is such that lines from S_1 and S_2 are essentially parallel, (7.2) may be simplified to

$$p(\theta) \frac{A}{r_0} e^{j(\omega t - kr_0)}(e^{j(kd/2)\sin\theta} + e^{-j(kd/2)\sin\theta}) \tag{7.3}$$

or

$$p = \frac{2A}{r_0} e^{j(\omega t - kr_0)} \cos \Psi \tag{7.4}$$

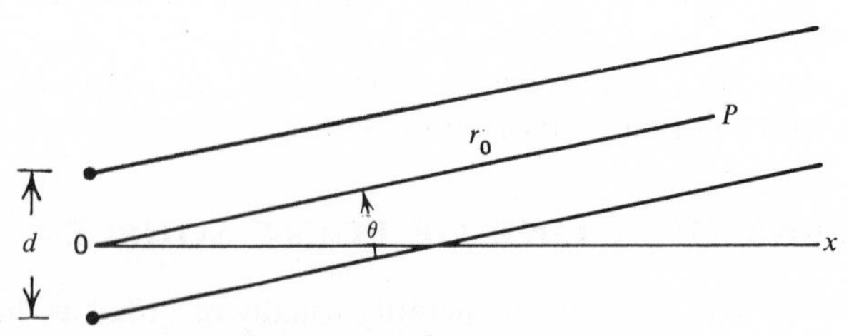

FIG. 7.2 Two simple sources.

where

$$\Psi = \frac{\pi d}{\lambda} \sin \theta \tag{7.5}$$

The pertinent question at this point involves what distance r_0 allows the approximation introduced by (7.3). To some extent the answer may be arbitrarily determined by the circumstances. The distance r_1 in the coordinate system of Figure 7.2 is

$$r_1 = \left[r_0^{\,2} + \left(\frac{d}{2}\right)^2 - r_0 d \sin \theta \right]^{1/2} = r_0 \left(1 + \frac{d^2}{4r_0^{\,2}} - \frac{d}{r_0} \sin \theta\right)^{1/2} \tag{7.6}$$

or

$$r_1 \simeq r_0 - \frac{d}{2} \sin \theta \tag{7.7}$$

The degree of approximation is determined by the ratio of d to r_0. We may use this simple two-source system to define a number of terms generally employed to describe radiation patterns. The locus of (7.4) is the directivity pattern of two point sources operating as described. The line OX normal to the line of the sources is the system axis. Along the axis responses from sources equidistant from the origin or acoustic center of the system will meet in phase and add to a maximum p_0 where

$$p_0 = \frac{2A}{r_0} e^{j(\omega t - kr_0)} \tag{7.8}$$

for the two points. The resultant pressure at the point P relative to the pressure along the axis is

$$\frac{p}{p_0} = \cos x = \cos \left(\frac{\pi d}{\lambda} \sin \theta\right) \tag{7.9}$$

Equation (7.9) defines the locus of the directivity pattern in all planes containing the line joining the two sources. So the pattern in space may be obtained by revolving the locus of (7.9) about the line S_1S_2. For convenience a discussion and also the measurement of a pattern is limited to planes containing the acoustic axis of the array. Also a polar plot will usually show the pressure as a function of the angle of deviation from the axis over the limited range from $-\pi/2$ to $+\pi/2$. The remainder of the pattern, which may be thought of as outlining the back radiation, may be a repetition, as in the case of a line of sources, or quite different when produced by a radiating surface.

The two patterns of Figure 7.3 illustrate the effect on the pattern of changing the ratio d/λ or the distance between sources in terms of the acoustic wavelength. For a very small ratio of d/λ the two points would appear as one and have no directivity, while for a ratio larger than $\frac{1}{2}$, the pattern

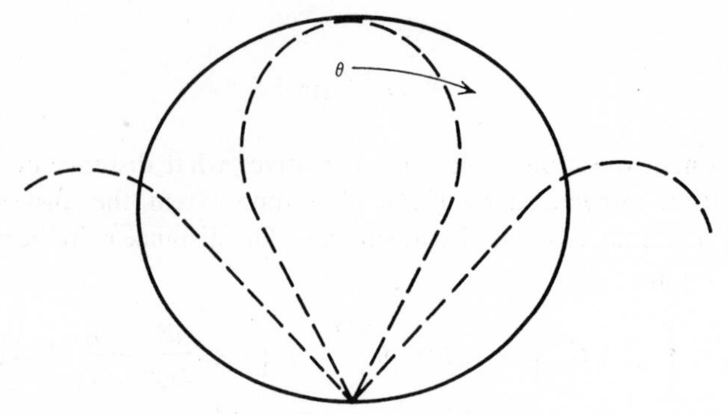

FIG. 7.3 Spacing two-point sources (——) $d = \lambda/2$; (- - - -) $d = \lambda$.

begins to repeat before the angle of deviation has reached $\pi/2$. If the spacing is one half wave the pattern is described by $P(\theta) = \cos(\pi/2 \sin\theta)$ and has nulls at $\theta = \pm\pi/2$. The section between the nulls is called the main beam or primary lobe. Additional beams resulting from greater spacing are called secondary lobes.

7.3.2 Multiple point sources

Let us assume equally spaced point sources arranged along a line as shown by Figure 7.4 and further that they are driven in phase with amplitudes that are symmetrical about a center point O. That is, points equally distant on either side of the center point are driven with equal amplitudes. Under the same conditions described in Section 7.3.1 the resultant pressure at the distant point P due to pairs of sources equally distant from O is given by (7.3):

$$p(\theta) = \frac{A}{r_0} e^{j(\omega t - kr_0)}\left[\exp\left(j\frac{\pi d}{\lambda}\sin\theta\right) + \exp\left(-j\frac{\pi d}{\lambda}\sin\theta\right)\right] \tag{7.3}$$

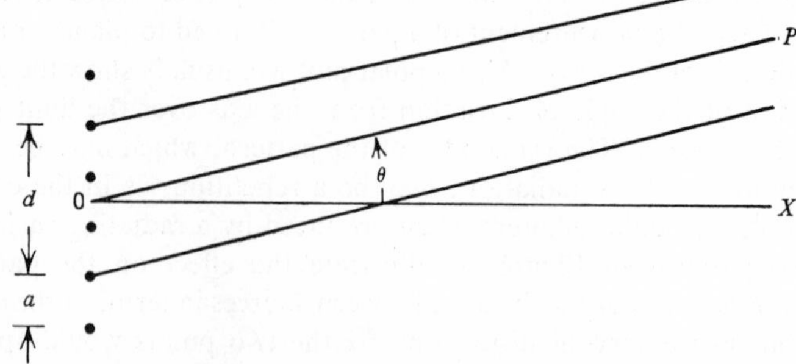

FIG. 7.4 A multiple point source.

where d is the distance between the chosen pair of points. A typical term for the contribution of the nth pair to the effective pressure at P is

$$p_n(\theta) = \frac{A_n}{r_0} e^{j(\omega t - kr_0)} \left\{ \exp\left[j\pi \frac{(2n-1)a}{\lambda} \sin\theta \right] + \exp\left[-j\pi \frac{(2n-1)a}{\lambda} \sin\theta \right] \right\} \tag{7.10}$$

where the total number of sources is $2n$, and a is the distance between adjacent points.

Following the example of (7.4),

$$p_n(\theta) = \frac{2A_n}{r_0} e^{j(\omega t - kr_0)} \cos(2n-1)\Psi \tag{7.11}$$

Under the assumption that the superposition principle holds throughout the sound field, we may sum the contribution of all sources to the pressure at P to get

$$p_t(\theta) = \frac{2}{r_0} e^{j(\omega t - kr_0)} \sum_{n=1}^{N/2} A_n \cos(2n-1)\Psi \tag{7.12}$$

for an even number of sources N. Before plotting the locus of (7.12) it is normalized by a division of its value at $\theta = 0$, where

$$p_t(0) = \frac{2}{r_0} e^{j(\omega t - kr_0)} \sum_{n=1}^{N/2} A_n \tag{7.13}$$

giving

$$P(\theta) = \left(\sum_{n=1}^{N/2} A_n \right)^{-1} \sum_{n=1}^{N/2} A_n \cos(2n-1)\Psi \tag{7.14}$$

For N an odd number of sources a similar development leads to

$$P(\theta) = \left[\sum_{n=0}^{(N-1)/2} \right]^{-1} \sum_{n=0}^{(N-1)/2} A_n \cos 2n\Psi \tag{7.15}$$

Baffles which may be either rigid or compliant surfaces adjacent to, or a part of sound sources are special boundary conditions which will affect to some extent the patterns demonstrated in the following discussion.

7.3.3 A continuous line source

A line of simple sources so densely assembled as to be considered a continuous line is shown by Figure 7.5. Let the total length of the line be l and its source strength per unit length be Q_s. Then the small section Δy

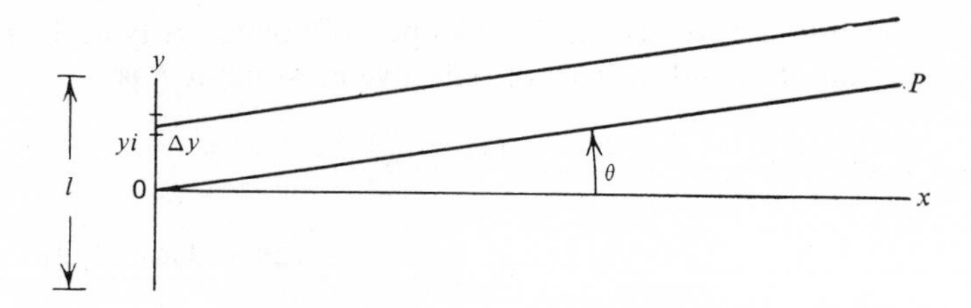

FIG. 7.5 A continuous line source.

at the point y_1 is developing at the distant point P a sound pressure

$$dp(\theta) = \frac{Q_s\,\Delta y}{r_0}\, e^{j(\omega t - ky_1 \sin\theta)} \tag{7.16}$$

where as before k is the wave number $2\pi/\lambda$.

Integration to get the total contribution of the line to the pressure at P

$$p_t(\theta) = \frac{Q_s}{r_0}\, e^{j\omega t} \int_{-l/2}^{l/2} e^{-jky\sin\theta}\, dy$$

$$= \frac{Q_s}{r_0}\, e^{j\omega t} \frac{l}{(kl\sin\theta)/2} \left[\frac{e^{-j(kl/2)\sin\theta} - e^{j(kl/2)\sin\theta}}{-2j}\right]$$

$$= \frac{Q_s l}{r_0}\, e^{j\omega t} \frac{\sin x}{x} \tag{7.17}$$

where $x = kl/2 \sin\theta$. Again it is necessary that the length of the line be quite small compared with the distance to P.

7.4 A SEGMENTED LINE SOURCE

If the source strength per unit length of the line of Figure 7.5 were a function of y, (7.17) would be written as

$$p_t(\theta) = \frac{1}{r_0}\, e^{j(\omega t - kr_0)} \int_{-l/2}^{l/2} Q_s(y) e^{-jky\sin\theta}\, dy \tag{7.18}$$

Equation 7.18 offers some interesting possibilities in shaping radiation patterns.

Figure 7.6 shows a segmented line. Segments may vary from each other in source strength per unit length and in length but for our purpose their arrangement must be symmetrical about the midpoint O, that is, segments at equal distances from either side of O must be identical. Selecting the segment of length S_i and source strength Q_i whose center is at y_i, its contribution to

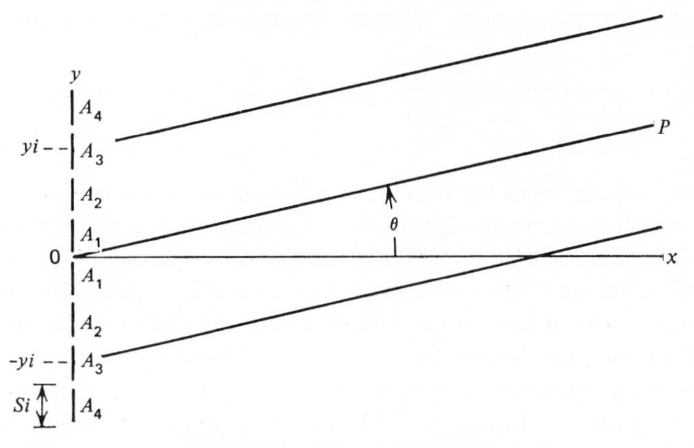

FIG. 7.6 A line of segments.

the acoustic pressure at far distant P is

$$
\begin{aligned}
p_i(\theta) &= \frac{Q_i}{r_0} e^{j(\omega t - kr_0)} \int_{(y_i - S_i/2)}^{(y_i + S_i/2)} e^{+jky \sin \theta}\, dy \\
&= \frac{Q_i}{r_0} e^{[\omega t - k(r_0 - y_i \sin \theta)]} \left[\frac{e^{jk(S_i/2)\sin\theta} - e^{-jk(S_i/2)\sin\theta}}{jk \sin \theta} \right] \\
&= \frac{Q_i S_i}{r_0} \exp\{j[\omega t - k(r_0 - y_i \sin \theta)]\} \frac{\sin x}{x} \qquad (7.19a)
\end{aligned}
$$

where $x = k(S_i/2) \sin \theta$. The identical segment centered at $-y_i$ will contribute a pressure:

$$
p_i(\theta) = \frac{Q_i S_i}{r_0} e^{j[\omega t - k(r_0 + y_i \sin \theta)]} \frac{\sin x}{x} \qquad (7.19b)
$$

the sum of the two being

$$
\begin{aligned}
2p_i(\theta) &= \frac{Q_i S_i}{r_0} e^{j(\omega t - kr_0)} (e^{jky_i \sin\theta} + e^{-jky_i \sin\theta}) \frac{\sin x}{x} \\
&= \frac{2}{r_0} e^{j(\omega t - kr_0)} Q_i S_i \cos(ky_i \sin \theta) \frac{\sin x}{x} \qquad (7.20)
\end{aligned}
$$

Let us call the distance between segment centers d and replace $Q_i S_i$ by A_i. Then $y_1 = d/2$, $Q_1 S_1 = A_1$, $y_2 = 3d/2$, $Q_2 S_2 = A_2$, etc. We may then express the total pressure at P as

$$
p_t(\theta) = \frac{2}{r_0} e^{j(\omega t - kr_0)} \frac{\sin x}{x} (A_1 \cos u + A_2 \cos 3u + \cdots + A_i \cos(2i - 1)u) \qquad (7.21)
$$

Equation 7.21 normalized for plotting a pattern in polar coordinates becomes

$$P(\theta) = \frac{1}{A_1 + A_2 + \cdots + A_i} \frac{\sin x}{x} [A_1 \cos u + A_2 \cos 3u + \cdots + \cos (2i - 1)u] \quad (7.22)$$

Equation 7.22 expresses the pattern of a line of segments as a product of the pattern of one segment by the pattern of points located at segment centers. To get the equation in this simple form requires that the array be symmetrical about its center and that the segments be of equal lengths. However, they do not have to be of equal source strength. If one wishes to use segments of unequal lengths, the factor $\sin x/x$ would not be common to all terms but could be included in the amplitude coefficients A_i. This is one method of shaping a pattern. Equation 7.22 defines a design principle frequently termed the first pattern theorem. It is of more practical importance than (7.14) because line sources and hydrophones consist of segments instead of points. With few exceptions the segments of a line almost fill the line and for such arrangements the following observations are clear:

1. If the segments are uniformly driven, the directivity pattern will be that of a continuous line source regardless of the lengths of the segments or the distance between their centers.
2. If the number of segments could be increased indefinitely so that their lengths became very small, the pattern would approach that expressed by

$$p_t(2) = \frac{1}{r_0} e^{j(\omega t - kr_0)} \int_{-l/2}^{l/2} Q_s(y) e^{-jky \sin \theta} \, dy \quad (7.18)$$

where the amplitude of each segment is a function of its distance from the center.

3. Since the segments are approximately as long as the distance between centers, (7.22) may be rewritten as

$$P(\theta) = \frac{1}{A_i + A_2 + \cdots + A_i} \frac{\sin x}{x} [A_1 \cos x + A_2 \cos 3x + \cdots + A_i \cos (2i - 1)x] \quad (7.23)$$

For a distance a between segments of one wavelength

$$x = \frac{\pi a}{\lambda} \sin \theta \rightarrow \pi \quad \text{as} \quad \theta \rightarrow \pm \frac{\pi}{2}$$

which means that the sum representing the pattern of point sources would again reach a maximum. However, the factor representing the pattern of a single segment is approaching zero, so the product of the two approaches zero rather than a maximum.

7.5 PATTERN CONTROL

The two radiation patterns shown by Figure 7.7 illustrate an effect gained by manipulating the amplitudes by which the segments of a line are driven. There are six segments in the line. The accompanying legend shows that the segments completely fill the line, that the amplitudes are all equal for the solid pattern, but varied in proportion to the coefficient of each term for the dashed pattern. The control of patterns by this and other methods is called *shading*. The unshaded array of segments has the pattern of the continuous line or

$$P(\theta) = \frac{\sin x}{x} \quad \text{where} \quad x = \frac{\pi l}{\lambda} \sin \theta \tag{7.24}$$

and l is the total length of the six segments. Given the ratio of l/λ, we may determine the width of the main beam, the number and relative height of the secondary lobes between $-\pi/2 < \theta < \pi/2$; for example, if the six segments of Figure 7.7 are each one half wave, we have

$$l/\lambda = 3 \qquad P(\theta) = 0 \quad \text{for} \quad x = \pm k\pi \qquad k = 1, 2, 3, \ldots$$

or

$$\frac{\pi l}{\lambda} \sin \theta = k\pi \qquad \sin \theta = \frac{k\lambda}{l} = \frac{k}{3}$$

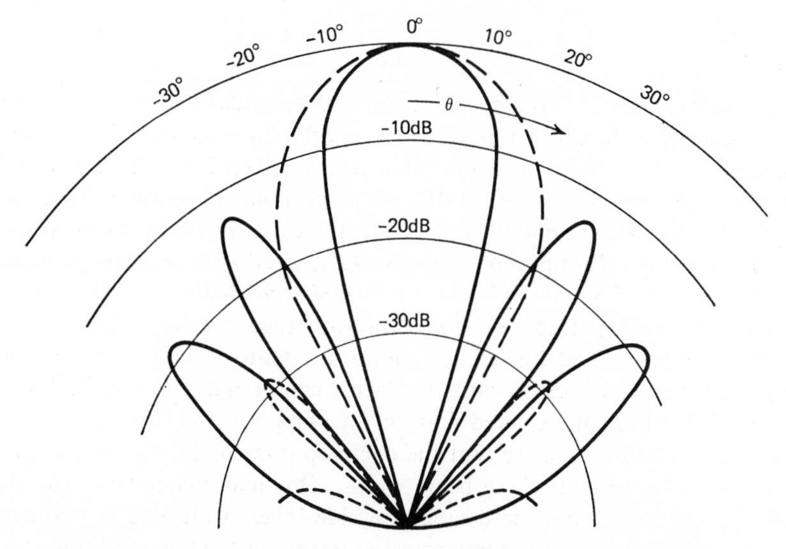

FIG. 7.7 Radiation patterns. (———) $P(\theta) = \frac{1}{3}(\sin x/x)(\cos x + \cos 3x + \cos 5x)$; (- - - -) $P(\theta) = (\sin x/x)(0.52 \cos x + 0.36 \cos 3x + 0.12 \cos 5x)$; $x = (\pi/2) \sin \theta$.

The nulls occur at $\pm 19.5°$, $\pm 41.8°$, $\pm 90°$. A more useful measure of the width of the main beam is given in terms of twice the deviation angle θ at a prescribed number of decibels below the peak. This number may be -3dB, the half-power point, -6dB, the quarter power point, or less often, -10dB. Since no one of these is universally accepted as a standard, one must take care in expressing or using beam width values to state which level is used as a reference. A convenient rule of thumb for estimating the transducer dimension required to produce a given beam width comes from the fact that at the -6dB point

$$\frac{\sin x}{x} = \tfrac{1}{2} \quad \text{at} \quad x \simeq 0.6\pi \tag{7.25}$$

and for the line of total length l

$$x = \frac{\pi l}{\lambda} \sin \theta = 0.6\pi$$

$$\theta = \sin^{-1} \frac{0.6\lambda}{l} \tag{7.26}$$

To locate the peaks of the secondary lobes we find the zeros of the derivative of (7.24)

$$\frac{d}{dx}\left(\frac{\sin x}{x}\right) = \frac{\cos x}{x} - \sin x \tag{7.27}$$

at the values of

$$x = \tan x \tag{7.28}$$

For a line of length 3λ, the peaks of the first secondary lobe occur at $\theta = \pm 28.4°$ corresponding to $x = 4.49^r$. This number inserted in (7.24) gives the height of this lobe relative to the main beam as 0.217 which expressed in decibels is $20 \log 0.217 = -13.3$ dB. Such prominent secondary lobes are undesirable for many applications; for instance, they could result in considerable bearing error in echo ranging or false steering in acoustic guidance. The dashed line of Figure 7.7 and the related legend shows a side lobe reduction from -13.3 dB to -30 dB by the amplitude shading indicated. The pattern of a segmented line (7.22) contains a finite Fourier series as one factor. By the use of transforms, this factor can be made to simulate to an approximate degree any desired pattern.[1] Comparing the two patterns and the related equations, we see that the decrease in secondary lobes has been accompanied by an increase in beam width. One may understand why this occurs by observing that the decrease in side lobes is affected by reducing the relative amplitudes of the outermost segments of the line, and that these segments contribute most to narrowing the main beam.

[1] Summary Technical Report of Division 6, NDRC, Vol. 13, pp. 117–19.

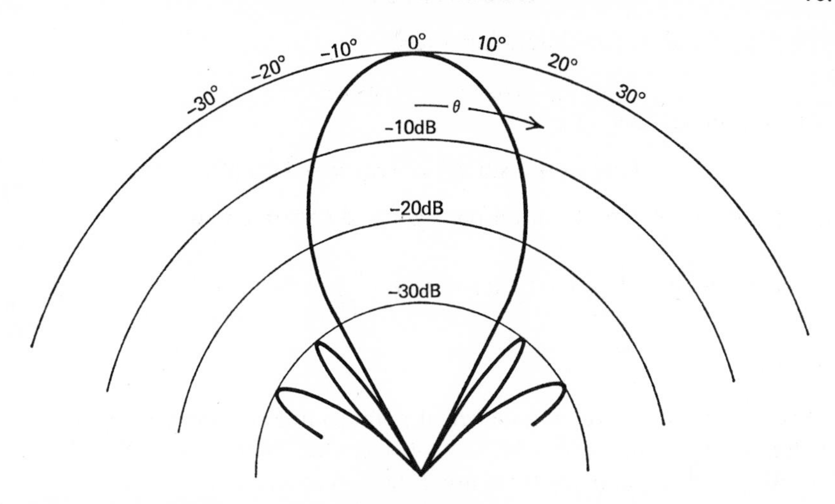

FIG. 7.8 A six-point radiation pattern for $P(\theta) = (1/1.901)(\cos x + 0.634 \cos 3x + 0.267 \cos 5x)$ and $x = (\pi/2) \sin \theta$.

The radiation pattern and amplitude shading of Figure 7.8 illustrates a superior arrangement of the segment amplitudes. The two advantages gained by this arrangement may be quite important where peak acoustic output per unit weight is critical. The increase in the main beam due to shading is a minimum and the array can handle more power.

The method of shading illustrated by Figure 7.8 was developed for antenna arrays.[2] It is based on the fact that the terms in an equation defining the pattern of a set of point sources such as:

$$P(\theta) = \frac{1}{A_1 + A_2 + \cdots}(A_1 \cos u + A_2 \cos 3u + \cdots)$$

for an even number of points and

$$P(\theta) = \frac{1}{A_0 + A_1 + \cdots}(A_0 + A_1 \cos 2u + A_3 \cos 4u + \cdots)$$

for an odd number can be expressed as polynominals in powers of cos u which by use of the definition

$$x = \cos u$$

become a set known as the Tschebyscheff polynomials; and that these polynomials have properties quite useful in pattern design.

[2] C. L. Dolph, "A Current Distribution of Broadside Arrays which Optimizes the Relationship between Beamwidth and Side Lobe Levels," in *Proc. IRE*, 335–348 (June 1946); 489–492 (May 1947).

First a procedure for converting a term:

$$\cos nu = f(\cos u)$$

uses De Moivres theorem:

$$(\cos nu + i \sin nu) = (\cos u + i \sin u)^n$$

By expanding the expression on the right and equating reals we get

$$\cos nu = \cos^n u - \binom{n}{2} \cos^{n-2} u \sin^2 u + \binom{n}{4} \cos^{n-4} u \sin^4 u - \cdots \quad (7.29)$$

where $\binom{n}{k} = \dfrac{n!}{k!\,(n-k)!}$

The pattern of an even number N of point sources is described by the sum of terms $A_1 \cos u$, $A_2 \cos 3u, \ldots, A_{N/2} \cos nu$, where $n = N - 1$. As Tschebyscheff polynomials there are $T_n(x) = \cos nu$:

$$\cos u = T_1 = x \qquad \cos 3u = T_3 = 4x^3 - 3x$$

$$\cos 5u = T_5 = 16x^5 - 20x^3 + 5x$$

$$\cos 7u = T_7 = 64x^7 - 112x^5 + 56x^3 - 7x, \text{ etc.}$$

An examination of the definitions

$$x = \cos u$$

and

$$T_n(x) = \cos nu$$

shows that x is limited to a region $-1 \leq x \leq 1$ and that $T_n(x)$ is bounded by the same limits, so that $T_n(x)$ oscillates between ± 1 over its area of definition. However, by taking a Tschebyscheff polynomial such as $T_7(x)$ previously shown as

$$T_7(x) = 64x^7 - 112x^5 + 56x^3 - 7x \quad (7.30)$$

which in the limited region of $-1 \leq x \leq 1$ oscillates between ± 1, we may extend the region to a point $x_0 > 1$; $T_7(xx_0)$ may take on a value as large as we please but still be limited in the region $-1 \leq xx_0 \leq 1$.

The locus of the function $T_7(x_0 x)$ plotted as a function of $u = \text{arc} \cos x$ is shown by Figure 7.9. The point

$$u = 0 \qquad x = 1 \quad \text{and} \quad T_7(x_0 x) = T_x(x_0)$$

is the peak value of the polynomial within the allowed values of $\cos nu$ bounded by ± 1.

This manner of pattern control may be applied to any number of point sources. For illustrative purposes let us consider an eight-point source that

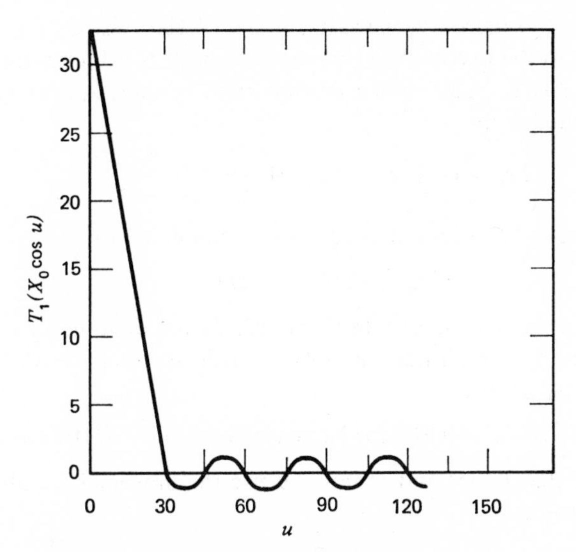

FIG. 7.9 The Tschebyscheff polynomial $T_7(x_0 \cos u)$ as a function of u for $x_0 = 1.18$.

has a pattern containing the series:

$$P_8(\theta) = A_1 \cos u + A_2 \cos 3u + A_3 \cos 5u + A_4 \cos 7u$$

which expressed in terms of cos u is

$$P_8(\theta) = 64A_4 \cos^7 u + (16A_3 - 112A_4) \cos^5 u$$
$$+ (4A_2 - 20A_3 + 56A_4) \cos^3 u + (A_1 - 3A_2 + 5A_3 - 7A_4) \cos u \quad (7.31)$$

or as polynomials in x

$$P_8(\theta) = 64A_4x^7 + (16A_3 - 112A_4)x^5 + (4A_2 - 20A_3 + 56A_4)x^3$$
$$+ (A_1 - 3A_2 + 5A_3 - 7A_4)x \quad (7.32)$$

The locus of (7.32) will duplicate that of Figure 7.9 if the coefficients of (7.32) are equated to those of

$$T_7(x_0x) = 64_0{}^7x^7 - 112x_0{}^5x^5 + 56x_0{}^3x^3 - 7x_0x$$

The solutions of these equations give the amplitudes as the following functions of $x_0 = 1.1807$

$$\begin{aligned} A_4 &= x_0{}^7 = 3.1988 \\ A_3 &= 7A_4 - 7x_0{}^5 = 6.3294 \\ A_2 &= 5A_3 - 14A_4 + 14x_0{}^3 = 9.9218 \\ A_1 &= 3A_2 - 5A_3 + 7A_4 - 7x_0 = 12.2451 \end{aligned} \quad (7.33)$$

Again referring to Figure 7.9, the value of $x_0 = 1.1807$ was chosen to give the ratio of $T_7(x_0)$ to the maximum value of $T_7(x)$ of 31.6, which ratio expressed in decibels is 30 dB. Calling the desired ratio r, the required value of x_0 may be computed from

$$2x_0 = (r + r^2 - 1)^{1/n} + \left(r + \frac{1}{r^2 + 1}\right)^{1/n} \tag{7.34a}$$

which for values of r usually needed may be simplified to

$$2x_0 \simeq (2r)^{1/n} + (2r)^{-1/n} \tag{7.34b}$$

Substituting the value of $x_0 = 1.1807$ into (7.33) and dividing by their sum in order to express the amplitudes in normalized form, we have for the pattern of an eight-point source

$$P(\theta) = 0.387 \cos u + 0.313 \cos 3u + 0.199 \cos 5u + 0.101 \cos 74 \tag{7.35}$$

We observe that (7.35) and Figure 7.9 use the universal variable u which has been defined as

$$u = \frac{\pi l}{\lambda} \sin \theta$$

where θ, the angle of deviation from the acoustic axis, is the variable describing the pattern of a set of points having the specific spacing to wavelength ratio of l/λ. To achieve the desired results from this type of shading the spacing ratio for point sources is restricted to $\frac{1}{2} \leq l/\lambda \leq 0.8$. However, for segments virtually filling the line, which is usually the case, the ratio is more flexible; for example, if, as could be the case at low frequencies, a half-wavelength segment became so large as to be unwieldy, it is obvious that it may be replaced by two quarter-wave segments of the same amplitude. It is also apparent that if segments approach a full wave in length the factor

$$\frac{\sin x}{x} \to 0 \quad \text{as} \quad \theta \to 90$$

and secondary lobes will not be of equal height.

For the eight-point source as described and a selection of $l/\lambda = \frac{1}{2}$ we may sketch the pattern of (7.35) rather well by locating the half-power points, the nulls, and the peaks of the secondary lobes at the half-power point x', $T_n(x_0 x') = 0.707 T_n(x_0)$ and

$$2x' x_0 \simeq (1.414r)^{1/n} + (1.414r)^{-1/n} \tag{7.36}$$

where $r = 31.6$, the preselected ratio of main lobe to side lobe. The value of x' from (7.36) inserted into the equation

$$\theta = \text{arc sin} \left[\frac{\lambda}{\pi l} \text{arc cos } x'\right] \tag{7.37}$$

yields the half beam width at −3 dB. It is interesting to compare the beam width resulting from the given data of 16.6° with that of a continuous uniform line source of the same length. At the half-power points at which

$$\frac{\sin x}{x} = 0.707, \qquad x = 79.8^\circ$$

and since in this case $x = (7\pi/2)\sin\theta$, the beam would be 14.6°. The increase in beam width is the price paid for reducing the side lobes from −13 dB to −30 dB.

To locate the nulls we must find those values of u that solve the equation

$$T_n(x_0 \cos u_k) = 0$$

For the case of eight-point sources in which

$$T_7(x_0 x_k) = 64x_0{}^7 \cos^7 u_k - 112x_0{}^5 \cos u_k{}^5 + 56x_0{}^3 \cos^3 u_k - 7x_0 \cos u_k \tag{7.38}$$

let us substitute into this equation

$$\cos u_k = \frac{1}{x_0}\cos\frac{k\pi}{2n} \tag{7.39}$$

to obtain

$$T_7(x_0 x_k) = 64\cos^7\frac{k\pi}{2n} - 112\cos^5\frac{k\pi}{2n} + 56\cos^3\frac{k\pi}{2n} - 7\cos\frac{k\pi}{2n}$$

which we recognize as the expansion of

$$T_7(x_0 x_k) = \cos 7\left(\frac{k\pi}{2n}\right) \tag{7.40}$$

for the eight-point source $n = 7$, and (7.40) has zero values for k equal odd integers. Therefore, from (7.39), nulls occur at

$$u_k = \text{arc cos}\left[\frac{1}{x_0}\cos\frac{k\pi}{2n}\right] \qquad k \text{ odd} \tag{7.41}$$

By the same reasoning it can be shown that secondary lobe peaks occur at

$$u_p = \text{arc cos}\left[\frac{1}{x_0}\cos\frac{k\pi}{n}\right] \qquad k = 1, 2, 3, \text{etc.} \tag{7.42}$$

All of this information adds up to the pattern of Figure 7.10. For a given assembly of points with l fixed, a change in the driving frequency changes the ratio of l/λ. The previously chosen dimension $l = 0.8\lambda$ as a maximum came from the point on Figure 7.9 marked with a cross where the pattern has already repeated itself to this point on the major lobe. With an assembly of points so spaced, an increase in frequency would tend to complete this lobe in the pattern.

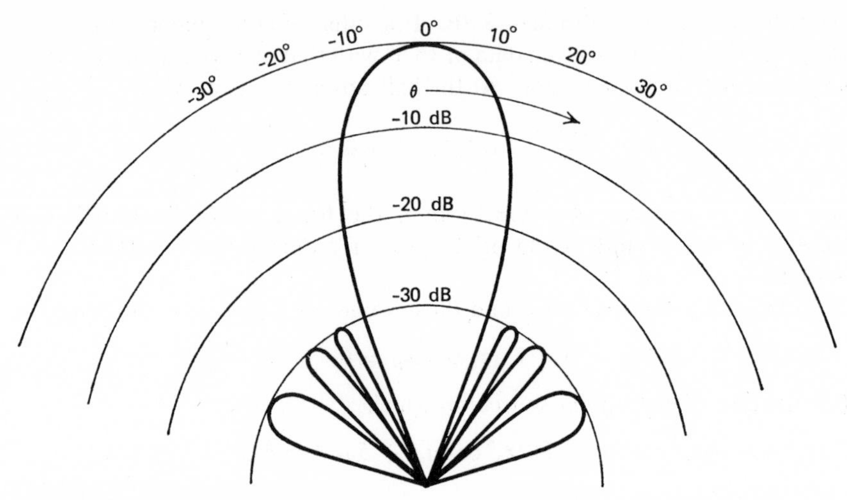

FIG. 7.10 An eight-point source with −30 dB side lobes. $P(\theta) = 0.387 \cos\theta + 0.313 \cos 3\theta + 0.199 \cos 5\theta + 0.101 \cos 7\theta$.

It is important to stress the fact that in these discussions the phase of signals arriving at a fixed distance from the source has been referred to that coming from the center point, that there is no change in phase within a lobe, and that there is a reversal of phase in passing from one to an adjacent lobe. This is shown in Figure 7.9 but of course cannot be shown in the logarithmic scale of Figure 7.10. The significance of these phase relationships will become obvious in our discussion of target tracking.

7.6 PLANE SURFACES AS SOUND SOURCES

The line sources that we have discussed are useful in many applications, but the addition of an orthogonal plane of directivity adds considerably to the versatility as a source, to signal to noise, and to target location as a receiver. As indicated by Figure 7.11, a variety of sources may be assembled on a plane. Shown there are points that may represent simple sources, line segments, or circular or rectangular surfaces.

In Chapter 6 we discussed the longitudinal resonator that radiates from one end only. A number of these assembled in a plane array provides a sound field which we may wish to control in any plane normal to the array and containing the acoustic axis. Such an array may be that of Figure 7.11 oriented in the y–z plane of the coordinate system chosen by the American Standards Association for pattern measurement. The method of recording patterns, described in Section 7.2, will assist in a simple explanation of patterns from plane surfaces. Because of the procedure used, the instrument measuring the sound field pressure is stationed in the xy plane at a distance r

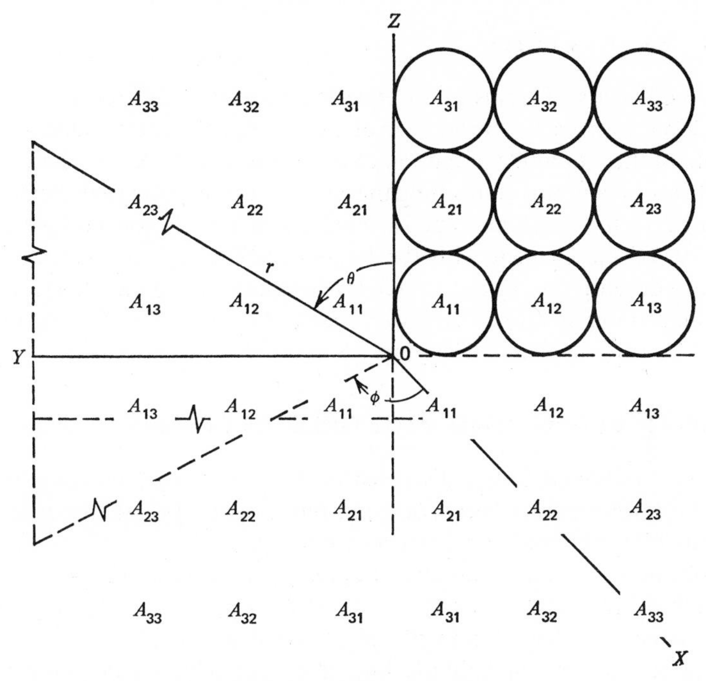

FIG. 7.11 A plane array.

such that lines drawn from all points on the array are essentially parallel. Variations in the pressure field are due strictly to phase differences which in the picture shown are function of φ, measuring the deviation angle from the acoustic axis which coincides with the x axis. Therefore there would be no effect on the pattern in the x–y plane if the acoustic output in each column were added and placed at the center of the column. To control this pattern as we would a line source by shading, the relative values but not the absolute values would be the same along each line. Since this does not control the shading along the columns, it illustrates the very important fact that pattern control in one plane is independent of that in an orthogonal plane. The pattern of a transducer is usually termed vertical or horizontal according to its position in operation. If the arrangement shown in Figure 7.11 measures the horizontal pattern, we may record the vertical pattern by rotating the transducer in its mounting fixture by 90°. Obviously such patterns for a square or circular surface would be identical. In addition, it may be desirable to take patterns in several other planes. A general statement can be made about the pattern in any one of these planes: it will be that of a line source where each section of this line has a source strength equivalent to the sum of strengths along a line perpendicular to it at that point.

7.6.1 Source control

The coefficients of terms in an equation such as (7.35) represent relative source levels from each element in an array. The coefficient is a function of the area and of the amplitude of motion of the element surface. One may conjure up a variety of ways for controlling the acoustic power from each element, but a limiting factor must be kept in mind; and that is the power-handling capability of the element with the largest coefficient. An unshaded plane array can develop the highest source level because it can handle more power and concentrates it into a narrower beam. For the same two reasons the manner of shading shown in Figure 7.8 is superior to others.

7.7 SPLIT BEAMS AND MULTIPLE PATTERNS

The type of sound source discussed in Section 7.6 was in common use in the early development of echo ranging, but as employed at that time it had little capability for tracking a fast-moving target.

Modifications of the plane array to enhance its usefulness have continued to a high degree of sophistication. As an introduction to the potentialities of the system, we shall begin with split beam designs.

If the sources of one-half the line array are reversed in phase but the driving amplitudes remain the same, the resulting pattern will be described by

$$P(\theta) = \frac{1}{A_1 + A_2 + \cdots + A_{n/2}} \times (A_1 \sin u + A_2 \sin 3u + \cdots + -A_{N/2} \sin (N-1)u) \quad (7.43)$$

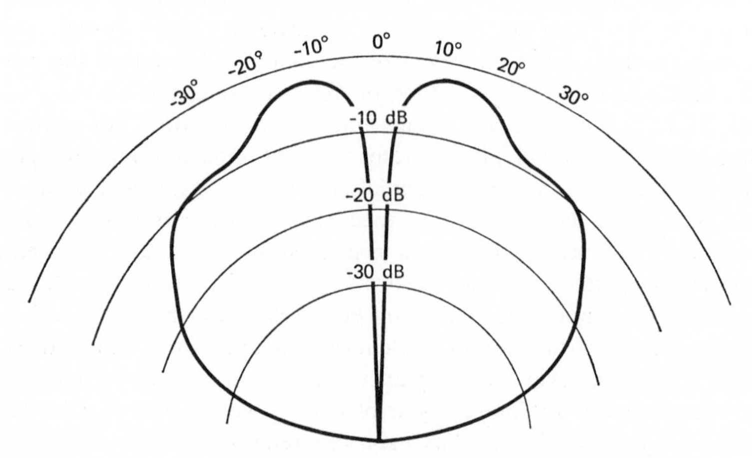

FIG. 7.12 A split beam pattern.

Illustrative of the locus of (7.43) is the pattern of Figure 7.12 which has the amplitude coefficients of Figure 7.7.

As an introduction to the use of split beams, let us consider the series of diagrams shown in Figure 7.13 illustrating responses from a plane array such as that of Figure 7.11. The individual elements of an electroacoustic transducer in a sound field develop electromotive forces that may be connected

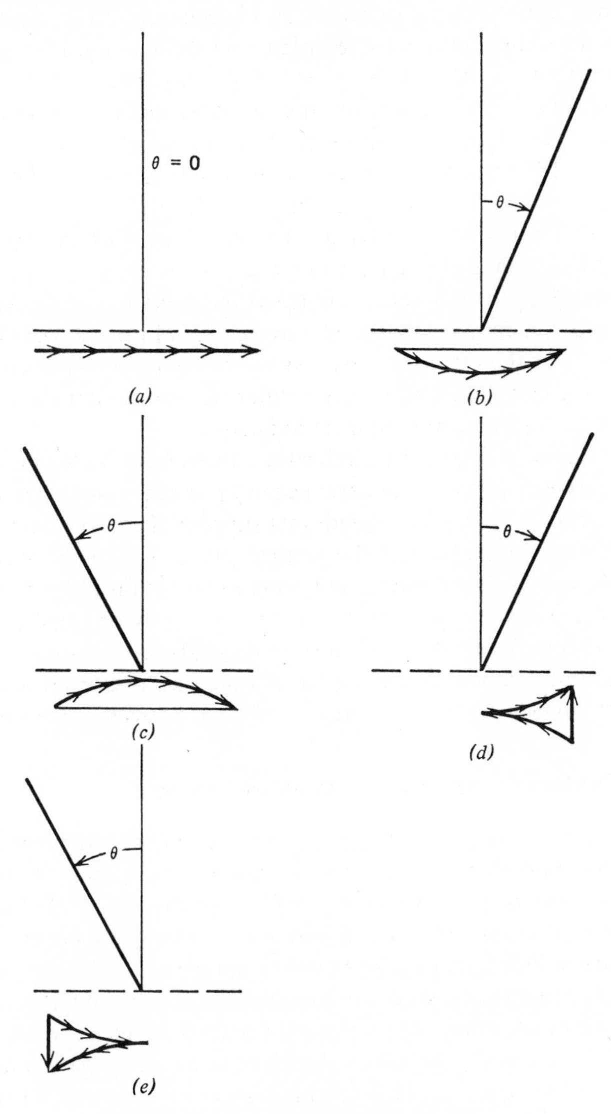

FIG. 7.13 Sum and difference signals.

in series or parallel, aiding or opposing. In each diagram an arrow is a phasor representing the sum of voltages from the elements in the column in which the arrow occurs. In diagram 7.13*a* the deviation angle θ is zero and these voltages are shown as being in phase. Diagram 7.13*b* shows that voltages to the right of center are advanced, and those to the left have been retarded as the sound field comes from a direction with a positive deviation angle. When this angle is negative as in diagram 7.13*c*, voltages from elements to the right of center are retarded, and those to the left are advanced. Only the amplitudes and not the phase of the resultant vary in these three cases. Diagrams 7.13*d* and *e* show the resultant voltage where emf's from elements to right of center oppose those from elements to left of center. We observe a 180° reversal of phase of the resultant as the deviation angle changes sign. No one of these resultant signals will supply a satisfactory instantaneous indication of bearing. The two halves of the array may be connected to furnish a difference voltage and simultaneously a sum voltage which leads or lags the difference voltage as the deviation angle changes sign. The sum resultant may be shifted 90° so that it is in phase with the voltage from one half, or the other, of the array as the deviation angle changes sign. The result is a conversion of a phase difference to an amplitude difference which gives an instantaneous bearing indication.

The split beam of Figure 7.12 indicates a transducer with top and bottom halves in reverse of phase. This arrangement provides tracking in the vertical plane. Dividing the array into quadrants supplies the information for tracking in both the horizontal and the vertical plane. Not only may the array connections and the amplitude coefficients in reception differ from those in transmission, but also the acoustic center of halves or quadrants may be shifted by adding the elements of one section to those of another. So long as the level of operation is in the region in which all responses are linear the principle of superposition allows any number of patterns simultaneously.

7.7.1 Conformal arrays and steered beams

A low-frequency source with the required bearing accuracy and directivity gain for discrimination against noise occupies a large area. In order to be mobile, an array of this sort must conform to the shape of the conveying vehicle or to a streamline form of minimum impedance to movement through water. It is possible here to present only a simple example, but one which is quite useful. This begins with the comparison of echo ranging problems of radar with those of sonar. The purpose of both is the location and identification of a distant object. Relative velocity of signal propagation places sonar under a great handicap. Instead of rotating an antenna, with a fixed pattern, sonar uses a fixed antenna with a rotating pattern.

The antenna may be an array of sources forming a cylinder. By the use

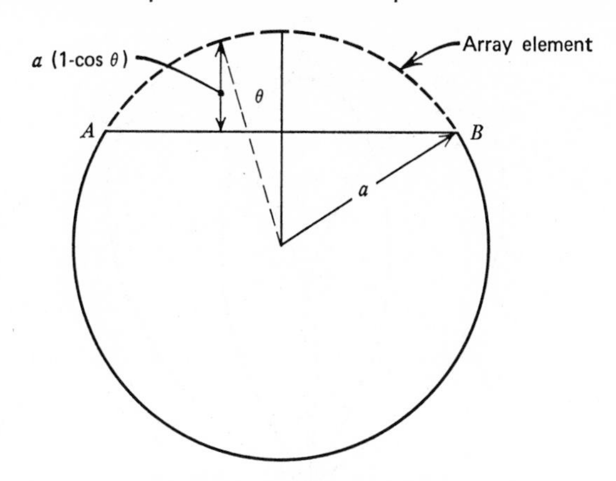

FIG. 7.14 A cylindrical array.

of delay lines a sector of this cylinder can be made to present a pattern closely simulating that of a line source. This set of delay lines may be rotated mechanically or electronically to give the required rotating pattern; for example, Figure 7.14 shows a sector from a cylindrical array of sound sources. By inserting time delays equivalent to that required for a signal to travel in water the distance between each element and the chord AB, the cylindrical sector approximates a plane array, and is responsive to similar methods of pattern control. Also, delay lines may be designed to steer the main beam of this or any plane array as illustrated diagrammatically by Figure 7.15. The sound beam is steered by rotating the acoustic axis of the line

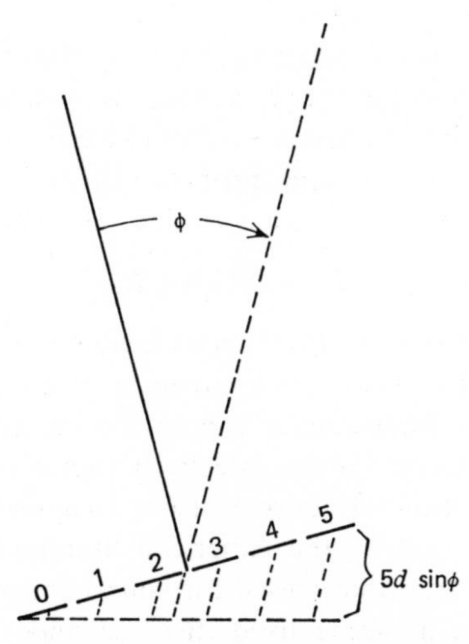

FIG. 7.15 A phase-shifted array.

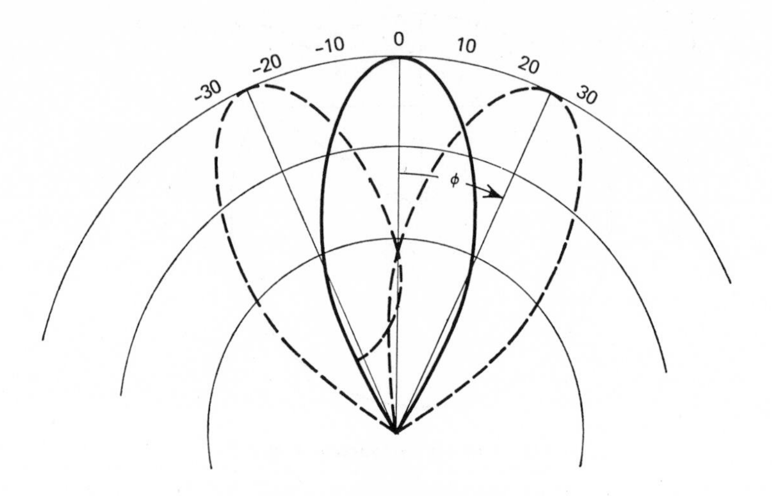

FIG. 7.16 Shifted lobe patterns.

through the angle φ. Numbering the elements from zero to N, a delay equivalent to a phase lag of $Nkd \sin \varphi$ is introduced to element N, where k is the wave number, d the distance between elements, and φ is the angle from the normal that the beam is steered. Figure 7.16 is experimental data showing the shift of a main beam between $\pm 22^\circ$. Definite limits are set to the amount of shift by the introduction of high secondary lobes and ambiguity of element location.

7.8 DIRECTIVITY INDEX

Our skill in shaping beam patterns results in many diverse forms which, to be compared in terms of efficiency as sources or in terms of discrimination against noise as receivers, require a determinable factor for relating patterns to total power. One is a unique property of each transducer called the directivity factor, R_θ, or the directivity index, D_i, where

$$D_i = 10 \log_{10} R_\theta \tag{7.44}$$

In simple terms the ratio of the total acoustic power from an omnidirectional projector to the total acoustic power from a given projector, required to develop equal acoustic intensities at a point on the acoustic axis in the far-field of the given projector, is the directivity factor of that projector. The reciprocal of this definition is frequently used. In agreement with it, however, is the following approved by the American Standards Association: "The directivity factor of a transducer used for sound emission is the ratio of the mean square pressure at some fixed distance and specified direction, to the mean square sound pressure at the same distance averaged over all directions." It is understood that the point of observation is in the far-field and in the direction of maximum response. This definition may be expressed concisely as

$$R_\theta = \frac{4\pi r^2 p_a^{\ 2}}{\int_s p^2(\varphi, \theta)\, ds} \tag{7.45}$$

where $p(\varphi, \theta)$ = sound pressure as a function of direction at some fixed distance,

p_a = sound pressure in the specified direction at the same distance,

r = the radius of a sphere whose center is the effective acoustic center of the source,

ds = the differential element of area on the surface of the sphere.

We may further develop the significance of the directivity factor by reference to Figure 7.17 showing a sphere of radius r in the sound field produced by a projector placed at the effective acoustic center. The total power radiated is found by an integration of the sound intensity $I(\varphi, \theta)$ over the spherical surface where $I(\varphi, \theta)$ has been defined as

$$I(\varphi, \theta) = \frac{p^2(\varphi, \theta)}{\rho c}$$

and p = mean square sound pressure on the spherical surface,

ρc = characteristic impedance of the medium.

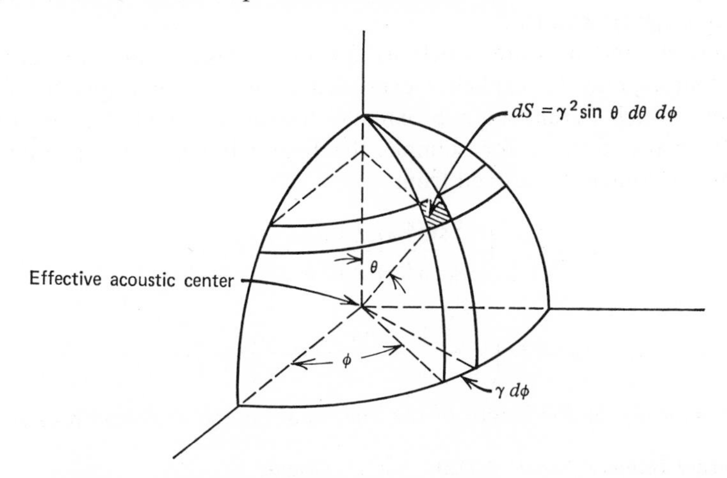

FIG. 7.17 Coordinate system for computation of the directivity factor.

Letting P_t represent the total power out,

$$P_t = \iint \frac{p^2(\varphi, \theta)}{\rho c} r^2 \sin\theta \, d\theta \, d\varphi \tag{7.46}$$

If an omnidirectional projector at the acoustic center develops a mean square pressure, p_a, uniformly over the spherical surface the integration indicated by (7.46) yields P_a where

$$P_a = \frac{4\pi r^2 p_a{}^2}{\rho c} \tag{7.47}$$

The ratio of (7.47) to (7.46) is the definition of the directivity factor given by (7.45).

7.8.1 Calculation of directivity factors

To find the directivity factor of a transducer, we must deal with the integral of (7.45) which by using the coordinates of Figure 7.17 becomes:

$$\int_s p^2(\varphi, \theta) \, ds = r^2 \iint p^2(\varphi, \theta) \sin\theta \, d\theta \, d\varphi$$

The limits of integration may be determined from the nature of the pattern and the orientation of the transducer. In most cases the function $p^2(\varphi, \theta)$ must be determined point by point from recorded patterns taken from one or more orientations of the transducer. From these the value of the integral is found by incremental summation. The latter chore may be avoided by the use of appropriate charts.[3,4]

Fortunately the pressure functions for lines, circular and rectangular piston sources may be explicitly expressed in integrable forms for ideal conditions which permit an approximate determination of the directivity factor for many sources; for example, the directivity factor of a continuous uniform line source is obtained from

$$R_\theta = 2\left[\int_{-\pi/2}^{\pi/2} \frac{\sin^2 \pi l/\lambda \sin\theta}{(\pi l/\lambda)^2 \sin^2\theta} \cos\theta \, d\theta\right]^{-1} \tag{7.48}$$

to give

$$R_\theta = 2x\left(\pi - \frac{1}{x} - \frac{\sin 2x}{x} + \frac{\cos 2x}{2x^3}\right)^{-1} \tag{7.49}$$

where $x = \pi l/\lambda$, l is the length of the line, and λ is the wavelength of sound

[3] Summary Technical Report, NDRC, Vol. 13, Chapter 5.

[4] P. M. Kendig, and R. E. Mueser, "A Simplified Method for Determining Directivity Index," in *J. Acoust. Soc. Am.*, **19**, 691 (1947).

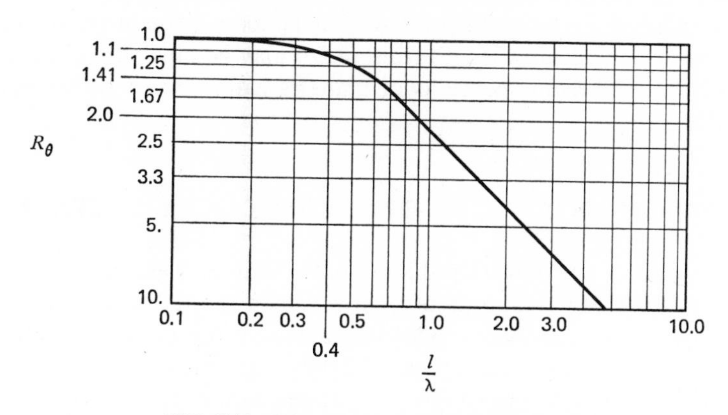

FIG. 7.18 Directivity factor of a line source.

produced in the medium by the line source. Figure 7.18 is the locus of (7.49) from which the directivity index may be computed. It may be observed that this locus with increasing length approaches

$$R_\theta = \frac{2l}{\lambda} \tag{7.50}$$

For a circular piston in an infinite baffle

$$R_\theta = (ka)^2 \left[1 - \frac{J_1(2ka)}{ka} \right]^{-1} \tag{7.51}$$

where $k = 2\pi/\lambda$, a = radius of piston source, and J_1 = Bessel function of first kind and order 1. As the radius exceeds the wavelength of sound being radiated, the directivity factor approaches a value given by:

$$R_\theta = (ka)^2 \tag{7.52}$$

which may be recognized as the product of $4\pi/\lambda^2$ by the piston area. Writing (7.52) as

$$R_\theta = \frac{4\pi A}{\lambda^2}$$

where A is the area of the radiating surface, we may use it for the directivity factor of a rectangular source that has both width and length greater than the wavelength of sound being radiated. When the effective radiating area or length differs from the measurable area, the former may be determined from the width of beam patterns; for instance, in (7.26):

$$\theta = \sin^{-1}\left(\frac{0.6\lambda}{l}\right)$$

the deviation angle θ is one-half the width of the main beam at points 6 db below its peak. An effective length from (7.26) may be used with (7.50) for computing R_θ. Effective dimensions of plane radiators may also be found from (7.26).

7.9 DIFFRACTION EFFECTS ON TRANSDUCER SURFACES

In the previous discussions of radiation patterns it has been assumed that the phase and velocity amplitude of each point on the radiating surface is controlled by the driving force. Since the radiation impedance varies from point to point, this assumption is successful only when the surface consists of elements of sufficient size and stiffness to respond to the average impedance. When the elements have dimensions considerably under a half-wave, some may be absorbing energy from rather than delivering energy to the sound field. This condition may produce amplitudes destructive to the transducer and the driver.[5] The problem has become more serious with the trend toward lower frequencies because of the difficulty of keeping element dimensions comparable to wavelengths.

PROBLEMS

7.1 Two simple sources radiating at 30 kHz are 2.5 cm apart. At what deviation angle is the resultant pressure down 6 dB from the maximum?

7.2 Sketch the pattern in polar coordinates of the radiation pressure versus deviation angle from $\theta = 0$ to $\theta = 90°$, for a line of six point sources equally spaced one-half wavelength apart. Express the relative pressure in decibels.

7.3 Sketch the pattern of a continuous line source of length 6λ from the peak of the primary lobe to the second null.

7.4 Verify the height shown for the first side lobe for the shaded pattern of Figure 7.7.

7.5 Required is a line source operating at 30 kHz with a main beam width of 10° at the −6-dB points. Find the length of the line. Where are the peaks of the first minor lobes?

7.6 Six-place log tables were used to compute the value of X_0 and the coefficients in the text. Compare with values obtained by the use of a slide rule.

7.7 A line of points using the amplitudes of Figure 7.8 is assembled with the points equally spaced one-half wave. Compare the −3dB beam width that of a similar unshaded source.

7.8 Find the directivity index for a circular piston of radius 3 in. when radiating at 60 kHz.

7.9 A piston radiating at 10 kHz has a D.I. at 18 dB. Find its effective area.

7.10 You are asked to design a line hydrophone having a beam width of 6° at the −6-dB points when receiving a signal at 1.5 kHz. Find the length of the line.

[5] D. L. Carson, "Diagnosis and Cure of Erratic Velocity Distributions in Sonar Projector Arrays," in *J. Acoust. Soc. Am.*, **34,** 1191–1196 (1962).

CHAPTER 8

Transducer Evaluation

8.1 INTRODUCTION

A transducer serves as a component in a system which may be quite simple, as in a depth sounder; or it may be in a system requiring the performance of a number of functions consecutively, simultaneously, or both. Each type will be designed to a unique set of specifications dictated by its one or many functions, and its calibration is a test of its approximation to these specifications. Therefore calibration procedures must be carefully planned to ensure that a given unit will perform as desired under the environmental conditions at which it is to operate. It was convenient in the early stages of transducer development to treat them as linear devices, for that simplified the theoretical discussion and reduced calibration problems. Actually this assumption was acceptable when power levels, measured by present-day standards, were quite low. The necessity for conceding a departure from linearity is due not only to the use of higher power levels but also to a change in materials. Calibration difficulties have been increased by the need for testing at levels at which the equipment is to work. Among the number of tests to be listed or described here only a directivity, an impedance, and one of the response measurements will be common to all types of transducers. One would not be interested, for example, in the power-output capability of a hydrophone or the receiving response over several octaves of a power projector. Because of the increasing variety of instruments and the many operations they must perform, it has been necessary to create an elaborate test program and standard procedures for instrument evaluation. For a better conception of the extent of this program one should consult the bibliography of the American Standard Procedures for Calibration of Transducers. In addition, there are published descriptions of the test facilities of the many government laboratories. It is this continued program of control that is largely responsible for the present advanced state of development of underwater sound instrumentation and the excellence of the facilities available for testing.

8.2 CALIBRATION PROCEDURES

The final calibration or performance check of an instrument for use under water is usually an expensive operation. Therefore every possible test should be made during the assembly period and before delivery to the final calibration station to ensure its proper performance there. At this station the degree of complexity of apparatus in use ranges from simple point-to-point voltmeter–ammeter readings to continuously controlled levels, passbands, and automatic recording devices. A detailed treatment of the instrumentation and its use is not within the scope of this discussion. Also, one may normally assume that personnel in charge of stations are thoroughly competent in maintaining proper test conditions. But in order to choose the type and place of test, the engineer should be aware in general of the problems involved. Measurements are made on the transducer while it is immersed in the equivalent of a boundless body of water that is relatively free of extraneous mechanical or electrical noise, uniform in composition and temperature, and stable with respect to these. The equivalent of a boundless body is one with surfaces from which reflected sound is attenuated 20 or more decibels either by absorption or by distance spreading. Absorption is achieved by anechoic tank linings while a sufficient spreading loss exists where the nearest reflecting surface is 10 or more times the pathlength of the acoustic signal being employed in the measurement.

The conditions just described provide what is termed a freefield. They may also exist in a limited medium over that time interval of a sound pulse lying between the arrivals of the leading edge and the first reflection. If this interval is sufficient to develop a steady state with enough time remaining for a measurement, pulsing may be used for response, impedance, and pattern measurements. If the value of the quality factor Q_m is approximately known, the number of required cycles in a pulse to represent the equivalent of a free-field condition may be computed from the transient of a harmonic vibrator driven sinusoidally. Given that ξ is the displacement from equilibrium of the radiating surface:

$$\xi = (1 - e^{-\alpha t})Ae^{-\omega t} \tag{8.1}$$

where from Chapter 1

$$Q_m = \frac{\pi f_0}{d} \qquad \alpha = \frac{\pi f_0}{Q_m} \tag{1.27}$$

From (8.1) and (1.27), one may deduce the fact that approximately Q_m cycles are required to reach 95% of the steady-state amplitude. In addition to at least three cycles for recording the amplitude, the path length of the pulse must cover the overall length of the transducer. As an example of the distances involved, the transient of the pulse from a 1000 Hz projector with a quality factor of 10 will cover a distance in excess of 60 ft. Of course the entire system involved in the measurements must have the bandwidth capability to handle these pulses without distortion.

The trend in sonar transducers has for some time been toward lower frequencies and larger sizes. And although there may be adequate facilities for the original calibration of these components, tanks or lakes of the required size do not exist at convenient repair depots. To resolve this problem, calibration tests conducted in the near field have been developed.[1] Because of the short distance from the radiating surface, measurement must be made near the leading edge of the pulse.

8.3 RECIPROCITY

The reciprocity theorem for electroacoustic devices has been discussed at several levels of complexity.[2] Our understanding of the ideas involved will probably be better served by deductions from less general principles. We may begin with a statement of the theorem in the following form[3]:

If a simple source of strength Q_1 at a point A produces a sound pressure p_2 at point B, a simple source of strength Q_2 at point B will produce a sound pressure p_1 at point A such that

$$\frac{Q_1}{p_2} = \frac{Q_2}{p_1} \tag{8.2}$$

Another statement of (8.2) is that given a linear electroacoustic network with an input branch A and an output branch B, the ratio of the output at B to the input at A is equal to the ratio of the input at B to the output at A. In order to apply this theorem to a reciprocity calibration, let us consider the system of Figure 8.1 consisting of a transducer at A and a simple source at P of strength Q. A current I into the transducer at A will develop an rms pressure p' at P. The simple source of strength Q located at P will develop at

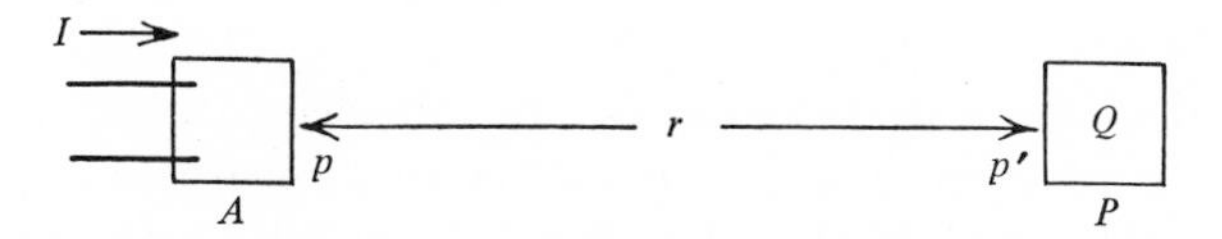

FIG. 8.1 Transducer and simple source.

[1] D. D. Baker, "Determination of Far-Field Characteristics of Large Underwater Sound Transducers from Near-Field Measurements," in *J. Acoust. Soc. Am.*, **34,** 1737 (1962).
[2] Lord Layleigh, *Theory of Sound*, Chapter V and Section 294 of Chapter XIV.
[3] L. E. Kinsler and A. R. Frey, *Fundamentals of Acoustics*, 2nd ed., Wiley, New York, 1962, p. 326.

A a pressure p of

$$p = \frac{j\rho c Q}{2\lambda r} e^{j(\omega t - kr)} \tag{3.47}$$

or

$$|p| = \frac{\rho c Q}{2\lambda r} \tag{8.3}$$

which produces an open circuit voltage of E across the terminals of the transducer at A. According to our statement of the reciprocity principle:

$$\frac{p'}{I} = \frac{E}{Q} \tag{8.4}$$

and using (8.3)

$$\frac{p'}{I} = \frac{\rho c}{2\lambda r} \frac{E}{p} \tag{8.5}$$

or as most frequently written

$$\frac{E}{p} = J \frac{p'}{I} \tag{8.6}$$

The symbol adopted for E/p the open circuit voltage developed by a hydrophone in an effective pressure field p is M_0. The symbol used for p'/I, the effective pressure developed by a transducer per ampere, is S_0. Using these symbols (8.6) becomes

$$M_0 = JS_0 \tag{8.7}$$

where

$$J = \frac{2\lambda r}{\rho c} \text{ m}^4 \text{ sec/kg} \tag{8.8}$$

and J is called the reciprocity parameter. Since J of (8.8) is expressed in the MKS system, the units of pressure in (8.7) must be N/m^2. If, as is most frequently the case, pressure is measured as dyn/cm^2,

$$J = \frac{2\lambda r}{\rho c} \cdot 10^{-7} \text{ cm}^4 \text{ sec/g} \tag{8.9}$$

The use of the reciprocity theorem may be extended to most of the linear electroacoustic transducers commonly used in sonar, provided measurements meet the free-field and far-field conditions. We may deduce from it that the transmitting and receiving patterns of a transducer are identical and that the efficiency as a receiver is also the efficiency as a transmitter.

8.3.1 Reciprocity calibration

Given a transducer that is linear, passive, and reversible, we may wish to measure its performance or response to electrical or acoustical excitation as a function of frequency under several conditions. The four responses usually taken are

M_0 = the free-field voltage response in volts per microbar is found from the voltage developed across the open terminals with the unit in a free sound field of a previously established effective pressure;
M_s = the free-field current response in amperes per microbar found from the current flowing in the shorted terminals with the unit in a free sound field of a previously established effective pressure;
S_0 = transmitting current response in microbars per ampere at a distance of one meter found from the input current and a measurement of the sound field at a proper distance, the pressure value being corrected for the deviation from one meter;
S_s = transmitting voltage response in microbars per volt at a distance of one meter, found in a manner similar to above, but using the input voltage rather than the input current for excitation.

With a standard hydrophone of the appropriate dynamic range, all of these responses may be measured by substitution methods. For very rapid work such methods are quite convenient. Fortunately, instead of periodically referring the standard hydrophone to some primary standard for recalibration, a reciprocity method may be used to check its calibration. Three linear transducers, one of which must be reciprocal, are required.

Using Figure 8.2 we may outline a series of measurements at a fixed frequency to illustrate a reciprocity calibration. An input current i into a transducer at A produces an effective pressure at a distance r of $S_0^A i/r$. A hydrophone placed in this pressure field will develop an open circuit voltage of $M_0^\beta S_0^A i/r$, where the superscripts identify transducers A, B, or C. We

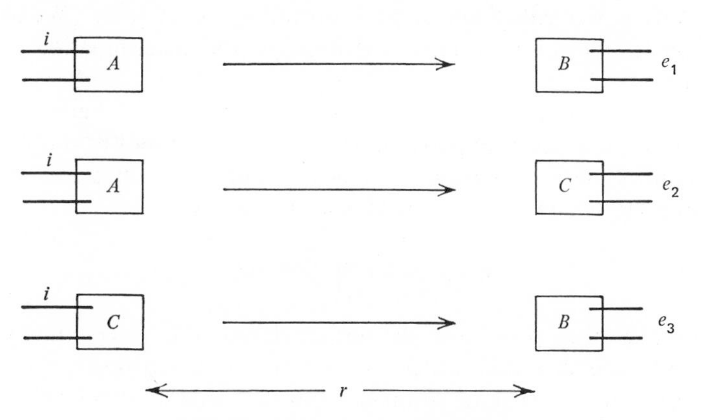

FIG. 8.2 Reciprocity calibration.

may write three equations for the open circuit voltages from Figure 8.2 as

$$e_1 = S_0{}^A M_0{}^\beta \tag{8.10}$$
$$e_2 = S_0{}^A M_0{}^C \tag{8.11}$$
$$e_3 = S_0{}^C M_0{}^B \tag{8.12}$$

where the values of e_1, e_2, and e_3 have been adjusted to a distance r of 1 m and current i of 1 A.

Dividing (8.11) by (8.10) and multiplying by (8.12), we get

$$M_0{}^C S_0{}^C = \frac{e_2 e_3}{e_1} \tag{8.13}$$

and combining with (8.7)

$$M_0{}^C = \left(J_1 \frac{e_2 e_3}{e_1}\right)^{1/2} \tag{8.14}$$

where the value of the parameter J_1 depends on the choice of units for pressure.

Additional results from the same data are

$$S_0{}^A = \left(\frac{e_1 e_2}{J_1 e_3}\right)^{1/2} \tag{8.15}$$

$$M_0{}^B = \left(J_1 \frac{e_1 e_3}{e_2}\right)^{1/2} \tag{8.16}$$

8.4 EVALUATION OF ACOUSTIC SOURCES

The information that one is asked to furnish with an electroacoustic source almost always includes its efficiency, its directivity pattern, its quality factor Q_m, and its power capability or peak operating source level. We shall use data from a typical transducer to illustrate the determination of these properties.

The transmitting efficiency η is expressed as the ratio of the acoustic power out, P_a, to P_e, the electrical power in. The acoustic power out is obtained by integration of the acoustic intensity over a sphere of radius r surrounding the source. In Chapter 7 a factor R_θ was defined as

$$R_\theta = 4\pi r^2 p_a{}^2 \left(\int_S p^2 \, dS\right)^{-1} \tag{7.45}$$

where p is the sound pressure at the surface element dS. p_a = sound pressure in the reference direction which is normally at the acoustic axis of the transducer, and at the same distance from the center. These pressures are measured in the far field and corrected for the distance r.

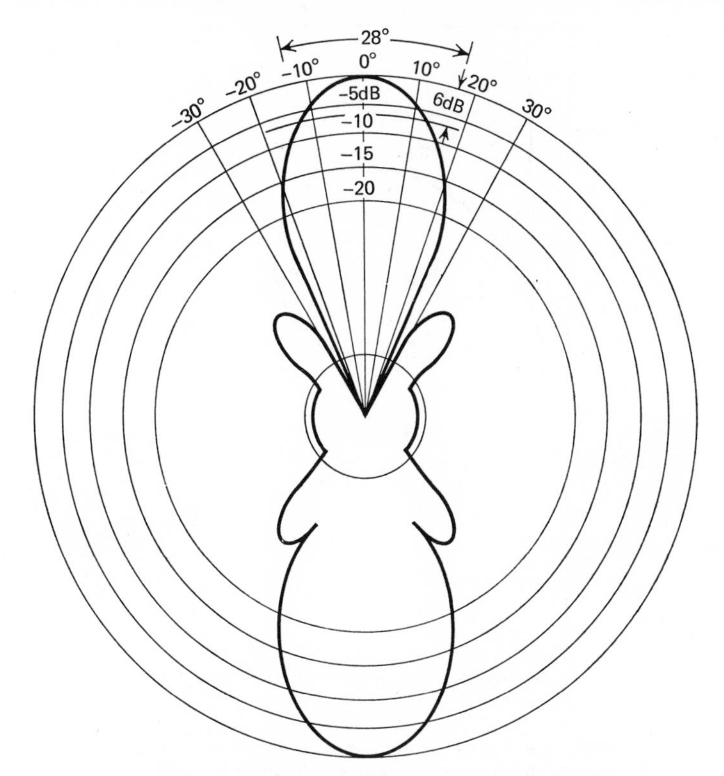

FIG. 8.3 Radiation pattern in vertical plane.

Since the total acoustic power P_a is $\int_S (p^2/\rho c)\, dS$, where ρc is the characteristic impedance of the medium, from (7.45)

$$P_a = \frac{4\pi r^2 p_a^{\,2}}{\rho c R_\theta} \tag{8.17}$$

We may compute R_θ for this particular transducer from Figure 8.3. The transducer is a free-flooding cylinder similar to those described in Chapter 7 having a radiation pattern of a continuous line. The effective length of the line may be found from the relation

$$\frac{\pi l}{\lambda} \sin x = 0.6\pi \tag{7.26}$$

where $2x$ is the width in degrees of the main beam between the -6dB points. Then from (7.50)

$$R_\theta = \frac{2l}{\lambda} \tag{7.50}$$

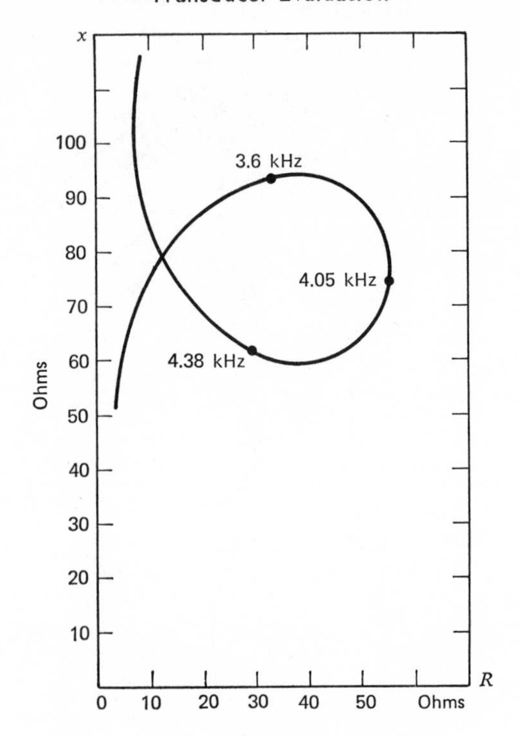

FIG. 8.4 Input impedance transducer in water. Frequency parameter in kHz.

From Figure 8.3 the angle $x = 14°$. Therefore

$$\frac{l}{\lambda} = \frac{0.6}{\sin x} = \frac{0.6}{0.242} = 2.48$$

$$R_\theta = 4.96, \qquad D_i = 10 \log 4.96 = 7 \text{ dB}$$

The pressure along the axis of the main beam at the frequency used in recording the radiation pattern is given by Figure 8.5 as 93.6 dB at 1 meter distance with reference to 1 dyn/cm^2. In MKS units this is 73.6 dB with reference to 1 N/m^2. Expressed in dB (8.17) is

$$10 \log P_a = 10 \log 4\pi + 73.6 \text{ dB} - 10 \log (1.5 \cdot 10^6) - 7 \text{ dB}$$

$$10 \log P_a = 11 \text{ dB} + 73.6 \text{ dB} - 61.8 \text{ dB} - 7 \text{ dB} = 15.8 \text{ dB}$$

The transmitting response of Figure 8.5 was recorded with a constant current input of one ampere. The real component of impedance at this frequency shown on Figure 8.4 is 55 ohms. Therefore

$$10 \log P_e = 10 \log I^2R = 10 \log 55 = 17.4 \text{ dB}$$

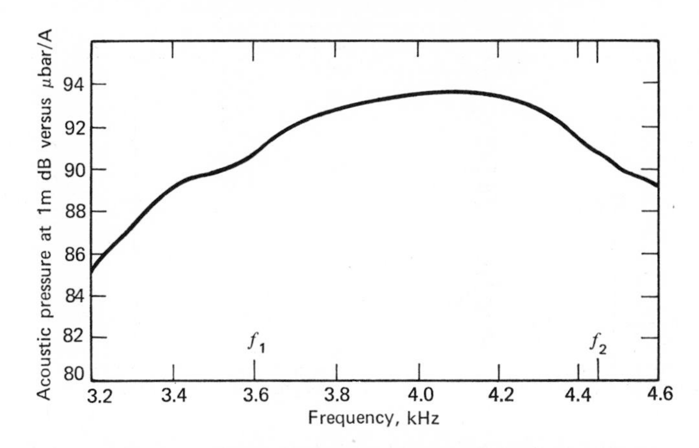

FIG. 8.5 Constant current transmitting response S_0.

The transmitting efficiency at this frequency and power level is

$$10 \log P_a - 10 \log P_e = -1.6 \text{ dB}$$

The quality factor Q_m of this transducer may be calculated from Figure 8.5 by using the definition

$$Q_m = \frac{f_0}{f_2 - f_1}$$

where f_0 is the frequency of peak response and f_2 and f_1 are the frequencies above and below f_0 where the response is 3 dB below the peak.

8.5 POWER LIMITS OF TRANSDUCERS

The problem of power limitation is closely related to the need for weight control. Minimum weight is a paramount issue with transducers used in missile guidance and with airborne equipment. Also, because power capability per unit weight varies inversely with frequency and bandwidth, sound sources designed for low frequency and heavy duty may become so massive as to be unmanageable. In such cases, it is necessary to design toward the power producing limits of the active material. These limits are fixed by magnetic saturation for magnetostriction transducers, by tolerable electric fields for the piezoelectric sources, and levels of temperature and mechanical strain for all types. With low duty cycles temperature poses no threat. Electroacoustic materials, however, are likely to be poor heat conductors, and since internal losses, the source of heating, increase with temperature an unstable situation could be reached in the center of large structures. The

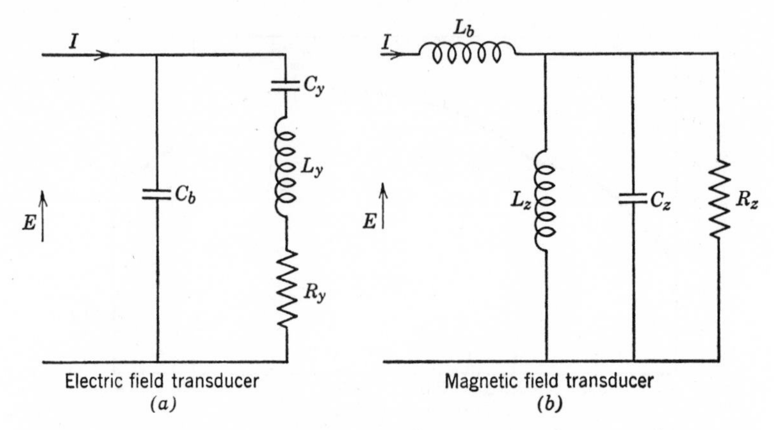

FIG. 8.6 Equivalent circuits. (a) Electric field transducer: $\omega_r^2 = 1/L_yC_y$; $Q_m = 1/\omega_r C_y R_y$; $k^2/(1-k^2) = C_y/C_b$; (b) Magnetic field transducer: $\omega_r^2 = 1/L_zC_z$; $Q_m = R_x/\omega_r L_z$; $k^2/(1-k^2) = L_z/L_b$.

following outlines a discussion by Woollett[4] of the electrical and mechanical power limits imposed by design and the properties of materials used. Figure 8.6 is a simplified version of equivalent circuits for electric and magnetic field transducers. Electrical losses are considered negligibly small and the mechanical arms have been transformed into equivalent electrical components. Figure 8.7 is a set of equations that are combined to express electrical power in terms of quantities applicable to all types of electroacoustic transducers. The last equation

$$P_r = \eta\omega_r \frac{k^2}{1-k^2} Q_m U_e \tag{8.18}$$

Peak stored energy, transducer blocked	
Electric field	Magnetic field
$U_e = \frac{1}{2}\lvert E\rvert^2 C_b$	$U_e = \frac{1}{2}\lvert I\rvert^2 L_b$
Radiated power at resonance	
$P_r = \dfrac{\frac{1}{2}\lvert E\rvert^2}{R_y\eta}$	$P_r = \frac{1}{2}\lvert I\rvert^2 Rz\eta$
$P_r = \eta\omega_r \dfrac{k^2}{1-k^2} Q_m U_e$	

FIG. 8.7 Electrical parameters of radiated power.

[4] R. S. Woollett, "Theoretical Power Limits of Sonar Transducers," in *IRE Conv. Record*, pt. 6, 90–94 (1962).

that applies to both electric and magnetic field devices gives the acoustic power radiated in terms of efficiency η, resonant frequency ω_r, electromechanical coupling coefficient k, the mechanical storage factor Q_m, and U_e, the energy that is storable in the electric or magnetic field of a mechanically blocked transducer. The first equations

$$U_e = \tfrac{1}{2}\,|E|^2\,C_b \quad \text{and} \quad U_e = \tfrac{1}{2}\,|I|^2\,L_b \tag{8.19}$$

express the peak energy stored in the electric or magnetic field of the transducer when no mechanical motion is allowed. At the resonant frequency the equivalent circuit of the electric field transducer becomes the capacitor C_b in parallel with the resistance R_y. At the resonant frequency the equivalent circuit of the magnetic field transducer becomes the inductance L_b in series with the resistance R_z. The power absorbed in the motional resistances R_y and R_z is partly useful radiated power and partly internally dissipated power. Since efficiency is defined as the ratio of the radiated power to the total power consumed across R_y or R_z, we get the two equations

$$P_r = \frac{\tfrac{1}{2}\,|E|^2}{R_y}\,\eta \tag{8.20a}$$

$$P_r = \tfrac{1}{2}\,|I|^2\,R_z\eta \tag{8.20b}$$

For a linear transducer the factor $\eta\omega_r(k^2/1 - k^2)Q_m$ is constant for all levels of input power. Therefore the maximum power available from the source may be limited by the maximum allowable value of U_e.

The following discussion of the mechanical limit is restricted by the assumption that no part of the vibrating structure other than the transducing material is involved. However, prestressing elements and radiating heads are vulnerable to fatigue, an effect difficult to define except by experience. Fatigue, however, is controlled by keeping peak strains within conservative limits, and our look at allowable strains in the active material may serve as an introduction to the overall problem. As in the electrical limit, the radiated power may be expressed in the general terms of efficiency, frequency, mechanical storage factor, and storable elastic energy.

In terms of a peak displacement ξ of the radiating surface, the radiation resistance P_r, and efficiency, η, the radiated power is

$$P_r = \tfrac{1}{2}\,|\omega\xi|^2\,R_r = \tfrac{1}{2}\,|\omega\xi|^2\,\eta R_z \tag{8.21}$$

and the peak elastic energy U_m, stored in the structure, is

$$U_m = \frac{1}{2}\frac{|\xi|^2}{C_m} \tag{8.22}$$

where C_m is the mechanical compliance of the transducer. The radiation resistance may also be expressed in terms of the mechanical storage factor from its definition as

$$Q_m = \frac{\eta}{\omega_r C_m R_r} \tag{8.23}$$

The elimination of displacement and compliance from these three equations yields

$$P_r = \frac{\eta \omega_r}{Q_m} U_m \tag{8.24}$$

Equations 8.18 and 8.24 relate the power limit of a transducer to the maximum permissible value of U_m or U_e, whichever is the lower. These values are characteristic of both the active material and the type of structure used. The one term common to both equations that may allow some freedom in design is Q_m. By equating them and solving for Q_m

$$Q_m = \left[\frac{1 - k^2}{k^2} \frac{(U_m)_{\max}}{(U_e)_{\max}}\right]^{1/2} \tag{8.25}$$

We will find a value for which the electrical and mechanical power limits are equal, an optimum situation.

Figure 8.8[5] is a table of estimated values for use in comparing power limits of common electromechanical materials. The referenced article presents an analysis of the ring transducer, the supported-edge flexural disk, and the

Materials Fig. 8-8	Coupling coefficient $k = k_{33}$	Estimated energies (J/m³) $(U_e)_{\max}$	$\frac{k^2}{1-k^2}(U_e)_{\max}$	$T_{\max}/2PC^2$ $(U_m)_{\max}$
Nickel	0.30	200	20	4000
Permenclur				
polarized	0.29	400	37	4000
remanenie	0.20	90	4	4000
Nickel				
ferrite	0.32	90	10	1500
ADP crystal	0.28	200	17	4500
Barium				
Titanate	0.48	400	120	2000
Leadtitanate				
zirconate	0.60	1200	470	3000

FIG. 8.8 Energy limits of transducer materials.

[5] *Ibid.*

Material	Optimum Q_m	Surface intensity, (W/cm^2)
Barium titanate	4	31
Leadtitanate Zirconate	2.1	36
Nickel	14	42

FIG. 8.9 Surface intensity limits of transducer materials.

equivalent half-wave longitudinal vibrator. Each has electrical and mechanical limits peculiar to the type with modifications to fit a specific model. It will suffice for our purpose to examine the performance of several materials as thin-walled ring transducers, a structure affording their best operation and one easily dealt with.

Equation 8.18 gives the radiated power at resonance per unit volume as directly proportional to the frequency. It is convenient to convert this to radiated power per unit area, or acoustic energy, as follows:

$$\frac{P_r}{S} = \eta \frac{k^2}{1 - k^2} (\mu_e)_{\max} Q_m \omega_r \frac{V}{S} \tag{8.26}$$

where S = radiating surface,
V = ring volume,
$(\mu_e)_{\max} V = (U_e)_{\max}$,
$(U_e)_{\max}$ = maximum storable electrical energy and since $(\omega_r V \rho / S \gamma_r)\eta = Q_m$,
where ρ = ring density,
γ_r = specific acoustic radiation impedance,

$$\frac{P_r}{S} = \frac{k^2}{1 - k^2} (\mu_e)_{\max} Q_m{}^2 \frac{\gamma_r}{\rho} \tag{8.27}$$

By introducing the appropriate data such as that from Figure 8.8 into (8.25) and (8.27), we may reach some conclusions which will serve as guides in selecting materials for specific purposes. Figure 8.9 lists the optimum Q_m and the maximum surface intensity available from a ring having a mechanical storage factor of that value. These data indicate the use of the piezoceramics for broadband operation, and the magnetostrictive metals where a narrow bandwidth is not objectionable.

8.6 PERFORMANCE LIMITS FOR HYDROPHONES

Certain classes of acoustic systems such as depth sounders, echo ranging, and missile guidance use the same transducer for transmitting and receiving sound energy. Such receivers cover only a limited frequency band in the vicinity of resonance, and are usually quite directional. They are not likely to be limited in measuring low levels because of their own characteristics and therefore are not the subject of discussion at this point.

Passive systems employing transducers strictly as receivers may serve in fixed arrays or as deployable arrays. Compared to active sonar devices or hydrophones in motion, they are working in a quieter environment limited only by ambient and thermal noise. Ambient sea noise results from a number of effects. Some of these cover a limited band, while others present a broadband spectrum. Some are directional and others isotropic. The differences call for specific solutions. Thermal noise is always present and since it can be reduced by good design it should not be the limiting factor at the usual sonar frequencies.

8.6.1 Thermal noise

After all other forms of excitation have been reduced to zero, there still exists in a material, energy resulting from a continual state of agitation of its elementary particles. In the case of a resistor, this energy will manifest itself as a voltage of broadband or random characteristics across the terminals. We may simulate this condition with an equivalent circuit shown by Figure 8.10. The available power over a bandwidth Δf is

$$P_{\Delta f} = kT\,\Delta f \tag{8.28}$$

where k = Boltzmann's constant,

T = temperature in degrees Kelvin.

If we equate this to the maximum power available from the circuit of Figure 8.10,

$$\frac{e_t^2}{4R} = kT\,\Delta f \tag{8.29}$$

$$e_t = (4RkT\,\Delta f)^{1/2} \tag{8.30}$$

Equation 8.30 defines the voltage across the open terminals of a hydrophone due solely to thermal agitation.

Another source of thermal noise is that due to agitation of the water

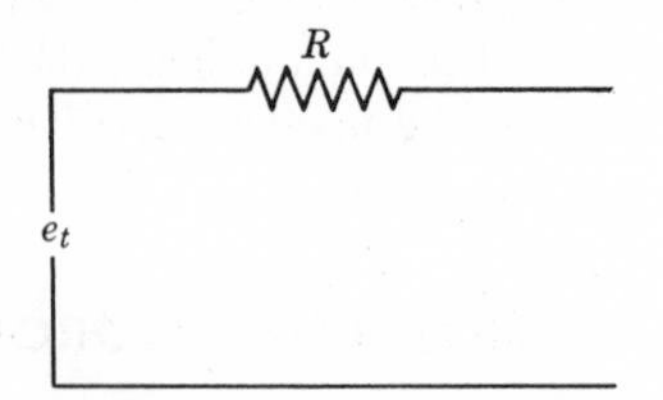

FIG. 8.10 emf from thermal agitation.

molecules bombarding the hydrophone. Its effect is defined in terms of an equivalent spectrum level which is the noise level that would be seen by an omnidirectional transducer of 100% efficiency. An analysis by Mellen[6] shows that this level is L_t

$$L_t = 20 \log \left(\frac{kT\rho\omega^2}{\pi c}\right)^{1/2} \tag{8.31}$$

or

$$L_t = 20 \log f - 194.8 \text{ dB ref N/m}^2 \text{ Hz}$$

8.6.2 Equivalent noise pressure level

Another useful concept is that of an acoustic wave in water which will develop a noise voltage across the electrical terminals equal to that generated by the noise of the hydrophone circuit. Obviously this sets a lower limit to a measurable signal level and is referred to as the threshold of the hydrophone. By definition

"... The equivalent noise pressure of an electroacoustic transducer or system used for sound reception is the root-mean-square sound pressure of a sinusoidal plane progressive wave, which, if propagated parallel to the principal axis of the transducer, would produce an open circuit signal voltage equal to the root-mean-square of the inherent open-circuit noise voltage of the transducer in a transmission bandwidth of 1 Hz and centered on the frequency of the plane sound wave."[7]

It is not a simple procedure to determine this value experimentally, but knowing the free voltage response and loaded impedance one may compute a theoretical value from

$$p_{en} = \frac{e_t}{M_0} \tag{8.32}$$

where p_{en} = the equivalent noise pressure,
M_0 = the free-field voltage response,
and from (8.30) the result expressed in decibels is

$$20 \log p_{en} = 10 \log R - 20 \log M_0 - 198 \text{ dB}$$

referred to 1 dyn/cm² at 15°C for a 1-Hz band (8.33)

An expression which shows explicitly the role of the hydrophone efficiency η and directivity, D, in fixing the value of the equivalent noise pressure level

[6] R. H. Mellen, "The Thermal-Noise Limit in the Detection of Underwater Acoustic Signals," in *J. Acoust. Soc. Am.*, **24,** 478 (1952).
[7] American Standard Acoustical Terminology.

is obtained from[8]:

$$M_0 = \lambda\left(\frac{R_T\eta D}{\pi\rho c}\right)^{1/2} \tag{8.34}$$

where λ = wavelength of sound in the medium,
R_T = series input resistance,
D = directivity factor,
ρc = characteristic acoustic impedance of the medium,
with (8.30) which yields

$$p_{en} = \left(\frac{4kT\pi\rho c}{\lambda^2\eta D}\right)^{1/2} \tag{8.35}$$

For an omnidirectional hydrophone with an efficiency of 100%, (8.35) reduces to (8.31) defining the thermal noise limit in the sea. The significance of the preceding expression may best be appreciated by referring to Figure 8.11. The curves lettered A to F are the Knudsen curves representing noise due largely to wind, wave motion, and other natural causes. These are averages of many measurements and fairly large random deviations from the curves may be observed. The dashed line shows the thermal noise limit defined by (8.31). An extension of the frequency range of Figure 8.11 and the noise curves would show the thermal noise curve intersecting the ambient noise curves, which means that ambient sea noise cannot be measured at these frequencies. An incoming acoustic signal lower than the equivalent pressure

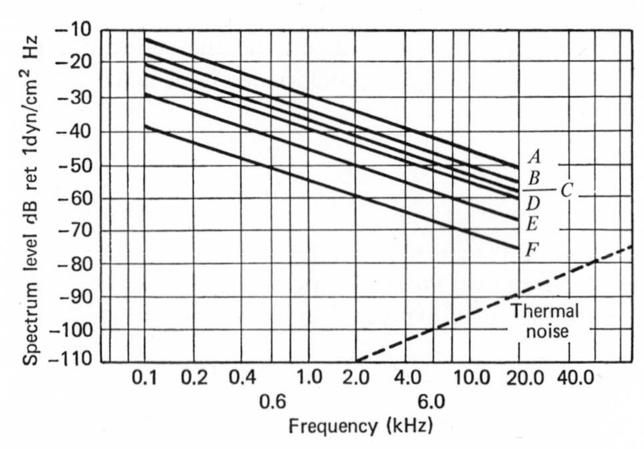

FIG. 8.11 Deep sea ambient noise (Knudsen). Noise sea state for ambient curve: $A = 6$, $B = 4$, $C = 3$, $D = 2$, $E = 1$, $F = 0$.

[8] P. M. Kendig, "Factors that Determine the Equivalent Noise Pressure, Free-Field Voltage Response, and Efficiency of a Transducer at Low Frequencies," in *J. Acoust. Soc. Am.*, **33,** 674 (1961).

level will be masked by the inherent noise of the hydrophone. Amplification does not improve this situation, but from (8.35) we can see that we can improve the signal advantage over noise by the number of decibels that can be gained by efficiency and by directivity.

8.7 HYDROPHONE SENSITIVITY

Although hydrophone sensitivity alone does not fix the threshold, we must know its value approximately in order to fit it into a system and, of course, accurately for its special use as a standard instrument for the measurement of sound fields. The following discussion suffices to predict the approximate value; a precise value must be determined by a reciprocity calibration or its equivalent.

For practical reasons we shall choose from among the many types and materials, the piezoceramic tube or cylinder because it fills a broader list of requirements than any other type. Figure 8.12 shows the active component of a piezocermaic hydrophone. It may be a continuous tube, as shown, or a cylindrical assembly of thin-walled rings. The electrodes may be arranged so that the mechanically induced electric field lies along any one of the three orthogonal axes. The interior of the tube may be isolated from the medium or it may be liquid filled for compensation against hydrostatic pressure. In the former case we must consider the effects of great static forces on the ceramic walls.[9]

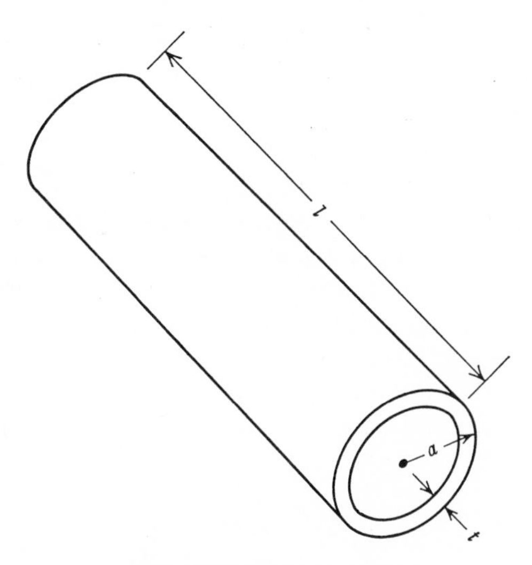

FIG. 8.12 Ceramic tube.

[9] Berlincourt, Curran, and Jaffe, in *Physical Acoustics*, Vol. 1, Part A, P. Mason Ed., Academic Press, New York, 1964, pp. 210–217.

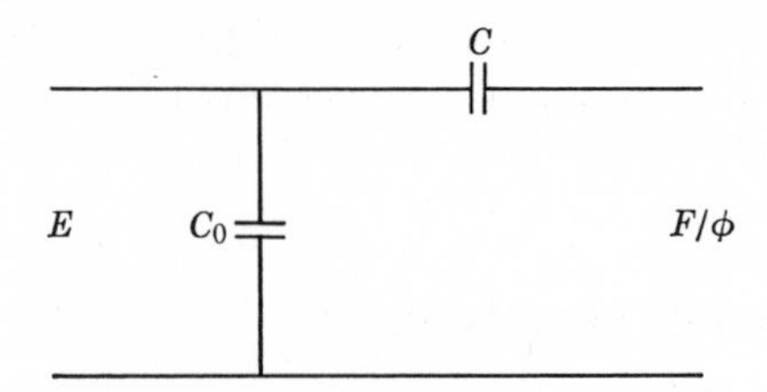

FIG. 8.13 Equivalent circuit for a piezoceramic hydrophone below resonance.

At frequencies well below the first fundamental resonance, the simple equivalent circuit of Figure 8.13 will apply to any one of the arrangements just mentioned. In the figure F represents the force applied to the hydrophone. It is the product of the rms acoustic pressure by the effective hydrophone area. By action of the transformation φ the force becomes a voltage which produces a current I through C_0 of

$$I = \frac{F}{\varphi} J\omega\left(\frac{CC_0}{C + C_0}\right) \tag{8.36}$$

and the current I develops an open circuit voltage across C_0 of

$$E = \frac{F}{\varphi}\left(\frac{C}{C + C_0}\right) = \frac{F}{\varphi} k^2 \tag{8.37a}$$

$$\frac{E}{p} = M_0 = \frac{k^2}{\varphi} S \tag{8.37b}$$

where S is the effective hydrophone receiving area.

To apply (8.37b) to a specific case the applicable values of k^2 and S must be determined. As an example, one of the most popular receivers is a barium titanate thin-walled tube of dimensions quite small compared to the wavelength of the sound signal being received. The interior of the tube is isolated from the medium either by sealing the ends or by the application of pressure release material to the interior wall. Approximate expressions for the parameters which determine the receiving sensitivity are

$$k_{31}^2 = \frac{d_{31}^2}{s_{11}^E \epsilon_{33}^T}$$

$$S = 2\pi a l$$

$$\varphi = \frac{2\pi l d_{31}}{s_{11}^E}$$

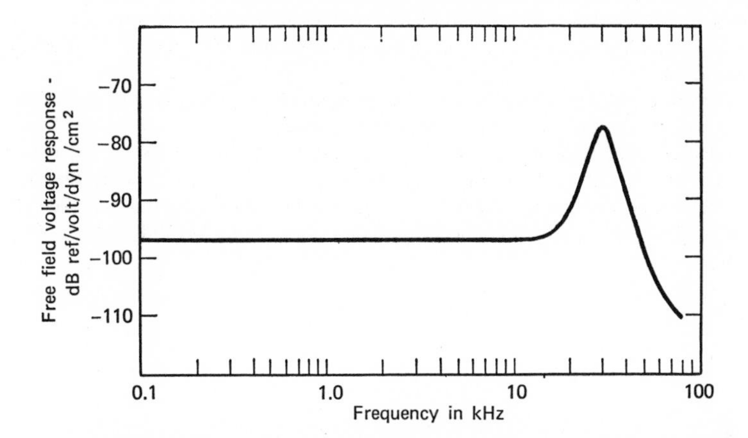

FIG. 8.14 A typical open circuit voltage response from a piezoceramic tube.

which substituted into (8.37b) gives

$$M_0 = \left(\frac{k_{31}{}^2 a^2 s_{11}{}^E}{\epsilon_{33}{}^T}\right)^{1/2} \tag{8.38}$$

Figure 8.14 is an example of the receiving response from a barium titanate thin-walled tube having a mean diameter of 5 cm. Variations from the values shown here result from the manner of polarization, the active material used, the mounting, and the ratio of wall thickness to diameter of the tube.[10] The latter ratio is involved because the piezoelectric radial and tangential strain constants are of opposite signs. Ratios of wall thickness to diameter between 0.2 and 0.5 must be avoided.

PROBLEMS

8.1 Given a projector with $Q_m = 10$, find the number of cycles required of a sinusoidal driving force to reach approximately 95 percent of the steady state. At a frequency of 5 kHz how far would the leading edge of the pulse have traveled?

8.2 In the MKS system J the reciprocity parameter has the units Volt Amperes m^2/N^2. There is a recommendation that the standard unit for pressure be 10^{-6} N/m^2. What change would this make in J?

8.3 Describe the reciprocity calibration of two identical transducers.

8.4 Let us assume that all three transducers used in a reciprocity calibration are linear, passive, and reciprocal. Write out the responses M_0 and S_0 for each in terms of J_1, e_1, e_2, and e_3.

[10] R. A. Langevin, "The Electro-Acoustic Sensitivity of Cylindrical Ceramic Tubes," in *J. Acoust. Soc. Am.*, **26,** 421 (1954).

8.5 Figures 8.3, 8.4, and 8.5 give the data for a free-flooding magnetostrictive scroll. Compute the quality factor Q_m, the efficiency at 4 kHz, and the open circuit receiving response at 4 kHz.

8.6 Check the values given in Figure 8.9 for the optimum Q_m for PZT, Batio$_3$, and nickel, and the power limits per unit area of these three materials.

8.7 Given a sealed barium titanate tube: o.d. $= 5$ cm, i.d. $= 4.5$ cm, $l = 10$ cm, $k_{31} = 0.19$, $s_{11}{}^E = 8.6 \cdot 10^{-12}$, $\epsilon_r/\epsilon_0 = 1200$, find M_0. Twenty of these tubes are assembled in a line and equally spaced to a total length of 15 m. Find the directivity index at 1000 Hz. Find M_0 and C_0 for a parallel connection of the tubes. Find M_0 and C_0 for a series connection.

CHAPTER 9

Listening and Echo Ranging

The science of underwater acoustics, while now embracing an ever-expanding number of engineering areas, had its origin in the development of the technology of ocean surveillance by the Department of Defense. It is most informative to follow this development from the early primitive attempts to the present.[1] It has long been known that water is quite superior to air as a medium for sound transmission, and recurrent investigations of possible methods have proven sound to be superior to other means for transmitting information through water. However, it was not until the submarine became an effective weapon for denying the use of the seas to shipping that underwater surveillance became a subject of paramount importance. In the beginning the interface between the two media of air and water was the barrier to an effective transfer of intelligence or power. This was removed by the development, which is also an interesting story,[2] of electroacoustic transducers matched to each medium with an electric cable serving as the connecting link.

9.1 SONAR

The growing technology of acoustics in a liquid medium rapidly found many applications[3,4] and also many misapplications. A useful designation for those systems relevant to our field of interest is the word, sonar, an acronym for *so*und *na*vigation and *r*anging. While the term may be applied to many nonmilitary functions, it does not include such things as sonic cleaning or sonic processing. A new science, ocean engineering, of great

[1] E. Klein, "Underwater Sound and Naval Acoustical Research and Applications Before 1939," in *J. Acoust. Soc. Am.*, **43,** No. 5, p. 931 (May 1968).
[2] F. V. Hunt, *Electroacoustics*, Wiley, New York, 1954, ch. I.
[3] L. Bergmann,
[4] Crawford,

military and commerical potential[5] is being formulated from the growing body of knowledge.

9.1.1 Passive sonar

For a necessarily limited survey we may consider the primary objectives of surveillance as being detection, classification, and tracking of underwater objects. Also, as in the early stage of sonar, that this is being done either by listening or by echo ranging systems.

The tactical advantage of the submarine is the possibility of coming within attack range without being observed. The target, if a surface ship, is noisy, and under favorable conditions is detectable by the use of listening devices at much greater ranges than those obtained by echo ranging. Therefore the submarine will use a passive system for detection, identification, and bearing, and send out a single pulse for range finding. Listening systems also have weight, cost, and power advantages over active systems. This, in addition to broad-band capability, recommends them also for coastal and harbor defense systems.

We may note that passive sonar systems are at a disadvantage in some respects: they are more easily decoyed; a series of echoes from a target may be processed to give range information quite superior to that from a single pulse; the self-noise of the system may resemble closely that of the target, causing a more serious masking problem.

9.2 NOISE

A discussion of noise must begin with the definition of a number of basic concepts that have become standardized by general acceptance. This will be done without going into an advanced analysis, a subject belonging to signal processing. We may classify the types as background noise, ambient noise, self noise, thermal noise, circuit noise, etc., any one of which may be the major problem in a given situation. For instance, the term background noise includes all sources of sound that tend to mask the signal; for sonar on helicopters, the airborne component of background noise not only limits aural reception, but also penetrates the air–water interface where it is picked up by the transducer.

The sound operator using the listening mode of target detection is receiving a signal with properties only slightly different from the background noise, but he must depend on these differences for recognizing and classifying the target. One of these is the character of the spectrum of the noise and of the

[5] J. P. Craven, "An Assessment of the Future of Deep Ocean Technology," in *Astronaut Aeronaut.*, 38 (July 1967).

signal. For either, the sound wave structure may be quite complex, consisting of a continuous distribution of components differing in phase, amplitude, and frequency. The wave spectrum is the description of these components as a function of frequency. Sound consisting of a number of discrete frequencies is said to have a line spectrum, whereas a sound resulting from a continuous distribution of frequencies has a continuous spectrum. A region between two frequencies f_1 and f_2 is called a frequency band and the numerical difference between the two is referred to as the bandwidth. A graphical definition of the spectrum of a sound is obtained by plotting its spectrum level as a function of frequency over the range of interest. The spectrum level at a given frequency is the level in terms of pressure, intensity, or power of that part of the sound within a band, 1 Hz in width, centered at the given frequency. Similarly we may define levels for bands of any chosen width.

Figure 8.11 shows two kinds of spectra of noise levels in the deep sea. One set illustrates how sea noise varies with sea state and with frequency. The other is the spectrum of thermal noise in the ocean. The spectra of other kinds of noise such as self noise and traffic noise are not so generally definable.

9.2.1 Masking by noise

The desired signal is more or less immersed in the noise background; for example, noise produced by the target. If its spectrum matches that of the background, detection will depend on a change in sound level only. An arbitrary but accepted criterion of audibility is recognition of the signal 50% of the time. The level at which it becomes audible against the background noise is termed the masked threshold. While the magnitude of the masked threshold is determined by the noise level, it is measured by the level of the signal. The masking effect is subject, to a large degree, to the character of the two spectra involved. There may be masking of one pure tone by another pure tone, a pure tone by complex sound, or of a complex sound by another complex sound. The first case would not be encountered in passive listening, but the second condition exists when the target is echo-ranging or producing a spectrum with some discrete frequencies. Aural listening is particularly useful for detection because the ear is a very effective analyzer in the audible range. As a result the masking effect, by the background, of a pure tone is limited to a narrow band, centered at the frequency of the tone, called the critical bandwidth. The Americal Standards Association defines "the aural critical band is that frequency band of sound, being a portion of a continuous spectrum noise covering a wide band that contains sound power equal to that of a pure tone centered in the critical band and just audible in the presence of the wide-band noise." To be just audible the pure tone must have a level

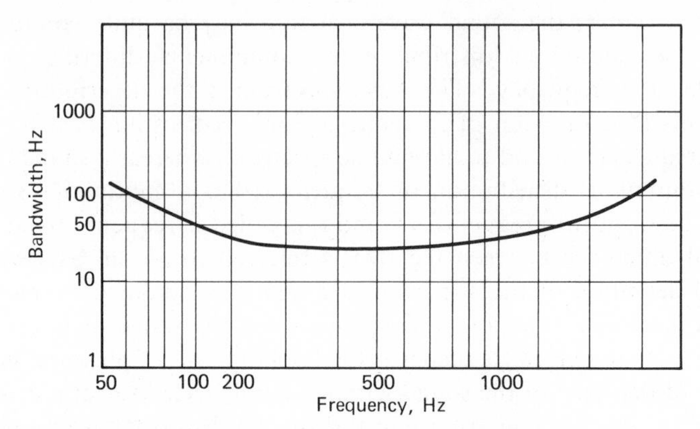

FIG. 9.1 Critical bandwidths for masking pure tones.

which exceeds the spectrum level of the continuous noise by 10 times the logarithm of the critical bandwidth.

The concept of critical bandwidths also applies to the situation where both the signal and the background are complex. Since the listener may choose the bandwidth and its location in the spectrum, he makes the most advantageous selection. Figure 9.1 shows how critical bandwidths for masking pure tones vary as a function of the tone frequency.

The level of a sonar field is almost invariably determined by a pressure measurement, and the spectrum described is a pressure spectrum. The use of the decibel in measuring pressure levels is justified by the assumption that the expression

$$L = 10 \log \frac{I}{I_0} = 10 \log \left(\frac{p^2/\rho c}{p_0^{\,2}/\rho_0 c_0} \right) = 10 \log \left(\frac{p^2}{p_0^{\,2}} \right)$$

is approximately true for waves in one direction within a single medium. If the sound field consists of a number of components differing in frequency and phase the resultant rms pressure squared is given by

$$p^2 = p_1^{\,2} + p_2^{\,2} + \cdots + p_n^{\,2}$$

and the sound pressure level for a band with a continuous spectrum is

$$L = 10 \log \int_{f_1}^{f_2} \frac{p^2(f)}{p_0^{\,2}}\, df \tag{9.1}$$

For a constant value of the pressure spectrum level L_s over the band (9.1) yields

$$L = L_s + 10 \log w \tag{9.2}$$

where $w = f_2 - f_1$. When the band is narrow, the spectrum level at the midfrequency may be used for L_s in (9.2). For a wider band a fair approximation to the overall sound pressure level may be obtained by using the spectrum level at the geometric mean of the upper and lower frequencies of the band.

In order to write the sonar equation for passive listening, we must define another level called the recognition differential and usually represented by the symbol M. Again quoting the American Standards Association, "the recognition differential for a specified aural detection system is that amount by which the signal level exceeds the noise level presented to the ear when there is a 50% probability of detection of the signal. The bandwidth of the system, within which signal and noise are presented and measured must be specified." Depending on the detection system and the physical conditions, this level may be zero, positive, or negative.

The first level involved in listening is the source level S of the target. It is measured at an appropriate distance and converted to a hypothetical value one yard from the apparent origin. At a distance r yards from the source, the signal level L_r will be

$$L_r = S - H(r) \tag{9.3}$$

where S is the overall noise level of the target from (9.1) and where $H(r)$ is the transmission loss in decibels. Since most acoustic measurements in water are made with pressure-sensitive instruments, sound pressure levels are appropriate for (9.1). These are absolute values and the reference level must be definitely understood. In line with our definition of the recognition differential, a 50% chance of detection requires that $L(r)$ be equal to, or greater than the sum of the background noise and the recognition differential. If the noise is isotropic, its effective level is the isotropic noise level, L_N, decreased by the directivity of the receiving hydrophone. Therefore, for signal recognition

$$L(r) \geq L_N - D + M_N \qquad \text{in decibels} \tag{9.4}$$

and solving the the maximum acceptable transmission loss

$$H(r) \leq S - (L_N - D + M_N) \qquad \text{in decibels} \tag{9.5}$$

Because of the complexity of the problem only a general discussion of the factors affecting listening ranges will be undertaken. These factors are the purpose of detection, the character of the target signal, the listening band, background noise, self noise, the receiving directivity, and signal processing. A harbor defense system might well use the same listening band as would a fixed system in the relative quiet depths of the sea, but its problem in recovering the signal would be quite different. Homing torpedoes aimed at submarines would not be the most efficient against commercial shipping. Two general principles apply in both cases: (a) over the chosen listening band ambient water noise should be audible above the self-noise of the system, and (b) the response curve of the receiving system should be such that all components of the ambient noise are presented at the same level. In meeting the first requirement, the design engineer has reduced the equivalent noise pressure level of his system below water noise, and in meeting the second, he will not be overloaded by some sections of the spectrum at the exclusion of other sections that could assist in identification.

9.3 DETECTION RANGES

The ship spectrum shown by Figure 9.2 is a greatly simplified version of a set of measurements which on any one occasion would show wide deviations from the average. It is presented to illustrate the 6–7 dB per octave slope common to ship's noise and the single frequency spikes that may be produced by some part of the ship's machinery. The general level will vary with type of ship, its tonnage, and its speed. For illustrative purposes this ship will serve as a target to be detected by a broad band listening system limited by ambient noise. The effective level of masking noise is

$$L_N = N_f + 10 \log w - D \tag{9.6}$$

where w = the listening bandwidth,
N_f = spectrum level of noise at center frequency of w,
D = directivity index of hydrophone.

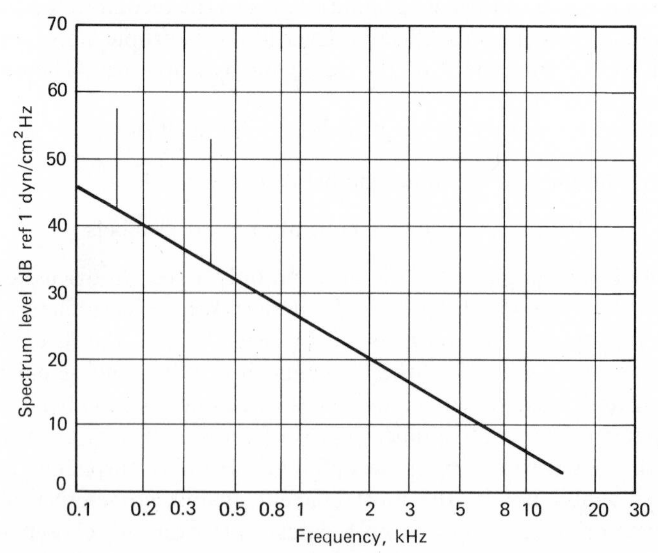

FIG. 9.2 A ship spectrum showing single frequency spikes.

By definition the required signal level for detection 50% of the time is

$$L_s = L_N + M_N \tag{9.7}$$

where M_N = recognition differential for noise. We also have for the same condition

$$S - H_m = L_s \tag{9.8}$$

where H_m = maximum allowable transmission loss for 50% detection,
S = source level.

Both transmission loss and source level must be clearly defined for a particular situation. At any appreciable distance from a ship, the sound from it may be treated as if from a point source. If the effective sound pressure adjusted to unit distance from the apparent point is p_0, then

$$p = \frac{p_0}{r} e^{-\alpha r} \tag{9.9}$$

where p = effective sound pressure at a distance r from the ship, or taking 20 log of (9.9)

$$L(r) = S - 20 \log r - ar \tag{9.10}$$

for

$$S = 20 \log p_0$$

and

$$a = 20\alpha \log_{10} e$$

The term $-20 \log r$ represents the spreading loss of a spherical wave. Because of refraction and reflection effects the transmission loss may vary considerably. Sound absorption, represented by the term ar has been the subject of a great deal of research in an effort to account for high values in seawater. Although this transmission loss is negligible in the low sonic range it becomes a limiting factor at the frequencies used for torpedo guidance, and has encouraged the use of lower frequencies for echo ranging.

The target source may produce a single frequency of level S or it may be a broad band source of level $S_f + 10 \log w$ where S_f is the spectrum level at the midfrequency of the band. Figure 9.2 was chosen to represent both cases. A relevant question is the importance of the single frequency spike for aural listening. Let there be a spike of intensity I_s:

$L_s = 10 \log I_s$ = level of spike

$L_f = 10 \log wI_f$ = level of a band w of sound in the absence of the spike with w centered at the frequency of the spike.

Then

$$\text{total intensity} \quad I_T = I_s + wI_f$$

and the resulting level S_T will be

$$S_T = 10 \log (wI_f + I_s)$$

an increase over L_f of

$$S_T - L_f = 10 \log \left(1 - \frac{I_s}{wI_f}\right) \tag{9.11}$$

For aural listening the bandwidth w is w_c selected from Figure 9.1 at the frequency of the spike. From the definition of w_c the spike will be heard 50% of the time if its intensity $I = w_c I_f$. With this condition satisfied, the increase in level given by (9.11) will be 3 dB. The ship's spectrum of Figure 9.2 shows a spike at 400 Hz, whereas Figure 9.1 indicates a critical bandwidth of 30 Hz at this frequency. Therefore for this example we have

$$L_s = 10 \log I_s = 52 \text{ dB} \qquad I_s = 10^{5.2}$$

$$L_f = 34 + 10 \log 30 = 48.8 \text{ dB} \qquad w_c I_f = 10^{4.88}$$

and substituting in (9.11)

$$S_T - L_f = 10 \log (1 + 10^{5.2-4.88}) = 4.9 \text{ dB}$$

If this spike is an intermittent effect, it may be readily detected.

Suppose that we are listening to the 400-Hz spike of Figure 9.2 with a hydrophone of directivity D of 15 dB located in a noise field corresponding to that of a sea state of 3. What is the maximum detection range? Combining (9.6), (9.7), and (9.8), we have

$$H_m = S - N_f - 10 \log w_c + D - M \tag{9.12}$$

The transmission loss is usually written to include a term A, called the anomaly, covering all deviations that cannot be accounted for, so

$$H_m = 20 \log r_m + ar_m + A \tag{9.13}$$

where r_m is the range for 50% detection. At low sonar frequencies the absorption term may be neglected, and omitting A we have

$$20 \log r_m = S - N_f - 10 \log w_c + D - M \tag{9.14}$$

From Figure 8.11 and sea state 3, $N_f = -30$ dB at 400 Hz. Also

$$w_c = 30 \text{ Hz}$$

$$D = 15 \text{ dB}$$

$$M_N = 0 \quad \text{for aural listening}$$

$$S = 10 \log (I_s + wI_f) \quad \text{where} \quad I_f = 10^{0.1S_f}$$

The insertion of these quantities in (9.14) yields

$$20 \log r_m = 53.6 + 30 - 14.8 + 15 = 83.8 \text{ dB}$$

or

$$r_m = 15{,}480 \text{ yd}$$

Considering the lack of precision of (9.14), one may wish to ignore the contribution of $w_c I_f$ to the source level S. In the above example this would lead to

$$20 \log r_m = 52 + 30 - 14.8 + 15$$

or

$$r_m \approx 13{,}000 \text{ yd}$$

Also, if the listening band includes no spike (9.14) becomes

$$20 \log r_m = S_f - N_f + D - M_N$$

9.4 ECHO RANGING

The major difference between passive and active sonar is one of tactics. Both are used defensively and offensively. Since a destroyer cannot conceal its presence, nothing is lost by its utilization of the advantage of echo ranging. On the other hand, the submarine may engage in offensive action without revealing its location, and therefore will limit its use to the minimum required for getting a range. Additional factors involved in active sonar are sound sources, target strength, Doppler, choice of frequency, ping length, and reverberation.

9.4.1 The range equation

We shall first introduce the equation for estimating the probable range and then discuss each of its terms. The most common systems will employ frequency modulated[6] signals or short bursts of constant frequency sound. In the latter case, a signal of level S undergoing a transmission loss, H, will be reflected by a target of strength, T, and after another transmission loss will return to the sender as an echo with a level L_E. We may express this event symbolically by

$$L_E = S + T - 2H \tag{9.15}$$

where identical transmission losses in the two directions are assumed.

For a 50% detection, L_E must have a level if noise limited of

$$L_E = N_f - D + M_N + 10 \log w \tag{9.16}$$

where

$$N_f = \text{noise spectrum level}$$

If reverberation limited

$$L_E = N_R - D + M_R + 10 \log w \tag{9.17}$$

[6] "Principles of Underwater Sound," Div. 6, Vol. 7, NDRC Summary Technical Reports, pp. 214–219.

where N_R = reverberation spectrum level and subscripts indicate that the recognition levels may not be identical. Equations 9.16 and 9.17 are equivalent to a combination of (9.6) and (9.7). From (9.15) and (9.17) for 50% recognition

$$S + T - 2H_M = N_f - D + M + 10 \log w \tag{9.18}$$

or

$$2H_M = S + T - (N_f + 10 \log w - D + M) \tag{9.19}$$

The resultant of terms on the right is the available signal output, and $2H_M$ is the allowable transmission loss for 50% recognition.

9.4.2 The echo ranging source

The design of the echo ranging source has been a compromise between directivity gain at high frequency with an attendant high transmission loss due to absorption, or negligible absorption at low frequency with very massive equipment required for directional gain. Increasing speed of targets forced a development in the direction of longer ranges, which in turn called for more acoustic power from the source at lower frequencies. Fortunately these two requirements are compatible, and transducer technology has successfully met the need. The subject of steerable arrays is introduced by Section 7.7.1. By using the type of construction illustrated by Figure 9.3, sound sources of any reasonable power level are feasible. This transducer consists of a cylindrical arrangement of columns, or staves of half-wave vibrators such as those described in Chapter 6. By driving all these in phase with equal amplitudes a sound field is produced that is uniform in the horizontal plane, but directional in the vertical plane. Hundreds of kilowatts are available from such sources. Directional patterns both for receiving and transmitting may be formed and rotated in the horizontal plane at great speed for scanning. Further details may be found in volume 16 of the Summary Technical Reports.[7] The accepted value for S is the equivalent pressure level one yard from the transducer when this unit is operating at its rated input power. Or

$$S = P + \eta + D + 71.5 \tag{9.20}$$

where P = input power in dB ref. one watt,
η = conversion efficiency,
D = directivity gain.

One may get some conception of source levels by observing that an omnidirectional source having a kilowatt power input will develop a source level near 100 dB, which transducers designed for directional gain may improve by as much as 25 dB.

[7] Div. 6, NDRC Summary Technical Reports; (a) p. 100; (b) p. 196.

FIG. 9.3 Echo-ranging sound source.

9.4.3 Target strength

An object in the path of a transmitted sound pulse intercepts a quantity of acoustic power and reradiates something less than that quantity in all directions. The effectiveness of that object as a target is determined by the fraction of intercepted power returning toward the receiver. The area of the target cross section and its reradiating efficiency in the desired direction are combined in a definition of the back scattering cross section, σ_b with

$$\sigma_b = \frac{4{p_{sb}}^2 r_0^2}{{p_i}^2} \tag{9.21}$$

where p_{sb} = backscattered sound pressure,
p_i = incident sound pressure,
r_0 = unit distance from the acoustic center of the object.

Because targets are so irregular in shape and in scattering behavior, any other but an empirical definition of backscattering cross sections is hardly feasible. However, the difficulty of measuring the backscattered sound pressure detracts from the usefulness of (9.21). This pressure must be measured at a considerable distance from the target with the resultant errors due to transmission losses and distance determinations. We may rewrite the equation as

$$\frac{{p_{sb}}^2}{{p_i}^2} = \frac{\sigma_{sb}}{4\pi r_0^2}$$

and

$$T = 10 \log \frac{{p_{sb}}^2}{{p_i}^2} = 10 \log \frac{\sigma_{sb}}{4\pi r_0^2} \tag{9.22}$$

A simple illustration of the use of (9.22) is its application to a sphere of radius a. This object will intercept an area of πa^2 and assuming perfect reradiation equally in all directions

$$T = 10 \log \frac{a^2}{4{\gamma_0}^2} = 20 \log \frac{a}{2} \tag{9.23}$$

referred to γ_0 as unit distance. For example a sphere with a 2-yd radius would show a target strength of zero dB, and one having a 10-yd radius would show a target strength of 14 dB, a level in the normal range of target strengths shown by submarines.

9.4.4 Reverberation

With the introduction of high-level sound sources as a means of increasing detection ranges, reverberation rather than noise becomes limiting. As a backscattering of the echo-ranging pulse, reverberation has a rather definite pitch and varies in level with that of the pulse. Both of these characteristics add to its masking effect on the echo. Backscattering results from bottom and surface irregularities and from echoes off bubbles, solid particles, and other inhomogeneities distributed throughout the ocean. There is evidence[8] that these scatterers may form layers in the deep ocean. By making the two assumptions that scattering particles are equally distributed throughout the sea and are of the same size, we may arrive at a qualitative but useful notion of reverberation levels. This follows conveniently from the discussion of target strength. Repeating (9.20) for $\gamma_0 = 1$ yd:

$$T = 10 \log \frac{\sigma_{sb}}{4\pi}$$

We must establish a value for σ_{sb}, the effective backscattering area. Figure 9.4 is a diagram indicating a section from the sound field producing reverberation. Given a sound pulse of duration to seconds: it will consist of a train of waves ct_0 in length for c, the velocity of sound. After a time, t, the unscattered part of the sound train will fill the outer spherical shell of radius, ct, and thickness, ct_0. If the spherical shell shown at the distance, r, represents the volume producing reverberation at the time t, the returning waves must have traveled the distance r and returned during that time, fixing r at $ct/2$ and Δr as $ct_0/2$. Thus the volume producing reverberation is

$$V = \frac{\Omega r^2 c t_0}{2} \tag{9.24}$$

where Ω represents the solid angle subtended by the shell. Under our assumption that N scattering objects each of uniform scattering area σ are uniformly distributed in V and that there are no secondary scattering effects,

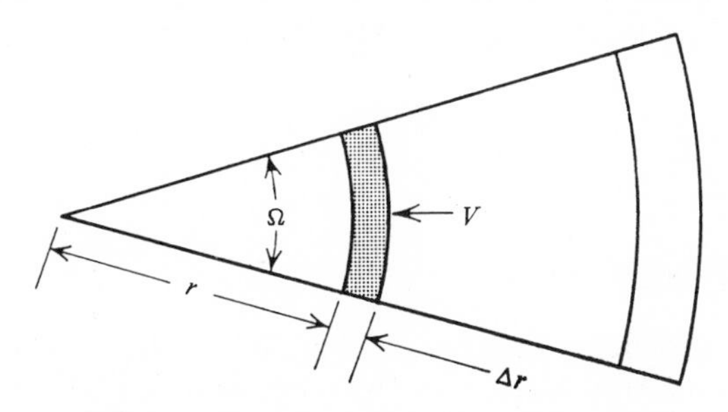

FIG. 9.4 Volume producing reverberation.

[8] *Ibid.*, p. 100.

the total backscattering effect of V will be that of a single target given by

$$T_R = 10 \log \frac{N\sigma V}{4\pi} = 10 \log \frac{N\sigma\Omega r^2 \Delta r}{4\pi} \tag{9.25}$$

Then corresponding to (9.15) which gives the returned echo level L_e as

$$L_e = S + T - 2H \tag{9.15}$$

We have a reverberation level

$$L_R = S + T_R - 2H \tag{9.26a}$$

or inserting the definition of T_R from (9.25) and twice the one-way transmission loss:

$$2H = 40 \log r + 2ar + 2A$$

we have

$$L_R = S + 10 \log N\sigma + 10 \log \frac{c\,\Delta t}{2} + 10 \log \frac{\Omega}{4\pi} - 20 \log r - 2ar - 2A \tag{9.26b}$$

The terms containing $N\sigma$, the number of scattering objects per unit volume, $2a\gamma$, the attenuation loss, and $2A$, the anomaly, experience large variations with place and time, and therefore are quite unpredictable. To assess the significance of the other terms, let us consider their weight in the equation for 50% recognition where the echo is masked by reverberation:

$$L_e = L_R + M_R \tag{9.27}$$

and L_R involves the source level, S, the pulse length $c\,\Delta t/2$, the directivity factor $\Omega/4\pi$, and the range r. Equating the L_e of (9.27) with that of (9.15) shows that an increase in source level does not improve recognition. Also we may observe that the reverberation level increases with the pulse length. Since the recognition differential M_R is also a function of pulse length, the two must be considered together. The echo level produced by a target decreases as 40 log r, whereas the reverberation decreases as 20 log r. Therefore if an echo is masked by reverberation at a given range, masking will be greater at longer ranges.

9.4.5 The Doppler effect

A familiar experience is the change in pitch that occurs when the distance between the sound and listener is changing at an appreciable rate. This change in pitch is known as the Doppler effect. A moving target intercepting sound pulses from a stationary sonar and reflecting an echo back presents two Doppler effects. The first is the shift in frequency when the receiver is in motion with respect to the fixed sound source. The second is the shift in frequency when the sound source—the echo—is in motion with respect to the receiver, which for echo ranging is the sonar sending out the original pulse. Let f_0 be the frequency of sound projected by the sonar and f_1, the frequency received by the moving target and transmitted as an echo. Then

$$\frac{f_1}{f_0} = \frac{c + u_t}{c} \tag{9.28}$$

where u_t represents a decreasing range rate of the target and c is the velocity of sound in water. Let f_2 be the frequency of the echo as received by the sonar. Then

$$\frac{f_2}{f_1} = \frac{c}{c - u_t} \tag{9.29}$$

From (9.26) and (9.27) we get

$$\frac{f_2}{f_0} = \frac{c + u_t}{c - u_t} \tag{9.30}$$

The total shift known as the target Doppler observed by a stationary sonar is the difference

$$f_2 - f_0 = f_0 \frac{2u_t/c}{1 - u_t/c} \simeq \frac{2f_0 u_t}{c} \tag{9.31}$$

If instead of being stationary the sonar is moving toward the target with a velocity resulting in a decreasing range rate, u_s, the resulting frequency shift known as own Doppler will be the same function of the range rate due to sonar motion. The combined Doppler effect will be the sum of target and own Doppler or

$$\Delta f = \frac{2f_0(u_t + u_s)}{c} \tag{9.32}$$

9.4.6 The effect of Doppler on aural recognition

During the transmission of the echo ranging pulse, the receiver is blocked so the operator hears only the reverberation and the echo. Since reverberation consists largely of echoes from fixed objects, the difference in frequency between the echo and reverberation is due to target Doppler. If ranging is reverberation limited, this frequency difference can effectively improve aural recognition. By way of explanation, let us recall (9.27):

$$L_e = L_R + M_R \tag{9.27}$$

and consider the effect of critical bandwidth, W_c, and pulse length on L_R, the reverberation level, and M_R, the recognition differential. If the target is

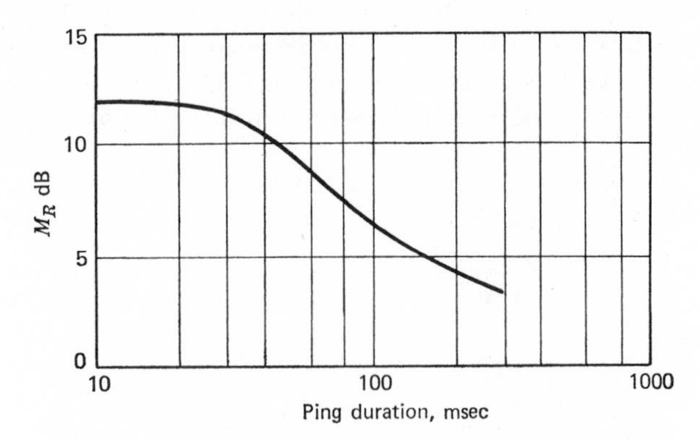

FIG. 9.5 Recognition differential as a function of ping length: zero Doppler.

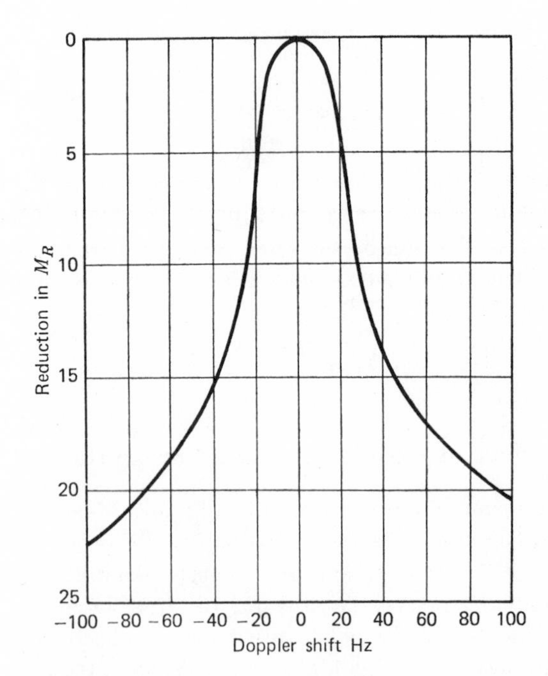

FIG. 9.6 Reduction in recognition differential for echoes with Doppler: —ping length = 114 msec.

showing zero range rate, the echo will be recognized only as an increase in sound level and M_R will be a function of ping length as shown by Figure 9.5.[9] If the target is showing a range rate, and the ping is of sufficient length to present a sound of definite pitch to the ear, its capability as a filter in discriminating against masking by reverberation is shown by Figure 9.6.[10] According to this curve, a Doppler shift of 40 Hz in a pulse 0.114 sec long will reduce the recognition differential 15 dB. The result is a considerable increase in a reverberation-limited range.

Figure 9.7 illustrates information to be gained by suitable processing of the return signal. The information in the unprocessed signal of Figure 9.7*a* is masked by reverberation. Figures 9.7*b* and 9.7*c* are the result of receiving the return signal through two channels; one showing the sound level and the other showing the Doppler shift. The latter is given by Figure 9.7*c* where the median line indicates no Doppler while a deviation up or down

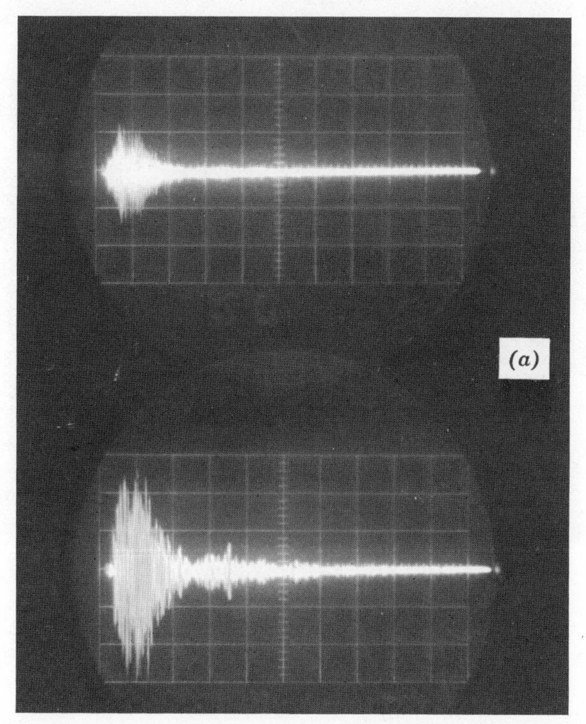

FIG. 9.7 (*a*) Reverberation-limited echo. (*b*) Sound level of unprocessed return signal. (*c*) Doppler shift of return signal. (*d*) Sound level of unprocessed return signal. (*e*) Showing no Doppler shift of return signal.

[9] *Ibid.*, p. 196.
[10] *Ibid.*

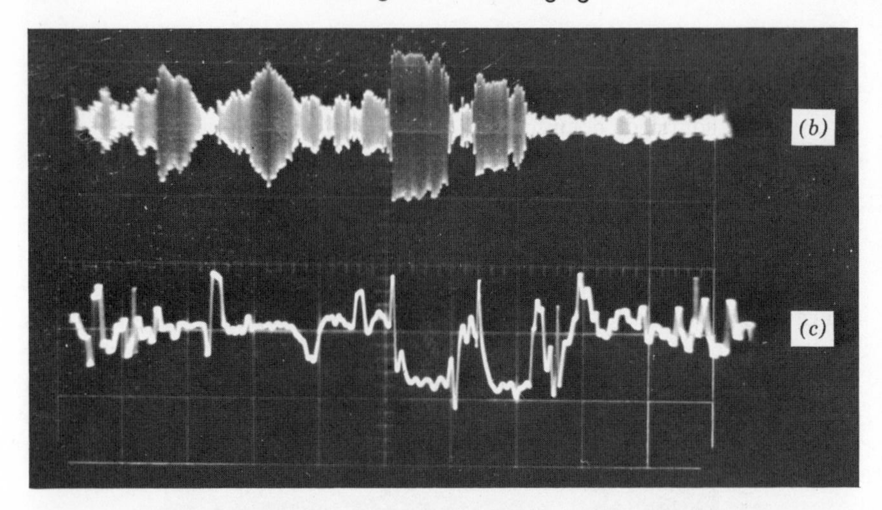

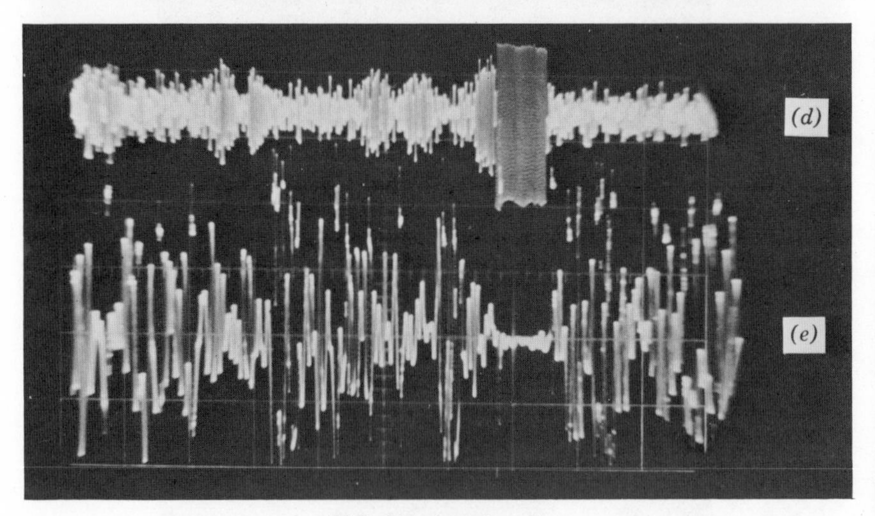

FIG. 9.7 (*continued*)

indicates a target approaching or moving away. Figure 9.7*d* shows a strong echo from a submarine while Figure 9.7*e* indicating no Doppler proves this echo to be off the target beam since this aspect presents no component of motion along the line of sight.

9.5 REVIEW OF SYMBOLS USED IN SONAR

A trying problem ever present in a discussion of sonar is a choice of symbols. The difficulty results not only from the large number necessary but also from the diversity of selection by current authors. The following is a review of those used in this chapter:

H Transmission loss, dB
H_M Maximum allowable loss, dB
I_f Intensity in a one-Hz band, W/cm²Hz
I_s Spike intensity, W/cm²
L Sound pressure level, dB/μbar
L_E Echo pressure level, dB/μbar
L_f Pressure level over a band, dB/μbar
L_N Background noise level, dB/μbar
L_S Signal level, dB/μbar
L_s Spike level, dB/μbar
M_N Recognition differential, noise dB
M_R Recognition differential, reverberation dB
N_f Noise spectrum level, dB/μbar in one Hz
S Source level, dB/μbar at one yard
S_f Source spectrum level, dB/μbar at one yard in one Hz
W Frequency band width, Hz
W_c Critical band for aural listening, Hz

PROBLEMS

9.1 The noise spectrum of an aircraft carrier may be represented by the equation $N_f = 55 - 20 \log f$, where f is frequency in kHz, and N_f is the acoustic pressure level in a 1-Hz band in microbars. (*a*) Using a coordinate system similar to that of Figure 9.2, plot the spectrum level from 100 Hz to 60 Khz. (*b*) Find the noise level for a 100-Hz band center at 1500 Hz. (*c*) Find the noise output over the band from 1 to 10 kHz using the spectrum level at the geometric mean frequency.

9.2 Figure 9.2 represents the pressure spectrum of a submerged submarine referred to an equivalent distance of one yard from its acoustic center. (*a*) Find the source level, S, for the noise band from 300 to 500 Hz including the spike. (*b*) Excluding attenuation and refraction effects, find the level over this band at 800 yards from the submarine. (*c*) How does this compare with the background noise over the same band in a sea of state 3? See Figure 8.11.

9.3 Given a detection system with a recognition differential of $M_N = 5$ dB, a passband w from 300 to 500 Hz, and a directivity D of 20, that is listening to a ship having the noise spectrum shown by Figure 9.2: (*a*) In a sea state of 3 find the effective level of noise masking the signal from the ship, (*b*) find the maximum allowable attenuation, H_m for 50% recognition, and (*c*) if there is an anomaly, A, of 10 dB due to refraction, and negligible attenuation by absorption, find the maximum range for 50% detection.

9.4 A stationary omnidirectional system employing aural listening is receiving a 4-kHz signal in background noise corresponding to a sea state of 4. At this frequency the critical listening band, w_c, for the ear is 200 Hz. The signal is being recognized 50% of the time. Give: (*a*) the signal to noise ratio, S/N, (*b*) the signal differential, $L_S - L_N = 10 \log (S/N)$, (*c*) the observed differential, $10 \log (S + N/N)$, and (*d*) the signal level. What advantage is gained by using a receiver having a 3 Hz filter and a recognition differential of 5 dB?

9.5 The available space and power allow a plane circular disc of 5-in. radius radiating an acoustic power of 1 kW. Absorption of sound by sea water is given approximately by $0.036f^2/(3600 + f^2)$ dB per yard where the frequency, f, is in kHz. Choosing 60 or 30 kHz as possible frequencies with which to drive the disc, compare the signal levels produced at 2000 yd.

9.6 A transducer with a source level of 140 dB/μbar at one yard is echo ranging on a submarine of target strength equivalent to that of a 120-ft diameter sphere. The transmitted frequency is 4 kHz. (*a*) What is the sound pressure level at 20,000 yd? What is the level of the echo returned to the source by the submarine at this distance? Assume the anomaly A to be zero.

9.7 The transducer of problem 9.6 has a receiving directivity of 20 dB. If operating in a sea state of 3, what is the signal differential for aural listening at 10,000 yd? Assume negligible absorption and no anomaly. What is the range for 50% recognition of the submarine?

9.8 A 4-kHz echo-ranging system provides a source level of 130 dB/μbar at 1 yd. Its receiver has a directivity of 20 dB, bandwidth of 100 Hz, and a recognition differential of 10. It is ranging on a submarine of target strength 15 dB. The limiting background noise pressure spectrum is -40 dB/μbar. Calculate the maximum range for 50% detection. What would be gained in range by increasing the source level to 150 dB/μbar at one yard? What will be the frequency of the echo if the range is decreasing at the rate of 20 knots?

CHAPTER 10

Sonar Signal Processing

10.1 INTRODUCTION

It is the job of the sonar system and its operator to determine the presence of targetlike signals in the acoustic fields near the receiving array, to aid in localization of the source of these signals, and to examine these signals to ascertain whether their source is a target of interest. The requirements for these three procedures are not necessarily met by the same signal processing and display equipment. In order to detect the presence of the targetlike signals when they are present at the greatest possible range, a system which makes use of all the signal energy seems intuitively to have an advantage over a system which looks for, let us say, spectral structure in the signal which represents only a fraction of the total signal energy. The classification procedure, that is, ascertaining whether the signal source is a target of interest, may succeed only if the structure of the signal is examined. Needless to say, a rather forbidding number of signal processing schemes have been developed for both active and passive systems for detection, localization, and classification. In the description and analysis of signal processing systems to be developed in the following pages, an attempt will be made to present a somewhat generalized view of signal processors with numerical relationships, describing their behavior, which can be used to determine the performance of specific processors. Emphasis will be placed upon systems for recognizing the presence of targetlike signals.

The generalized view is built around the concept of the resolution cell. It is assumed that all the space surrounding the sonar can be divided into cells, bins, or elements of volume, not necessarily the same size, each having the largest volume within which the sonar is not capable of distinguishing the presence of more than one acoustic source. That is, two sources within the same resolution cell will appear as a single source. In order for the sonar to distinguish two sources as separate sources, the two sources must be located in two different resolution cells.

The resolution cells are not really geometrical volumes in the ocean in the most general view, as will be seen presently. The simplest set of resolution cells which will be encountered are those associated with a wide band passive sonar. This sonar is capable of distinguishing acoustic sources as multiple sources provided the sources are separated in azimuth by an amount greater than φ, the angular resolution of the sonar. In this case the resolution cells are sectors of azimuth or bearing having angular width φ. A diagram of the azimuthal resolution cells is shown in Figure 10.1*a*. It is clear that the

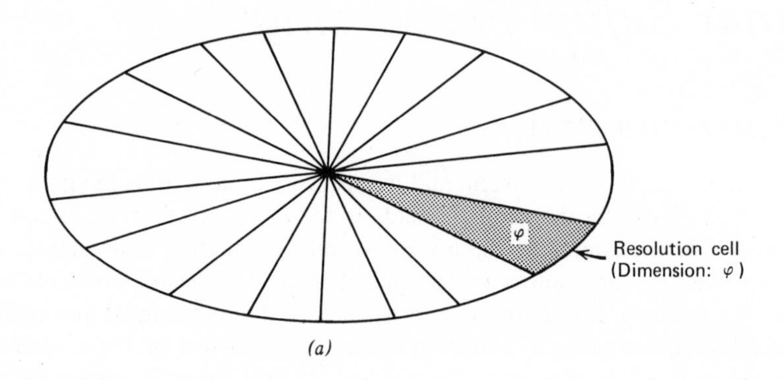

(*a*)

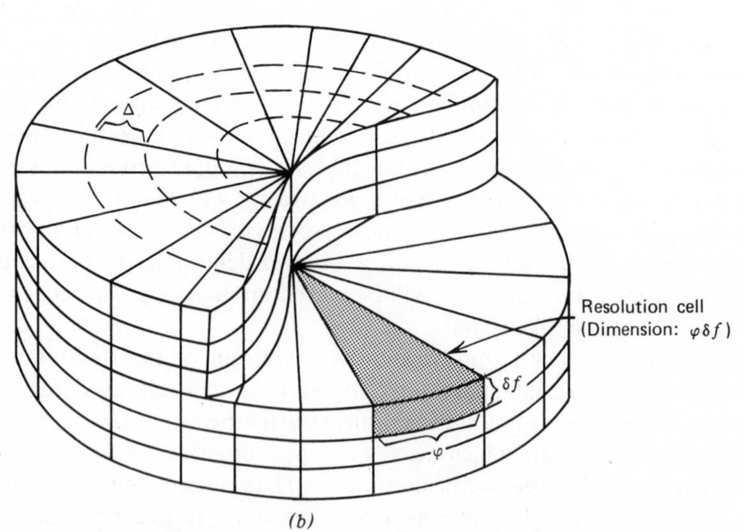

(*b*)

FIG. 10.1 The passive resolution cell structure: (*a*) bearing resolution. $k_p = 360/\phi$. (*b*) bearing and frequency resolution.

number K_p of resolution cells associated with this sonar is

$$K_p = \frac{360}{\varphi}. \tag{10.1}$$

The number of bearing cells[1] usually lies between 20 and 50 for sonars which search the entire 360°. In the implementation of this sonar it is necessary to examine the signal available from each resolution cell in turn and decide whether or not there is a signal present. This examination may be carried on simultaneously, as may be done in a preformed beam sonar receiver, or sequentially, as in a scanning receiver. Either type of receiver can be made to operate at maximum effectiveness provided care is taken not to destroy data useful in making the required decision.

The passive sonar could be implemented in another way. For example, the sonar could be fitted with a bank of contiguous band pass filters with bandwidth B_1 operating with each bearing resolution element. Such a system would be effective if the target signal contained one or more spectral lines with spectral density significantly higher than the continuous signal spectrum. If B is the entire bandwidth admitted to the sonar system, the number of spectral resolution cells will be B/B_1 and the total number of resolution cells in the spectrum analyzer, passive sonar is

$$K_p = \frac{360}{\varphi} \cdot \frac{B}{B_1}. \tag{10.2}$$

Such a sonar, if implemented, could contain more than 200,000 resolution cells. This implementation is clearly more complex than the first system described. Instead of having to examine the signal characteristics of perhaps 50 different cells, the sonar operator would have to examine the signals of perhaps 200,000 resolution cells. This increased complexity would be worthwhile only if, somehow in the process, the signal were made more easily recognizable. (In fact, there would be no point in using the system with azimuthal resolution unless doing so made the signal more easily recognized.)

A simple example comparing three types of passive sonars will indicate the kind of improvement which might be expected. The three sonars are passive sonars with sensor bandwidth $B = 3000$ Hz. The signal is a narrowband waveform with bandwidth $B_1 = 40$ Hz, having power S at the receiving array. It comes from a single source on a particular bearing. The noise sources are uniformly distributed in bearing and range about the receiving array. They produce acoustic power N at the array within the bandwidth B.

[1] If the sonar is capable of searching with depressed beams, K_p must be defined in terms of solid angle.

The characteristics of the three sonars follow:

Sonar No. 1: Omnidirectional hydrophone admits signal power S and noise power N. Signal is to be detected by observing the change in the power output of the array when the signal is turned on. This power changes from N to $S + N$. The ratio of the two powers is

$$\frac{S}{N} + 1.$$

This processor will be successful for a given averaging time for a specific value of S/N.

Sonar No. 2: The beamformer of this sonar provides azimuthal or bearing resolution φ, a total of 20 resolved beams. At the output of the beam aimed at the target there will appear all the signal power S and approximately $\frac{1}{20}$ of the noise power N since only those sources within the azimuthal resolution φ will contribute appreciably to any one beam output. All the other beams will contain $\frac{1}{20}$ of the noise N. In the beam containing signal, the power will rise from $N/20$ to $S + (N/20)$ when the signal is turned on. The ratio of these two powers is

$$20\frac{S}{N} + 1.$$

In this system it is clear that the same integration time used with the first sonar will provide success with a signal 13 dB lower ($10 \log \frac{1}{20}$) than in the previous system.

Sonar No. 3: If the sonar designer knows enough about target signals, he will know the expected bandwidth of any spectral lines which will be present. He will know to build in 40-Hz comb filters into the processing system of each beam. There will then be $\frac{3000}{40} = 75$ filters/beam. If the noise power is uniformly distributed in frequency space $\frac{1}{75}$ of the noise power N at one beamformer output ($\frac{1}{75} \times \frac{1}{20}$ of the noise power in the water) will be present at the output of each filter in the filter bank in each beam. All of the signal power S will appear at the output of the filter in the beam containing signal which is tuned to the signal band. When signal is turned on the power at the output of this filter changes from $N/1500$ to $S + N/1500$. The ratio of these two powers is

$$1500\frac{S}{N} + 1.$$

In this system it is clear that the same integration time used in the first sonar will provide success with signals 31.7 dB ($10 \log \frac{1}{1500}$) below that required in the first sonar and 18.7 dB ($10 \log \frac{1}{75}$) lower than that required in the second sonar.

These examples agree very well with our expectations concerning ability to "filter out" noise. In one case the filtering was done by taking advantage of the fact that the acoustic energy arrived on a single bearing. In the other, additional advantage was taken of the narrowband characteristics of the signal. It was possible to reject the noise which did not come from the same direction as the signal and the noise which did not lie in the same band as the signal.

The improvement in signal-to-noise ratio in this example was designed to make the result impressive. This kind of result has mislead beginners in signal processing for many years. The comparison of two signal processors on the basis of the signal-to-noise ratio improvement in the resolution cell containing signal is not adequate. Allowance must be made for the fact that although an additional 31.7 dB signal-to-noise ratio is available, the operator must now look at some 1500 cell outputs in making his decision. We will find that this increase in number of output channels (if you prefer, output data rate) costs in equivalent signal-to-noise ratio—in this case an estimated 4 dB. In addition, the problem of making the filter bands electrically equivalent plus the expense of such a complex system make it necessary to learn in detail something about the comparative performances of sonar systems before striking out to construct complex systems.

One other fact should also be inserted into the discussion: if the signal-to-noise ratio S/N is great enough that the spectrum-analyzing system would detect the signal reliably in 100 msec after the signal is turned on, a man listening to the output of a single trainable beam would detect the same signal with about the same reliability within 100 msec after the beam was trained "on target." Historically, this surprising result has been an important factor in keeping the operator as an essential part of the sonar. Much effort has been spent designing sonars which will perform with minimal operator involvement and perform as well as a simple sonar with an operator as an integral part of the system. The human operator is also rather versatile in recognizing patterns on graphic recorders. Except for the problem of maintaining alertness and concerned attention to the events which appear on his display or in his headphones, a man plus a relatively simple sonar can perform as well as rather sophisticated equipment.

Active sonar systems can also be described in terms of a set of resolution cells. In general, both bearing and spectral resolution are also available to the active sonar. In addition the expected signals have a finite duration τ so that resolution in time or its equivalent $\Delta = c\tau/2$ will also be available to the active sonar. In this expression Δ is the range equivalent of the expected signal length τ. Two targets separated in range less than Δ will not be observed as two targets. Δ is the range resolution of the sonar. The active sonar resolution cell structure is like the passive cell structure shown in Figure 10.1, but a series of concentric circles drawn symmetrically about the center of the diagram form the boundaries of the range dimension of the cells. These boundaries are indicated by dashed lines in Fig. 10.1b. The separation of the circles is Δ, the range resolution.

If R_{max} is the maximum range at which the sonar is designed to operate, the number of range cells or bins which will be associated with the sonar is R_{max}/Δ. If the active sonar has no spectral resolution, the number of resolution cells belonging to it will be

$$K_a = \frac{360}{\varphi} \cdot \frac{R_{max}}{\Delta}. \tag{10.3}$$

Sonars in existence have K_a values of 10,000 or more. If, in addition, spectral resolution is available to the sonar, the number of cells will be

$$K_a = \frac{360}{\varphi} \cdot \frac{B}{B_1} \cdot \frac{R_{max}}{\Delta}. \tag{10.4}$$

Active sonars with spectral resolution could make use of 100,000 or more resolution cells.

The task of the sonar operator is to examine the properties of the waveform (data) available from each of the resolution cells of his sonar and decide which of the cells, if any, contain a signal which can be associated with a target of interest. Furthermore, he must recognize the presence of a target signal with high probability when a target is present. Failure to recognize the presence of a target is a high cost error. The other type of error, calling "target" in a resolution cell which contains no target, is relatively a low cost error, but this error requires further investigation, delay in ship's mission, perhaps the expenditure of a weapon, or misdirecting the sonar operator's attention sufficiently that he misses the presence of a real target.

The specification of the performance of a sonar and its operator needs to include a statement of the probability of detection (recognition of the presence of a target) as a function of something like target range or signal-to-noise ratio in the water for a specified *rate* of calling "target" incorrectly. This latter quantity is called the false target rate or the false alarm rate.

The description of the waveforms which appears at the output of a signal processor is relatively easy in comparison to the description of the decision-making process. It should be pointed out that the waveforms in the water, as well as those at the output of a signal processor or depicted on a display, represent data. The information received by the brain of the operator depends upon his decision processes. In this discussion of signal processing we will spend a great deal of time showing how to make the "data" more accessible to the operator and will leave the discussion of the rate of "information" transfer to the brain of an operator to those who claim to understand the process.

The methods for setting up such performance specifications will be described in the remainder of this signal-processing discussion. Both active and passive sonars will be included in the discussion. Coherent and incoherent temporal processors will be included. These descriptions will be based upon somewhat idealized models of the signal and noise structure. Section 10.3, entitled "Real World Considerations" will describe deviations from the ideal assumptions and their effects on the performance predictions.

10.2 THE IMPLEMENTATION OF SIGNAL PROCESSORS

The block diagram of a sonar signal processor is shown in Figure 10.2. It consists of a number of hydrophones in an array in the water connected to a display through a spatial processor (beamformer) and a temporal processor. The connections from the hydrophones to the beamformer circuitry are indicated as a simultaneous or parallel system. The beamformer sums the hydrophone outputs with proper relative delay to form beams. The beam outputs are connected to the temporal processor by simultaneous or parallel circuitry. In the temporal processor the waveform is modified in one of a number of ways which will be discussed in detail presently. The output of the temporal processor is transferred to the display by means of a serial or multiplex technique. That is, a single circuit is time shared to get data from individual

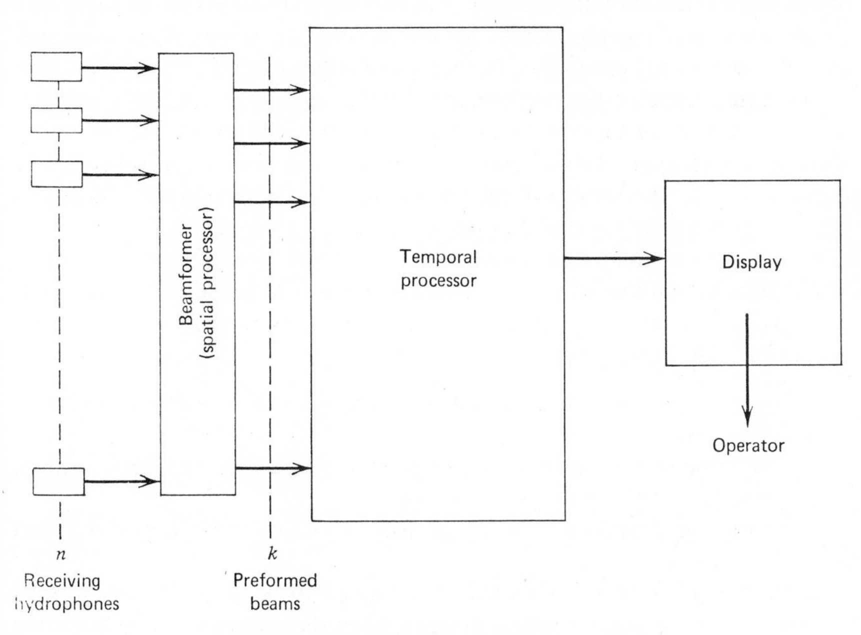

FIG. 10.2 The generalized signal processor.

beams to preassigned locations in the display. This description of the generalized signal processor is not meant to solidify our concepts around any specific processor form. It would be possible to use the simultaneous or parallel structure from the hydrophones to the display. This procedure requires the duplication of many parts of the system and usually this requires amplifiers, detectors, and thresholders which are electrically identical to each other to high precision. Implementation of this type of system is technically difficult because of the "electrically identical" requirement. On the other hand, it is possible to multiplex at the hydrophone output and insert waveform information serially by scanning across the array, beam form, scan the beams, and introduce the outputs serially into the temporal processor where waveform modification takes place. A multiplex system is then used to transfer data to the display.

The point of this description is that we may think of the system as a completely parallel system or an equivalent scanned system. It is possible to get the same data on the display using either system. In one case many complete parallel systems are required, each handling data at a relatively low rate. In the second case, a simpler processing system is required but it must handle data at a much higher rate. There is a significant advantage to multiplexing all the channels through the same amplifiers, detectors, and thresholders because variations in the circuits, such as bias shifts which would be disastrous in the parallel thresholding circuits, would treat all channels in the same way in the serial implementation. As the state-of-the-art permits there is a tendency to push signal processing further and further toward serial techniques.

The equivalence of the performance of the parallel and serial systems is guaranteed if all of the data available are processed by both. If the multiplexing rate is improperly chosen, so that waveform samples obtained are introduced into the serial system too seldom, the performance will be degraded. It is important that this not be allowed to happen.

In order to simplify the mathematics involved in calculating processor performance a number of specific assumptions will be made. These assumptions will allow the complete description of the performance of a particular processor. The list of assumptions follows:

1. Input spectra will be assumed flat over a band of width B and zero outside the band.

2. The nonlinear or demodulation process will either be multiplication, linear detection, or square law detection.

3. Averaging processes will be perfect, running, time averagers with averaging time T.

4. Input waveform statistics will be assumed to be Gaussian noise[2] with variance σ_0^2 except when signal is present, when the statistics will be Gaussian

[2] See Appendix 1, (A1.1).

signal plus Gaussian noise, with variance $\sigma_0^2 + \sigma_1^2$. σ_0^2 and σ_1^2 may be functions of time.

5. When the input waveform is narrowband, that is B/f_c (f_c is the band center frequency) $\ll 1$, then the envelope statistics will be assumed to be described by the Rayleigh distribution.[3]

6. Except where expressly stated otherwise the number of independent samples included in averaging calculations will be considered large enough that the central limit theorem[4] applies.

7. In the examination of each temporal processor it will be assumed that the resolution cells are examined serially in a prescribed order at a rate that each time an independent data sample becomes available in a particular resolution cell it will be examined. In this model each processor has a single terminal at which all the information appears.

8. The output statistics of the temporal processor multiplexed as described in 7 will be described as a function of signal-to-noise ratio at the beamformer output.

9. The statistical description will be used with a few known display characteristics to show how overall system performance may be predicted.

10.2.1 The incoherent temporal processor

A typical incoherent detector is the detector averager shown in Figure 10.3. It consists of a number of identical processors shown processing the outputs of k preformed beams simultaneously. At point a in each channel the waveform bandwidth is B, the input bandwidth. The absolute value of the waveform in each channel is produced[5] by the nonlinear element. This element is called a linear detector. The rectified waveform appears at point b. A running average of the waveform from time $t_1 - T$ to time t_1 appears at the output of each averaging integrator. As time passes and new waveform data are added to the integrator, a corresponding amount of the oldest waveform data in the integrator must be deleted. The integrator computes $1/T$ times the area under the latest T seconds of the waveform curve observed at point b, that is, the average of the latest T seconds of the point b waveform, to form the current integrator output. The output scanner scans all of the integrator outputs somewhat oftener than every T seconds, say 3 or 4 times every T seconds, to obtain the processor output waveform. An interpolation network at the outputs will provide a smooth rather than a discontinuous waveform at the scanner output. Independent output samples, of course, can occur only every T seconds, but some redundancy in sampling is necessary

[3] See Appendix 1, (A1.6) and (A1.7).

[4] See Appendix 1, (A1.12), (A1.13,) and (A1.14).

[5] A squaring device can also be used as the nonlinear element. The performance using a square law detector differs from that using a linear detector by perhaps 0.1 dB.

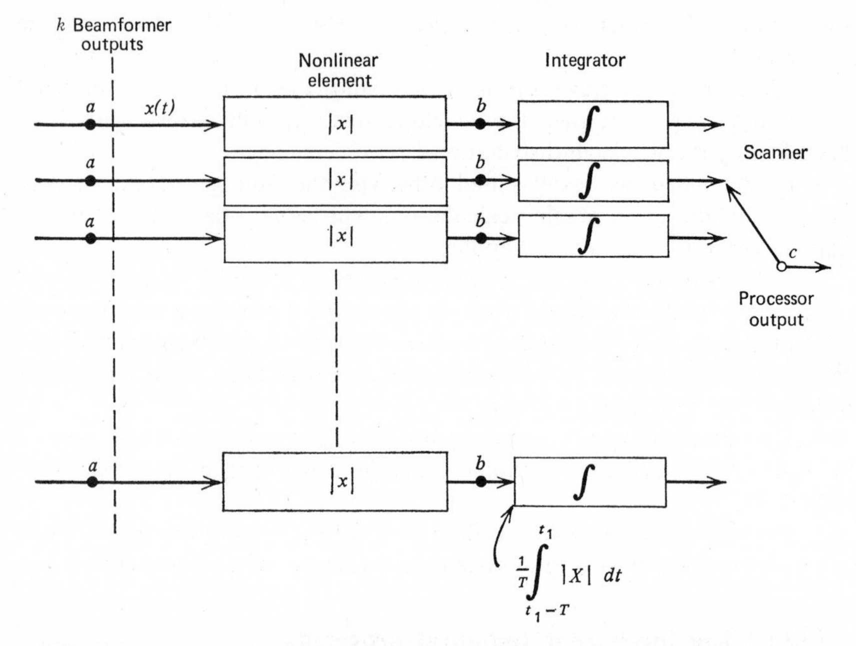

FIG. 10.3 A preformed beam receiver without frequency resolution. *B*, waveform bandwidth at *a*; *S*, average signal power in bandwidth *B* (Gaussian signal) in one channel; *N*, average noise power in bandwidth *B* (Gaussian noise).

in order to avoid complexities in waveform reconstruction. (See Appendix 2.)

According to Appendix 2, there are $n = 2BT$ independent input samples averaged into each independent output sample. It follows from (A1.12), (A1.13), and (A1.14) of Appendix 1 that the mean of the waveform at point c is the same as the mean of the waveform at point b:

$$\mu_c(N) = \mu_b(N)$$
$$\mu_c(S + N) = \mu_b(S + N) \tag{10.5}$$

and the variance at point c is $(2BT)^{-1}$ times the variance at point b.

$$\sigma_c^2(N) = \frac{\sigma_b^2(N)}{2BT}. \tag{10.6}$$

If $2BT > 20$, the central limit theorem allows approximation of the point c waveform statistics with a Gaussian density function:[6]

$$G_c'(z) = (2\pi\sigma_c^2)^{-1/2} e^{-(z-\mu_c)^2/2\sigma_c^2}. \tag{10.7}$$

[6] See (A1.14) of Appendix 1.

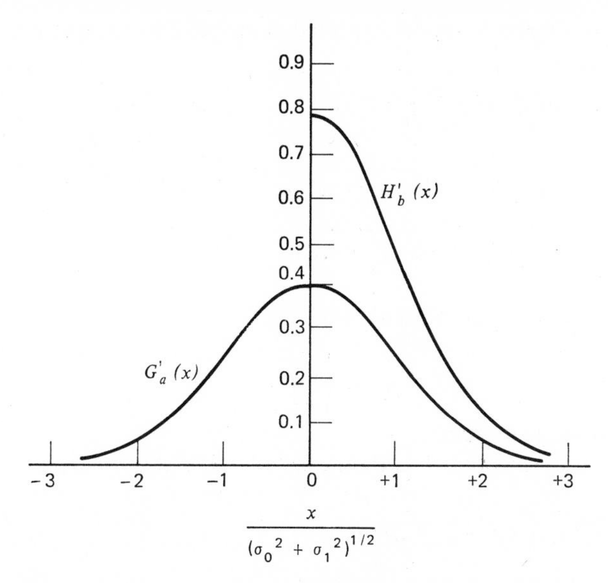

FIG. 10.4 The waveform density functions at points *a* and *b*.

The complete statistical behavior of the output waveform can therefore be specified provided the mean and variance of the waveform at point b can be calculated. This can be easily done by the methods outlined in Appendix 1. [See (A1.3) and (A1.4), for example.] The waveform at point a is Gaussian with zero mean. The density function for this waveform is

$$G_a'(x) = [2\pi(\sigma_0{}^2 + \sigma_1{}^2)]^{-1/2} e^{-x^2/2(\sigma_0{}^2+\sigma_1{}^2)} \tag{10.8}$$

This density function is shown in Figure 10.4 as $G_a'(x)$. The absolute value box in the processor (Fig. 10.3) makes all the negative samples positive. The density function describing the point b waveform must indicate zero probability for negative x. Furthermore all the formerly negative values now appear in the positive x region, hence the distribution in the positive x region is double in size but has the same shape in this region as $G_a'(x)$. The required density function is $H_b'(x)$ defined by

$$\begin{aligned} H_b'(x) = 2G_a'(x) &= \left[\frac{2}{\pi(\sigma_0{}^2 + \sigma_1{}^2)}\right]^{1/2} e^{-x^2/2(\sigma_0{}^2+\sigma_1{}^2)} \qquad 0 \le x < \infty, \\ &= 0 \qquad x < 0. \end{aligned} \tag{10.9}$$

$H_b'(x)$ is shown also in Fig. 10.4. This distribution can be used to obtain the mean and variance at point b.

The mean μ_b at point b is the expected value[7] of the waveform at point b:

$$\begin{aligned}\mu_b = \langle x\rangle_b &= \int_0^1 x\, dH_b(x)\\ &= \left[\frac{2}{\pi({\sigma_0}^2 + {\sigma_1}^2)}\right]^{1/2} \int_0^\infty x e^{-x^2/2({\sigma_0}^2+{\sigma_1}^2)}\, dx\\ &= \left(\frac{2}{\pi}\right)^{1/2} \sqrt{({\sigma_0}^2 + {\sigma_1}^2)} \end{aligned} \quad (10.10a)$$

when signal and noise are present and

$$\mu_b = \left(\frac{2}{\pi}\right)^{1/2} \sigma_0 \quad (10.10b)$$

when noise alone is present.

The variance ${\sigma_b}^2$ at point b is the expected value of $(x - \mu_b)^2$:

$$\begin{aligned}{\sigma_b}^2 &= \int_0^1 (x - \mu_b)^2\, dH_b(x)\\ &= \int_0^1 x^2\, dH_b(x) - 2\mu_b \int_0^1 x\, dH_b(x) + {\mu_b}^2 \int_0^1 dH_b(y) \quad (10.11)\\ &= \int_0^1 x^2\, dH_b(x) - {\mu_b}^2.\end{aligned}$$

The center integral in (10.11) is simply $2\mu_b$ times the integral (10.10) defining μ_b. The last integral in (10.11) is ${\mu_b}^2$ times the area under $H_b'(x)$ which is unity, hence the simple form for ${\sigma_b}^2$. The remaining integral[8] in (10.11) is

$$\begin{aligned}\int_0^1 x^2\, dH_b(x) &= \left[\frac{2}{\pi({\sigma_0}^2 + {\sigma_1}^2)}\right]^{1/2} \int_0^\infty x^2 e^{-x^2/2({\sigma_0}^2+{\sigma_1}^2)}\, dx\\ &= {\sigma_0}^2 + {\sigma_1}^2 \end{aligned} \quad (10.12a)$$

when signal and noise are present and

$$\int_0^1 x^2\, dH_b(x) = {\sigma_0}^2 \quad (10.12b)$$

when noise alone is present. The variance ${\sigma_b}^2$ at point b is therefore

$${\sigma_b}^2 = ({\sigma_0}^2 + \sigma_1^2)\left(1 - \frac{2}{\pi}\right) \quad (10.13a)$$

[7] This integral is evaluated by substituting φ for $x^2/2({\sigma_0}^2 + {\sigma_1}^2)$. The resulting integral has the form $\sqrt{(2/\pi)({\sigma_0}^2 + {\sigma_1}^2)}\int_0^\infty e^{-\varphi}\, d\varphi$.

[8] Dwight, *Tables of Integrals*, No. 861.5, Macmillan, New York, 1947.

when signal and noise are present, and

$$\sigma_b^{\,2} = \sigma_0^{\,2}\left(1 - \frac{2}{\pi}\right) \tag{10.13b}$$

when noise alone is present.

Equations 10.10a, 10.10b, 10.11a, and 10.11b give the means (signal plus noise and noise alone) and the variances at point b in terms of the variance σ_0^2 of the noise and the variance σ_1^2 of the signal at point a. It is easy to show that these quantities are the mean of x^2, that is $\langle x^2 \rangle$, at the temporal processor input. [See Appendix 1, (A1.4), where the variance of a Gaussian distribution with zero mean is calculated.] Since the mean of x^2 represents the average power associated with the waveform in some units $\sigma_0^2 = N$ is the average noise power at each point a and $\sigma_1^2 = S$ is the average signal power at one point a. The means and variances at point b can therefore be written explicitly in terms of N and S, the average input noise and signal power:

$$\mu_b(S + N) = (2/\pi)\sqrt{S + N} \tag{10.14a}$$

$$\mu_b(N) = (2/\pi)^{1/2}\sqrt{N} \tag{10.14b}$$

$$\sigma_b^{\,2}(S + N) = (S + N)(1 - 2/\pi) \tag{10.15a}$$

$$\sigma_b^{\,2}(N) = N(1 - 2/\pi) \tag{10.15b}$$

Equations 10.5 and 10.6 can now be used to find μ_c and σ_c^2 in terms of the input signal and noise power.

$$\mu_c(S + N) = \mu_b(S + N) = \left(\frac{2}{\pi}\right)^{1/2}\sqrt{S + N} \tag{10.16}$$

$$\sigma_c^{\,2}(S + N) = \frac{\sigma_b^{\,2}(S + N)}{2BT} = \frac{(1 - 2/\pi)(S + N)}{2BT} \tag{10.17}$$

Substitution of these values into (10.7) provides an explicit expression for the density function at point c in the processor. This substitution is cumbersome and does not help specifically in understanding the processor. The point is that we know the output signal-plus-noise statistical distribution. For fixed S and N the distribution is Gaussian (10.7) with μ_c and σ_c^2 values given by (10.16) and (10.17). The density functions may be plotted. In preparing the plots, it is convenient to start with the noise-alone plot. It is shown in Figure 10.5, a copy, except for scale, of the normal curve of error shown in Figure 10.4 but translated out from the origin to the noise mean at point c.

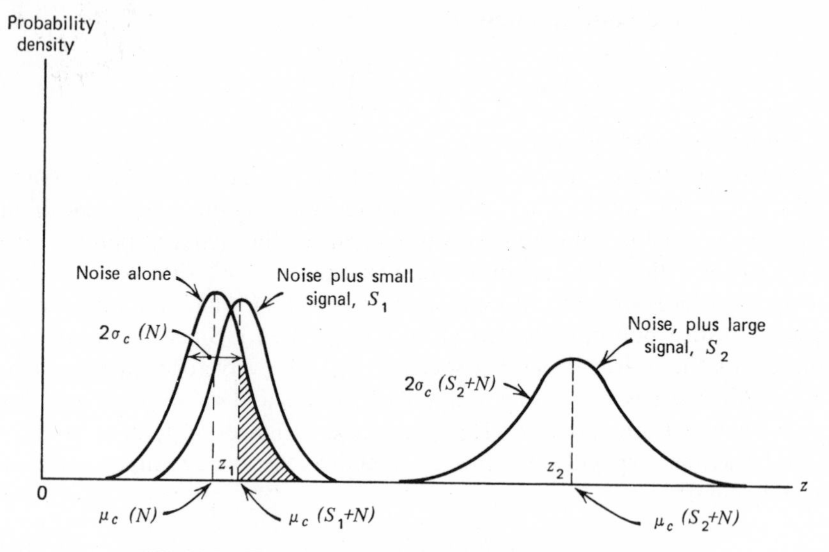

FIG. 10.5 The waveform density function at point c.

The width of this curve at 0.6 of its peak value represents $2\sigma_c(N)$.[9] The noise mean is

$$\mu_c(N) = \left(\frac{2}{\pi} N\right)^{1/2} \tag{10.16a}$$

and the noise variance is

$$\sigma_c^2(N) = \frac{(1 - 2/\pi)N}{2BT} \tag{10.17a}$$

and $2\sigma_c(N)$, the marked width at 0.6 peak of the plot is

$$2\sigma_c = 2\left(\frac{(1 - 2/\pi)N}{2BT}\right)^{1/2} \tag{10.17b}$$

When signal is added to the noise, the mean shifts to $\mu_c(S + N)$:

$$\mu_c(S + N) = \left[\frac{2}{\pi}(S + N)\right]^{1/2} = \left[\frac{2}{\pi}(N)\right]^{1/2}\left(1 + \frac{S}{N}\right)^{1/2} \tag{10.16b}$$

the variance shifts to

$$\sigma_c^2(S + N) = \frac{(1 - 2/\pi)(S + N)}{2BT} \tag{10.17c}$$

[9] See table of the normal curve of error, *CRC Handbook of Probability and Statistics*, Chemical Rubber Publ. Co., Cleveland, Ohio, 1966, p. 23.

and the width of the plot at the 0.6 of peak value is

$$2\sigma_c(S + N) = 2\left[\frac{(1 - 2/\pi)(S + N)}{2BT}\right]^{1/2}$$

$$= 2\left[\frac{(1 - 2/\pi)(N)}{2BT}\right]^{1/2}\left(1 + \frac{S}{N}\right)^{1/2}. \qquad (10.17d)$$

It appears that μ_c increases by the same factor as σ_c when signal is present. When noise alone is present, the waveform fluctuates symmetrically about $\mu_c(N)$. When a small signal S_1 is present, the signal-plus-noise waveform fluctuates about $\mu_c(S_1 + N)$. The small signal-plus-noise distribution is also shown in Figure 10.5. If a threshold is set at $z_1 = \mu_c(S_1 + N)$, so that the only parts of the waveform passed on to the display are required to exceed z_1, the signal-plus-noise waveform would do so with probability 0.50. Half the area under the signal-plus-noise distribution is the the right of z_1. At the same threshold setting the noise-alone waveform would exceed z_1 with probability ~0.10 because approximately 10% of the area under the noise curve lies above z_1. The noise-alone waveform would result in a display response 10% of the time. The larger signal with mean, $\mu_c(S_2 + N)$ centered at z_2, would exceed the z_1 threshold with probability near 1.00 because nearly all of its area is to the right of z_1. If the threshold is raised to $z_2 = \mu_c(S_2 + N)$, the probability that the $S_2 + N$ waveform will exceed the threshold is 0.50 because half of the area of the $S_2 + N$ density function lies above z_2. The noise-alone waveform would exceed the threshold z_2 with extremely low probability.

The statistical behavior shown in Figure 10.5 is illustrated in another way in Figure 10.6. Here a typical output waveform $u_c(t)$ is shown at the processor

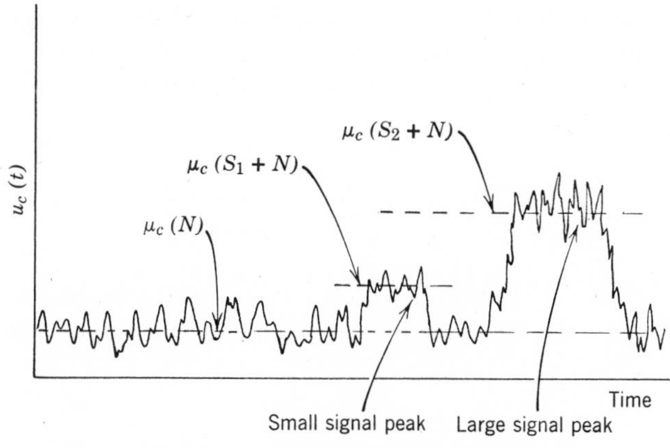

FIG. 10.6 Noise and signal behavior at point c.

output. When noise alone is present, the waveform fluctuates more or less symmetrically about $\mu_c(N)$, the noise mean. When the small signal S_1 is present, the waveform fluctuates about $\mu_c(S_1 + N)$, and when the large signal S_2 is present, the wave fluctuates about $\mu_c(S_2 + N)$. If the threshold is set at $z_1 = \mu_c(S_1 + N)$, the noise rises above the threshold a few times, about half the S_1 signal peak lies above the threshold and the S_2 peak is well above the threshold. If the threshold is raised to $z_2 = \mu_c(S_2 + N)$, the S_2 signal peak lies about half above the threshold; neither the noise nor the S_1 signal come close to z_2.

Since the areas under the density functions from the threshold up determine the probability that the waveform will exceed the threshold, a plot of the area under $G'(z)$ computed from z to infinity is the probability that a measured waveform sample will exceed z. This area is $1 - G(x)$ listed in the tables of area under the normal curve and in the plots of Appendix 1 (Fig. A1.4). A special grid, called Gaussian or probability grid, is available on which $1 - G(x)$ plotted as a function of x will be a straight line. For convenience in Figure A1.4 the origin of the x axis is placed at $\mu_c(N)$. This is equivalent to subtracting the noise mean from the waveform in Figure 10.6 and making the noise waveform fluctuate symmetrically about zero. The unit of the x axis is made the noise standard deviation $\sigma_c(N)$. The tabulated values of the area under normal curve of error provide points which lie on the curve marked 0 (or $-\infty$) in Figure A1.4, Appendix 1. The slope of this curve is such that $1 - G(x) = 0.841$ at $[x - \mu_c(N)]/\sigma_c(N) = -1$ and $1 - G(x) = 0.159$ at $[x - \mu_c(N)]/\sigma_c(N) = +1$. This slope will be referred to as the noise slope. It should also be noted that this curve passes through $1 - G(x) = 0.50$ at $[x - \mu_c(N)]/\sigma_c(N) = 0$, that is, at $x = \mu_c(N)$. The 0.50 probability point coincides with the peak of the density function. Half the area under the density curve is to the right of $x = \mu_c(N)$.

The plot for the waveform with signal present shifts to the right, because of the change in mean. Precisely the $1 - G(x) = 0.50$ point of this curve must lie to the right by an amount equal to the difference in the signal and noise means expressed in units of noise standard deviation:

$$\frac{\Delta\mu_c}{\sigma_c(N)} = \frac{\mu_c(S + N) - \mu_c(N)}{\sigma_c(N)} \tag{10.18}$$

The square of this quantity is called the output signal-to-noise ratio $(S/N)_{\text{out}}$. The position of the $1 - G(x) = 0.50$ points for an output signal-to-noise ratio $(S/N)_{\text{out}}$ is

$$\frac{\Delta\mu_c}{\sigma_c(N)} = \frac{\mu_c(S + N) - \mu_c(N)}{\sigma_c(N)} = \left[\left(\frac{S}{N}\right)_{\text{out}}\right]^{1/2}. \tag{10.19}$$

If we choose[10] $(S/N)_{out} = 10$, $\sqrt{(S/N)_{out}} = 3.16$ and the $1 - G(x) = 0.50$ point for this output signal-to-noise ratio needs to be located at

$$\frac{\Delta\mu_c}{\sigma_c(N)} = 3.16\,. \tag{10.20}$$

If the standard deviation of the signal-plus-noise distribution is identical to that of the noise-alone curve, the plot will be a straight line through the point just determined, parallel to the noise curve. Since the ratio $\sigma_c(S + N)/\sigma_c(N)$ is dependent on the input signal-to-noise ratio $(S/N)_{in}$,

$$\frac{\sigma_c(S + N)}{\sigma_c(N)} = \left(\frac{S + N}{N}\right)^{1/2} = \left[1 + \left(\frac{S}{N}\right)_{in}\right]^{1/2} \tag{10.21}$$

as can be seen from (10.17a) and (10.17c), the slopes are not equal unless $(S/N)_{in} \ll 1$. In many cases this condition is satisfied for large BT products. The curves in Figure A1.4 have been plotted with slopes equal to the noise slope for convenience when the approximation is valid. An auxiliary plot, shown dashed, in Figure A1.4 is used to determine what slope correction needs to be made. For wide band processors with $BT > 1000$ little if any slope correction needs to be made.

The curves in Figure A1.4 are, in effect, universal curves, applicable to all averaging processors, and can be used to determine their performance in terms of their output signal-to-noise ratios. The corresponding input signal-to-noise ratios can be found using the definition of output signal-to-noise ratios stated immediately after (10.18), that is, the square of the output mean shift caused by signal, expressed in units of noise standard deviation. Substitution of $\mu_c(S + N)$, $\mu_c(N)$, and $\sigma_c(N)$ from (10.16), (10.16a), and (10.17a) gives

$$\left(\frac{S}{N}\right)_{out} = \frac{4BT}{\pi - 2}\left\{2 + \left(\frac{S}{N}\right)_{in} - 2\left[1 + \left(\frac{S}{N}\right)_{in}\right]^{1/2}\right\} \tag{10.22}$$

as the relationship between the input and output signal-to-noise ratios for the linear, detector averager. It is not convenient to solve this expression for $(S/N)_{in}$, but a plot has been made showing output $S/N(\text{dB})_{out}$ as a function of input signal-to-noise ratio in decibels (Fig. 10.7). In order to make this plot universal the output signal-to-noise ratios have been reduced by $10 \log BT$. This plot is known as the processing gain curve.

All the information needed is now available to make a prediction concerning the probability that signal will exceed a threshold and mark the display

[10] If $(S/N)_{out} = 10$, $S/N\,\text{dB}_{out} = 10$ dB.

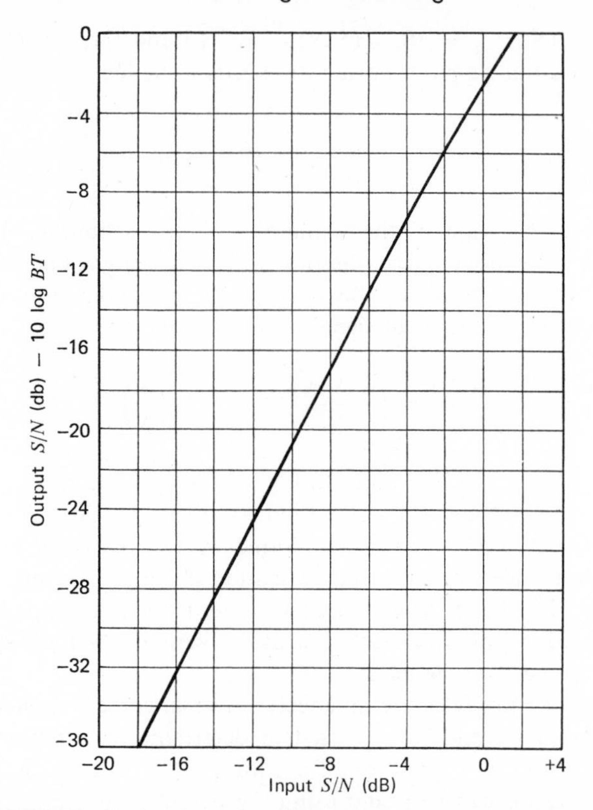

FIG. 10.7 Processing gain of linear detector averager (Gaussian signal in Gaussian noise).

as a function of input signal-to-noise ratio with a specified rate for noise marking of the display. Since the output averaging time is T, there is a new opportunity to mark the display every T seconds in each resolution cell. Since there are K_p cells and each must be scanned every T seconds, there are K_p/T opportunities for cells to mark on noise alone each T seconds. The expected rate of noise marking is therefore

$$\dot{F} = \frac{\text{number of marks per } T \text{ sec}}{T} = \frac{K_p P(N)}{T} \tag{10.23}$$

where $P(N)$ is the probability that a single-cell, noise sample will mark in one opportunity. $P(N)$ is a number which can be read from Figure A1.4, if the threshold setting is known.

Application to a passive sonar. It is instructive to prepare a performance prediction graph for a specific processor. A passive sonar will be examined.

The following specific sonar system will be used:

1. 30 preformed beams, 12° beam width.
2. 4000 Hz $= B$, input noise bandwidth.
3. 4000 Hz $= B_s$, input signal bandwidth.
4. Linear detector with 10-sec averager.
5. Scanner scans every 5 sec, independent output samples occur each 10 sec in one beam.
6. Display: an A-scan, a plot of the output wave-form as a function of time.
7. Threshold line drawn on the A-scan so that output signal-to-noise ratio of 12 dB marks the display with probability 0.50.

With this information and the characteristics that have been described for this processor a number of facts can be generated:

1. From Figure A1.4 it can be seen that the threshold must be set at 4.00 to make $1 - G(x) = 0.50$ when the output signal-to-noise ratio is +12 dB.
2. With this threshold $P(N)$, the probability that a noise sample will exceed the threshold is $P(N) \simeq 3.5 \times 10^{-5}$.
3. With this noise probability the rate at which noise will exceed the threshold is [(10.23)]

$$= \frac{K_p P(N)}{T} = \frac{(30)(3.5 \times 10^{-5})}{10}$$

$$\simeq 1.05 \times 10^{-4} \text{ marks/sec}$$

4. With this same threshold a table (Table 10.1) can be prepared showing the probability $P(S + N)$ that signal-plus-noise will exceed the threshold as a function of output signal-to-noise ratio. The first two columns have been read from Figure A1.4.

TABLE 10.1

Threshold $= 4.00$; $P(N) \simeq 3.5 \times 10^{-5}$; $\dot{F} \simeq 1 \times 10^{-4}$/sec

$(S/N)_{out}$(dB)	$P(S + N)$	$(S/N)_{in}$(dB)
+6	0.025	−20.4
+8	0.072	−19.4
+10	0.205	−18.3
+11	0.330	−17.9
+12	0.500	−17.2
+13	0.685	−16.6
+14	0.845	−16.0
+15	0.948	−15.5
+16	0.990	−14.9

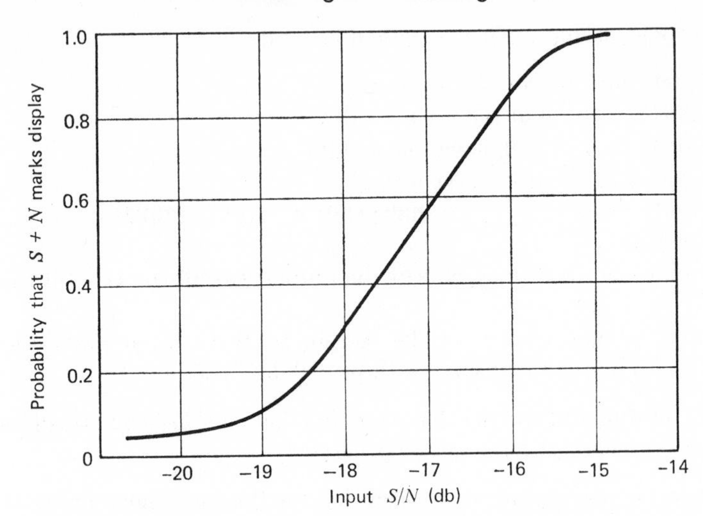

FIG. 10.8 Passive sonar performance curve (noise marking rate = 1×10^{-4}/sec = 9/day); B = 4000 Hz; 30 Beams; T = 10 sec.

5. The BT product for this processor is (4000)(10) = 40,000. If each output signal-to-noise ratio is decreased by 10 log 40,000 = 46 dB and used to enter Figure 10.7, the third column in Table 10.1, containing the input signal-to-noise ratios, can be filled in.

The results in Table 10.1 constitute a performance description for this processor. A plot of the probability that the signal will cause the output waveform to exceed the threshold as a function of input signal-to-noise ratio is shown in Figure 10.8.

Certain typical results appear on the graph. The system causes the signal to exceed the threshold about 50% of the time when the input signal-to-noise ratio is ~ -17.3 dB. Increasing the input signal-to-noise ratio by less than 2 dB causes the signal exceeding probability to be in excess of 0.98, while decreasing the ratio 2 dB lowers the signal exceeding probability to less than 0.08. A 4-dB interval of input signal-to-noise ratio changes the system from a condition where the signal will seldom be seen (when it is present) to a condition where it will be seen almost continuously.

It may also occur to the reader that nine events per day in which noise alone exceeded the threshold may be less than would be annoying to an operator. Could we perhaps lower the threshold so that the noise exceeded the threshold 900 times/day? This would still be only 6×10^{-3} noise exceedings per second or about 20/hr. In order to have this rate the threshold would have to be lower so that $P(N) = 3.5 \times 10^{-3}$ (100 times the value used earlier in step C). The required threshold, according to Figure A1.4, is 2.85. The output signal-to-noise ratio at which $1 - G(x) = 0.50$ at this threshold is 8.4 dB. Figure 10.7 along with $9.2 - 10 \log BT = -36.8$ dB indicates that the corresponding input signal-to-noise ratio to be -19.1 dB. This means that if the operator raises the number of noise threshold exceedings by a factor 100, he will improve his ability to see a signal by 1.9 dB, the decrease in input signal-to-noise ratio required to obtain 0.50 probability that signal exceeds the threshold. (-17.3 dB $-$ (-19.1 dB) $=$ 1.8 dB.) This amounts to less than 1 dB improvement per decade of clutter (threshold exceedings by noise). Another factor 100 increase in clutter will bring the noise exceeding events up to 200/hr ($\sim$3.3/min). Under these conditions about 20% of the noise marks exceed the threshold. This high rate of threshold exceeding by noise is excessive when the output waveform is observed with an A-scan.

Observation of the passive sonar output on an area display. The processor output can be observed on an area display, such as a graphic recorder, or on a long persistence or storage-driven cathode ray tube employing intensity modulation. Each output scan is presented as an intensity-modulated sweep on the display. Subsequent sweeps appear displaced downward from the previous one. The resolution cell containing target is associated with a position located at the same lateral distance from the left edge of the display in every sweep. In this type of display, if the operator waits until he has perhaps 20 sweeps on the screen, he can observe the presence of target signals (as vertical lines on the display) under threshold conditions when the noise marks *half* the resolution cells on the display and signal-plus-noise marks the cell to which it belongs about *three quarters* of the time. According to Figure A1.4 the threshold must be set at zero to achieve this noise marking density. The output signal-to-noise ratio which is required to exceed this threshold with probability 0.75 is -3.5 dB. The required input signal-to-noise ratio read from Figure 10.7 entering with $-3.5 - 46$ dB $= -49.5$ dB is -25.5 dB.

The additional performance obtained with the area display arises from the ability of the operator to average the result of many sweeps by recognizing the line patterns on the display. His ability to do this increases as the number of sweeps on the display containing a target trace increases. By waiting for 20 traces to accumulate on the screen, he requires 10 sec/sweep $\times$ 20 sweeps $=$ 200 sec integration time. This additional integration time has roughly the effect of increasing T by a factor 20. Increasing T by a factor 20 changes $10 \log BT$ by $10 \log 20 = 13$ dB. A change of 13 dB along the vertical axis in Figure 10.7 results in a change of 6.5 dB on the input signal-to-noise ratio axis. As a rough rule of thumb, area displays of this type are assumed to improve the processor input signal-to-noise required by $5 \log m$, where m is the number of independent traces on the display.

Choice of parameters for the sonar receivers. The results obtained in the previous example show that the input signal-to-noise ratio required to make the signal easily observable at the passive sonar output depends upon the input bandwidth, the integration time, and the type of display employed. The example shows that increasing the input bandwidth improves the performance of the sonar provided the input signal-to-noise ratio is constant. This condition can only be satisfied as long as signal fills the input band. If increasing the input bandwidth admits more noise power and not more signal power, input signal-to-noise ratio will decrease more than the expected increase in and compensate for expected gain from the increased bandwidth.

It should be remembered that it is the bandwidth of the input waveform which appears in the equations which have been set up. The claims which have been made depend upon having noise which is spectrally flat in the bandwidth B. If this is not the case in the acoustic fields in the water, it must be made the case by prewhitening filters associated with the beam formers.

Increasing the averaging time is a distinct advantage in passive processors although there are limiting considerations here as well. If the integration time is too long, the target will not remain in one resolution cell long enough to accumulate an observable output mean shift.

It has also been assumed in preparing the passive sonar example that the acoustic noise sources are isotropically distributed in the ocean. If this is not the case (for example, the noise in the aft beams is usually dominated by screw noise), some kind of bearing normalization is required in presenting the sonar output on the display. This normalization makes the detectability of the signal, as seen at the sonar output, depend only upon the signal-to-noise ratio at the beamformer output and not upon the noise level in the water.

Active sonars. The principles which have been applied in the passive sonar example apply equally well in an active sonar receiver. The signal for the active sonar is an echo from a target submarine which is formed when acoustic energy is transmitted from the active sonar. It has the advantage that the signal form can be altered (its spectral content, its amplitude and its duration) to take advantage of special characteristics of submarines and the medium. It has the disadvantage that echoes from the ocean surface, the bottom, and inhomogeneities in the ocean volume produce echoes which primarily raise the level of the interfering acoustic fields but which also produce distinct echolike signals which may be confused with echoes from targets of interest. The induced interfering fields are called reverberation. The level of reverberation changes with time after transmission until it finally falls below the level of the acoustic noise present in the absence of transmissions. Usually, the reverberation is statistically Gaussian with time-dependent standard deviation. The discrete reverbs (the echolike signals) constitute real "false" targets which must be weeded out by classification procedures. In addition to the bearing normalization problems which arise in displaying the sonar receiver output, the active sonar has time variable normalization problems because of variable reverberation level which is present for a time after transmission. For the present it will be assumed that AGC (Automatic Gain Control) or TVG (Time Variable Gain Control) overcomes the problem of the variable background level.

The sonar transmits a pulse (pings) of length τ. This pulse may be a constant-frequency sinusoid, frequency-modulated sweep, or some type of noiselike waveform such as a pseudorandom pulse. (A constant-frequency sinusoid is usually referred to as a CW pulse. This is an unfortunate usage because of the original and generally accepted meaning of CW, "continuous wave.") The spectral content of the signal which reaches the water may or may not reflect the own-ship Doppler shift depending upon the availability of ODN (own Doppler nullifier).

The echo which is returned contains spectral content characteristic of the structure producing the echo. It reflects the gross range rate, Doppler shift of the target hull, the differential Doppler shift if the hull is turning, and an echo with nearly zero Doppler shift associated with the wake of the target.

The sonar receiver will respond to this echo provided its input bandwidth is sufficiently wide to accept all the spectral components of the signal. A number of different situations may arise depending upon whether or not the echo and the reverberation are spectrally separable or whether or not the target echo occurs late in the ping cycle when the background is noise. If either of these situations occurs, it is possible to search for the presence of an echo in a noise background; this condition of operation is known as the *noise-limited operation*. Noise-limited operation is available for high Doppler targets at all ranges and low Doppler targets at sufficiently long range. When the echo is not spectrally separable from the reverberation and the target echo occurs at a time when the background is dominated by the reverberation, the situation is said to be *reverberation limited*. The nonstationarity of the background level and bandwidth introduces a complication not found in the passive systems.

Sonar designers have approached this situation from several points of view depending upon the performance required:

1. The reverberation and noise are not spectrally altered; AGC methods are used to hold the background level nearly constant. The background bandwidth changes abruptly when the reverberation level falls below the

noise level. Targets at all range rates are sought in the entire background of reverberation plus noise.

2. Two pulse forms are used, one specifically for searching for high Doppler targets, the other for low Doppler targets. A sinusoidal pulse, used in connection with a reverberation notch filter, makes it possible to search for targets outside the notched band in the noise background. A wideband pulse (bandwidth approximately equal to the receiver input bandwidth) is used to search for low Doppler targets.

3. Two pulse forms are used as in 2, but the equivalent of comb filter processing techniques are used to search for high Doppler targets.

There is an important difference between the active sonar and the passive sonar. Target signals (echoes) occur only at a specific time, when the transmitted pulse travels outward, reflects from the target, and returns. The echo is never shorter than the transmitted pulse length τ and it may be as long as $\tau + 2L/c$, where L is the projection of the target length on the range vector and c is the speed of sound in the water. The term $2L/c$ is called the *target elongation* of the signal. It may be 5 msec up to 120 msec. Pulse lengths τ between 5 msec and 1 sec are feasible, although attempts to keep reverberation down at short range have kept pulse lengths near the low end of this interval. The limit on signal length imposed by the pulse length and the expected target elongation prevent arbitrarily increasing the integration time T of the processor averager. A compromise must be made concerning the averaging time which should be built into an active processor. If $\tau < 2L/c$, little increase in target strength occurs because of multiple arrival overlap from the target and an averaging time approximately equal to the pulse length is about optimum. The echo has the appearance of separated arrivals from the target structural detail. If $\tau \sim 2L/c$, multiple arrivals from the target overlap, there is a significant increase in echo level and the echo is continuous. In this case there is some advantage in choosing an averaging time T greater than the pulse length τ.

The performance of active sonars with incoherent processors is describable using the statistical plots already set up. In order to prepare an example of the performance of a particular active sonar, one follows the steps outlined in Section 10.2.1-1, substituting the parameters for the active sonar. Such a procedure will be illustrated.

An active sonar example. The characteristics of the active sonar to be used in the example follow:

1. 30 preformed beams.
2. 330 Hz = B_n, input noise bandwidth.
3. 25 Hz = B_r, input reverberation bandwidth.
4. 8 Hz = B_s, input signal bandwidth.

5. 125 msec = τ, transmitted pulse length, constant frequency sinusoid.
6. Linear detector with 175-msec averager.
7. Scanner scans all detected beam outputs every 75 msec. Independent samples occur each 175 msec. The sonar is on the 10-kyd range scale.
8. Display, a plan position indicator, PPI. Shows intensity modulation when output waveform exceeds a preset threshold.
9. It will be assumed that AGC maintains the background level constant and that the threshold is set so an output signal-to-noise ratio of 10 dB marks the display with probability 0.50.

With this information and the characteristics which have been described for this type of processor and the statistical information in Appendix I, specific facts concerning the processor performance can be generated. Two performance examples are required: one for the reverberation-limited condition of operation and one for the noise-limited condition of operation. The calculation of the performance follows:

1. From Figure A1.4 it can be seen that the threshold must be set at 3.16 to make $1 - G(x) = 0.50$ when the output signal-to-noise ratio is 10 dB.
2. With this threshold $P(N)$, the probability that a noise sample will exceed the threshold is $P(N) \simeq 8 \times 10^{-4}$.
3. With this noise probability 8×10^{-4} times the total number of resolution cells in one ping cycle will mark on the average. The total number of bearing cells is 30. The 10-kyd range ping cycle is completed in about 12 sec and the range dimension of the resolution cell corresponds to an elapsed time of 175 msec, the output averaging time. There are therefore $12/0.175 = 68.5$ range resolution cells. The total number of cells is $68.5 \times 30 = 2055$ and the average number of cells which mark each ping cycle is $8 \times 10^{-4} \times 2055 = 1.64$.
4. With this same threshold a table (Table 10.2) can be prepared showing the probability $P(S + N)$ that cells containing signal-plus-noise will mark the display as a function of output signal-to-noise ratio. This is done using Figure A1.4 to obtain the first two columns. The product $n = BT = (330)(0.175) \simeq 58$ for the noise-limited condition must be used to make the slope corrections in Figure A1.4.
5. The BT product for this processor is $B_rT = (25)(0.175) \simeq 4.4$ when reverberation limited and $B_nT = (330)(0.175) \simeq 58$ when noise limited. Only one of these satisfies the criterion that BT must be greater than 20 if the curves in Figure A1.4 are to be useful. The noise-limited input signal-to-noise ratio may be obtained by subtracting $10 \log B_nT = 17.6$ dB from each output signal-to-noise ratio and entering Figure 10.7 with the result. The required input signal-to-noise ratios found in this way have been inserted in column 4 of Table 10.2. (A similar process with the reverberation

TABLE 10.2

Threshold = 3.15

	$P(S + N)$		
$(S/N)_{out}$(dB)	No slope correction	Slope correction	$(S/N)_{in}$(dB)
0	0.016	0.028	−8.5
+3	0.034	0.048	−6.9
+6	0.112	0.150	−5.1
+8	0.250	0.280	−4.1
+10	0.500	0.500	−2.9
+12	0.790	0.760	−1.9
+14	0.961	0.910	−0.8
+15	0.992	0.955	−0.3

B_rT product for the 0.50 signal marking probability would give $10 - 10 \log B_rT = 3.56$ as the entry into Figure 10.7 and +4.0 dB as the required input signal-to-noise ratio at 0.50 probability. This prediction is better than the actual performance of the processor under these conditions. Even so, the estimated 4.0 dB as the required input signal-to-noise ratio is 7 dB poorer than the −2.9 dB required at $P(S + N) = 0.50$ for the noise-limited condition.)

A plot showing the performance of the active processor in the noise-limited condition using the results in Table 10.2 is given in Figure 10.9. Again it is noteworthy that although the signal will mark half the time when the input signal-to-noise ratio is −2.9 dB, an increase of 2 dB makes the marking probability about 0.15. This range of 4 dB carries the processor from a condition where signal marks would seldom occur to a condition where they would seldom fail to occur.

The operator might choose to operate with somewhat more noise marking. If he felt he could tolerate ten times as many marks on the screen (about 16), he could lower the threshold until the probability that a particular noise-containing cell would mark is 8×10^{-3} (10 times the value in B). The required threshold is 2.4 (from Fig. A1.4). With this threshold, the output signal-to-noise ratio for 0.50 signal-plus-noise marking probability is +7.5 dB. $7.5 \text{ dB} - 10 \log BT = 7.5 \text{ dB} - 17.6 \text{ dB} = -10.1$ dB is used to enter Figure 10.7 to find the corresponding input signal-to-noise ratio $(S/N)_{in}(\text{dB}) = -4.5$ dB. The threshold decrease associated with the factor 10 increase in clutter makes it possible for the signal plus noise to mark with probability 0.50 at $(S/N)_{in}(\text{dB}) = -4.5$ dB, 1.6 dB lower. The factor 10 in clutter provided only 1.6 dB improvement in signal marking performance.

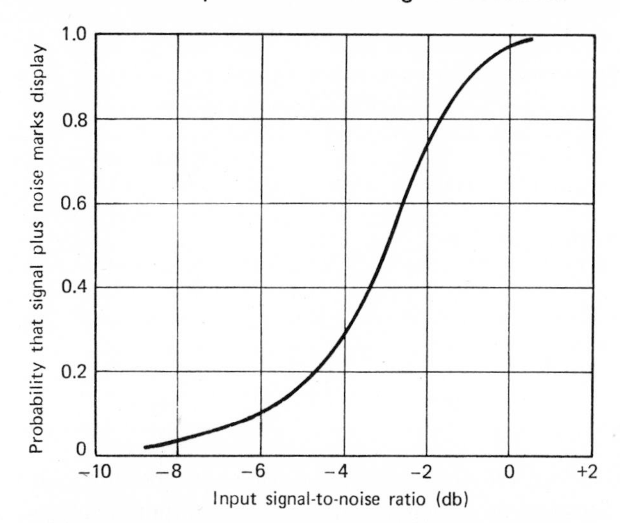

FIG. 10.9 Performance of active sonar ($T = 0.175$ sec; $B_n = 330$ Hz; 30 preformed beams; 2055 resolution cells; linear detector). (Average noise marking = 1.6 resolutions cells/range sweep.)

To this point plots like Figure 10.8 and Figure 10.9 have been constructed to show how a processor performs. A complete graph of this type will consist of a family of curves, one curve for each condition of noise marking, that is, the average number of the resolution cells which mark per scan divided by the time per scan. In addition, if comparison is to be made among processors with different BT products, it will be necessary to have a family of graphs with different BT products and different conditions of noise marking. Getting all this information in a form which is easily used is rather difficult.

A simple representation is possible if the 0.50 signal-plus-noise marking probability is used as the point of comparison. The simple representation of processor performance is a plot showing input signal-to-noise ratio required for 0.50 probability that signal-plus-noise will mark plotted as a function of noise marking rate in resolution cells which mark per second. There will be one such curve for each bandwidth B, for each averaging time T, and for each detector type.

For the example to be prepared, we will use the active sonar which has just been discussed and fill in Table 10.3. The first step in filling the table is to assume suitable values of the noise-marking probability. The seven values listed were arbitrarily selected to be approximately $\frac{1}{2}$ decade apart. The next step is to enter Figure A1.4 with each $P(N)$ value, determine the threshold

TABLE 10.3

Processor Performance as a Function of Clutter Rate
(10 log $BT = 17.6$)

$\dot{F}$(mark/sec)	$P(N)$	$(S/N)_{\text{out}}$(dB)	$(S/N)_{\text{out}}$(dB) $-10 \log BT$	$(S/N)_{\text{in}}$(dB)
1.71×10^{-3}	10^{-5}	12.7	−4.9	−1.5
5.1×10^{-3}	3×10^{-5}	12.0	−5.6	−1.8
1.71×10^{-2}	10^{-4}	11.4	−6.2	−2.3
5.1×10^{-2}	3×10^{-4}	10.9	−6.7	−2.6
1.71×10^{-1}	10^{-3}	9.8	−7.8	−3.1
5.1×10^{-1}	3×10^{-3}	8.9	−8.7	−3.5
1.71	10^{-2}	7.2	−10.4	−4.4

required, and read the output signal-to-noise ratio required for 0.50 signal-plus-noise marking probability. The required output signal-to-noise ratios are shown in the table. To this point the procedure is the same for all incoherent processors. Now $10 \log BT = 17.6$ dB is then subtracted from each output signal-to-noise ratio. To this point the procedure is the same for all incoherent processors with the same BT product. At this point one enters the processing gain curve for the processor under study with the results of the subtraction and determines the required input signal-to-noise ratio. Since the processor being used is one with a linear detector, we can use Figure 10.7. The values in the right column are the result. These results could also be computed using (10.22) by solving for $(S/N)_{\text{in}}$. The final step is to work left from the $P(N)$ column. The probability that a particular resolution cell will mark is $P(N)$. The scanner samples each resolution cell, 2055 of them in 12 sec; that is, it scans $2055/12 = 171$ of them per second. It follows that $\dot{F} = 171P(N)$ is the average number which mark per second. This multiplication gives the $\dot{F}$ column.

The plot of $(S/N)_{\text{in}}$(dB) as a function of $\dot{F}$ provides a plot suitable for comparing this processor with other processors. It is shown in Figure 10.10.

Treatment of other incoherent processors. It is a simple matter to prepare curves like Figure 10.10 for other processors because the performance is predicted at only the 0.50 probability that signal-plus-noise will mark the display. In the steps carried out in setting up Table 10.3 and in the construction of Figure 10.10, only the input signal-to-noise ratios and $\dot{F}$ are processor dependent and $\dot{F}$ depends only upon the resolution of the processor since $\dot{F} = P(N)$ times the number of resolution cells scanned each second. This leaves only the relationship between the input and output signal-to-noise ratios as the controlling difference between processors.

In general, the connection between input and output signal-to-noise ratio

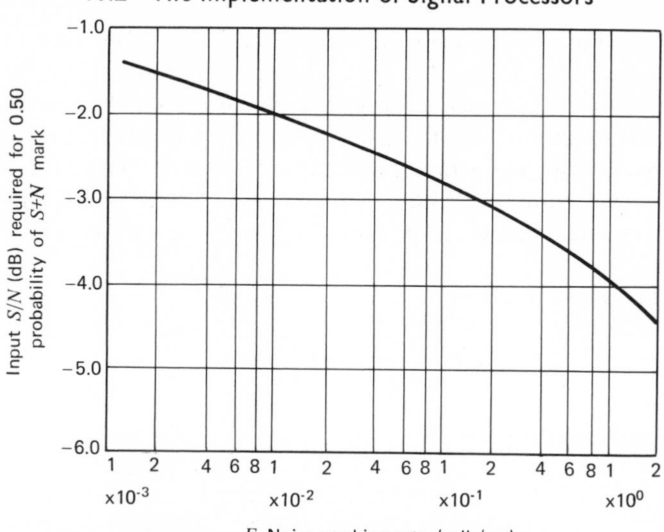

FIG. 10.10 Detection performance curve: linear detector averager ($B = 330$ Hz; $T = 175$ msec).

for incoherent temporal processors can be computed by carrying out a procedure for finding μ_b, σ_b, μ_c, and σ_c for the incoherent processor under study similar to the example built around the Figure 10.3 processor. Among other processors which have been used extensively is the square law detector. In this processor, the absolute value box, $|x|$, in Figure 10.3 is replaced by a squaring box, x^2. The waveform parameters for this processor are computed in Appendix 1 (A1.8),

Square law detector parameters:

$$\mu_b = \left\langle \frac{r^2}{2} \right\rangle = N\left(1 + \frac{S}{N}\right) = \mu_c \tag{10.24a}$$

$$\sigma_b{}^2 = \left\langle \left(\frac{r^2}{2}\right)^2 - \left\langle \frac{r^2}{2} \right\rangle^2 \right\rangle = 2N^2\left(1 + \frac{S}{N}\right)^2 = 2BT\sigma_c{}^2 \tag{10.24b}$$

$$\left(\frac{S}{N}\right)_{\text{out}} = \left(\frac{\mu_c(S+N) - \mu_c(N)}{\sigma_c(N)}\right)^2 = BT\left(\frac{S}{N}\right)_{\text{in}}^2 \tag{10.24c}$$

Processors using split arrays and correlation of the waveform obtained from the two halves have a pair of input filters ahead of the nonlinear element. The nonlinear element is a box in which the product of the two waveforms is computed instant by instant to form the point b waveform. Such "cross correlators" are either linear cross correlators in which amplitude information is preserved, or "clipped" cross correlators in which only the sign of the product is used as the point b waveform. Processors of this type are automatically degraded 3 dB because the signal-to-noise ratio at the half array beam outputs is 3 dB lower than would be obtained from the corresponding full array beam outputs. Furthermore, it is usually assumed that the noise-alone outputs of the half array beams are uncorrelated. This assumption is usually not met entirely. With these two assumptions Faran and Hills[11] give the parameters needed to prepare the processing gain curves for these processors.

The linear cross correlator:

$$\left(\frac{S}{N}\right)_{\text{out}} = \tfrac{1}{2}BT\left(\frac{S}{N}\right)_{\text{in}}^{2} \tag{10.25}$$

The clipped cross correlator:

$$\left(\frac{S}{N}\right)_{\text{out}} = 1.56BT\left[\sin^{-1}\frac{\frac{1}{2}\left(\frac{S}{N}\right)_{\text{in}}}{1+\frac{1}{2}\left(\frac{S}{N}\right)_{\text{in}}}\right]^{2} \tag{10.26}$$

The processing gain curves for the four processors are shown in Figure 10.11. The plot has been made independent of BT by subtracting 10 log BT from the output signal-to-noise ratio.

These curves can be used along with the knowledge that all these processors have the same output noise and signal statistics for equal output signal-to-noise ratios to show which is the better processor for a given BT product. To do this one merely needs to enter the processing gain curves with an output signal-to-noise ratio (reduced by 10 log BT) and determine the input signal-to-noise ratio. It is clear that the square law detector outperforms the linear detector by perhaps 0.2 dB, the linear cross correlator by 1.8 dB, and the clipped cross correlator by 2.8 dB over the range of input signal-to-noise ratio where the plots are linear.

It would be hasty to say that the choice of incoherent processor should be the square law detector at this point. Consideration of practical problems, which will be undertaken in a subsequent section, must be included before making this choice.

10.2.2 Coherent processors

The incoherent signal processors considered in the previous sections made no use of knowledge of the signal form in their operation. All the noise in

[11] J. J. Faran, and R. Hills, Jr., "Correlators for Signal Reception," Tech. Memo. 27, Acoustic Research Laboratory, Harvard University, September 15, 1962.

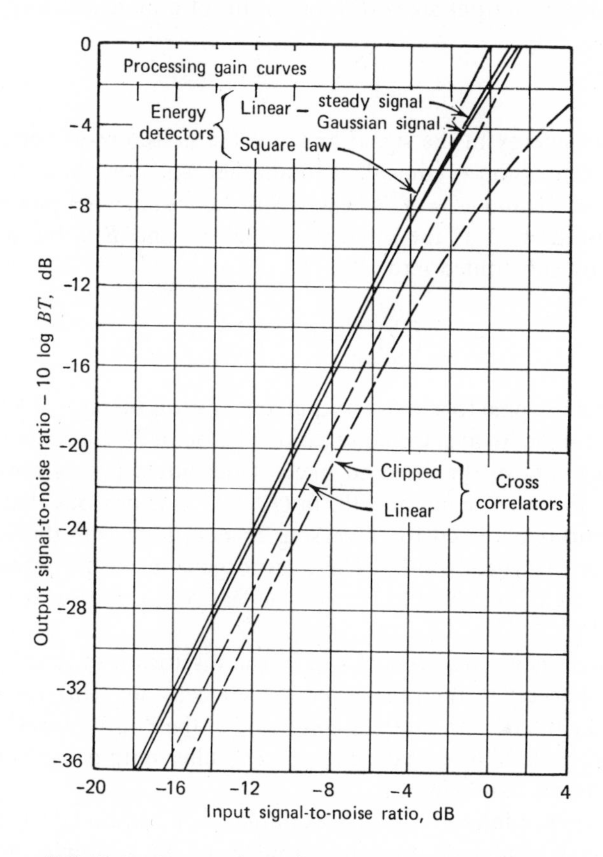

FIG. 10.11 Processing gain curves for four processors.

the input passband of the receiver plus the signal when present was passed on to the detector-averager. Enhancement of the signal-to-noise ratio occurred because of the smoothing of the noise in the averager. Such systems have also been called postdetection integrators. There is another important class of signal processors known as the *matched filter*. They are also known by the names optimum filter and coherent processor.

Signal processors based upon this concept reject as much of the noise energy as possible before detection. The signal-to-noise ratio of the waveform produced by the matched filter and provided to the detector input cannot be further improved by post detection integration.

A *matched filter* is a processor which maximizes the output signal-to-noise ratio. The processor which accomplishes this maximization must be selected with detailed knowledge of the input signal waveform and the noise waveform. For spectrally flat, Gaussian noise, the optimum filter is one whose spectral response is a constant times the complex conjugate of the spectrum of the signal.[12] The peak output signal-to-noise ratio of a matched filter is

$$\left(\frac{\hat{S}}{N}\right)_{\text{out}} = \frac{2E}{\eta} \tag{10.27}$$

where E is the energy in the signal and η is the noise power per unit bandwidth. This expression is equivalent to another which is obtained when the quantities $E = ST$ and $\eta = N/B$, where S is the input signal power and T is the signal duration, N is the input noise power, and B is the input noise bandwidth, are substituted into (10.27).

$$\left(\frac{\hat{S}}{N}\right)_{\text{out}} = 2BT\left(\frac{S}{N}\right)_{\text{in}} \tag{10.28}$$

The peak-output-signal to average output noise ratio is therefore $2BT$ times the input average-signal to average noise ratio. The factor $2BT$ is sometimes called the processing gain of the matched filter. One should not be misled by the factor B, the input noise bandwidth, in (10.28). It is obvious that increased processing gain is obtained by increasing B, but along with this increase, N also increases with bandwidth. There is no gain in output signal-to-noise ratio as a result of bandwidth increases unless S also increases with bandwidth.

Two characteristics are brought out in the discussion of matched filters[13] which are important to the prediction of their behavior: (*a*) The frequency response of the filter is the complex conjugate of the Fourier transform of the signal to which the filter is matched; (*b*) The filter output bandwidth is the same as the signal bandwidth.

A simple matched filter. A simple example will illustrate how the gain of a matched filter is achieved. For the purposes of this example the situation shown in Figure 10.12 represents the idealized spectral structure of the waveforms involved. At a single beamformer output, the reverberation spectrum is flat over a bandwidth $B_R = 25$ Hz; its spectrum level will be used as the reference level 0 dB in this example. The noise is flat over a bandwidth $B_N = 330$ Hz and its spectrum level is -10 dB, relative to the reverberation spectrum level. The signal spectrum is displaced relative to the reverberation spectrum. Its spectrum is flat over a band $B_S = 8$ Hz; its spectrum level is -6 dB relative to the reverberation spectrum level. The numerical values corresponding to the decibel levels are shown in Table 10.4. In this table the power per unit bandwidth in the reverberation is taken as unit power in agreement with the 0 dB reverberation spectrum level. The

[12] M. Schwartz, *Information Transmission, Modulation and Noise*, McGraw-Hill, New York, 1959, p. 282.

[13] *Ibid.*

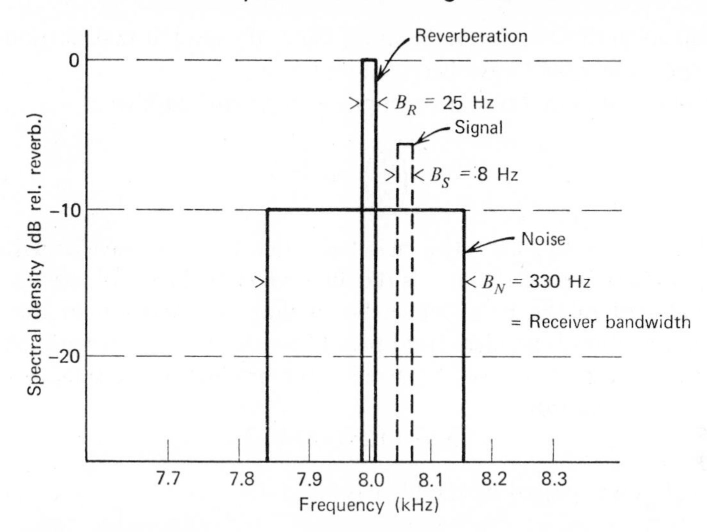

FIG. 10.12 Idealized receiver input spectrum (approximate pulse length = 125 msec; center frequency = 8 kHz).

spectrum levels (dB) listed in the second column are 10 log (spectrum level, third column). The product of the spectrum level and the bandwidth is the band level given in the fifth column.

Examination of Figure 10.12 and Table 10.4 reveals that the conditions for the validity of Eqs. (10.27) and (10.28) are not met because the interference (reverberation plus noise) spectrum is not flat across the receiver pass band. The requirement of spectral flatness can be met by reducing the power gain of the input circuits by $\frac{1}{11}$ in the reverberation band. If this is done, the reverberation plus noise power (at power level 1.1, before the gain change) in the 25-Hz band centered at 8.0 kHz will be reduced to 0.1. The resulting interference spectrum will be flat after this change. Had there been signal

TABLE 10.4

Waveform Characteristics: Simple Matched Filter[a]
(Power Unit: Power Per Unit Bandwidth in Reverberation)

	Spectrum level (dB)	Spectrum level (power)	Bandwidth, B	Band level (power)
Reverberation	0	1.00	25 Hz	25
Noise	−10	0.100	330	33
Signal	−6	0.251	8	2

[a] Signal duration = approximately 125 msec

components in the 25-Hz reverberation band, the level of that portion of the signal spectrum would have been reduced also.

The signal-to-noise ratio at the input to the matched filter is

$$\left(\frac{S}{N}\right)_{\text{in}} = \frac{2}{33} \tag{10.29}$$

or $10 \log \frac{2}{33} = -12.2$ dB. The matched filter for this waveform in spectrally flat Gaussian interference is the filter which will pass all the signal and only that part of the noise which lies within the spectrum of the signal. The required filter is an ideal band pass filter which coincides with the signal spectrum. The signal power S passed is the product of the signal spectrum level and its bandwidth:

$$S = (0.251)(8) = 2. \tag{10.30}$$

The noise power passed by the filter is

$$N = (0.1)(8) = 0.8. \tag{10.31}$$

The matched filter output signal-to-noise ratio is therefore

$$\left(\frac{S}{N}\right)_{\text{out}} = \frac{2}{0.8} = 2.5 \tag{10.32}$$

or $10 \log 2.5 = +3.98$ dB. The matched filter has provided a gain in signal-to-noise ratio amounting to $3.98 - (-12.2) = 16.2$ dB. The gain for this example is simply $10 \log B_{\text{out}}/B_{\text{in}}$, where $B_{\text{out}}/B_{\text{in}}$ is the ratio of the output and input noise bandwidths.

This simple example has many of the features of more involved situations but does not have mathematical complexity. Usually a calculation of the noise power passed by the filter involves the square of the integral of the complex conjugate of the signal spectrum, rather than just the product of the signal bandwidth and the noise level.

The energy in the signal is

$$E = ST = (2)(0.125) = 0.250.$$

The noise spectrum level is $\eta = 0.1$. According to (10.27) the peak output signal-to-noise ratio at the output of the matched filter is

$$\left(\frac{\hat{S}}{N}\right)_{\text{out}} = \frac{2E}{\eta} = 2\frac{(0.250)}{(0.1)} = 5 \tag{10.34}$$

or $10 \log S = 6.98$ dB, just 3 dB higher than the average output signal-to-noise ratio obtained with Eq. (10.32).

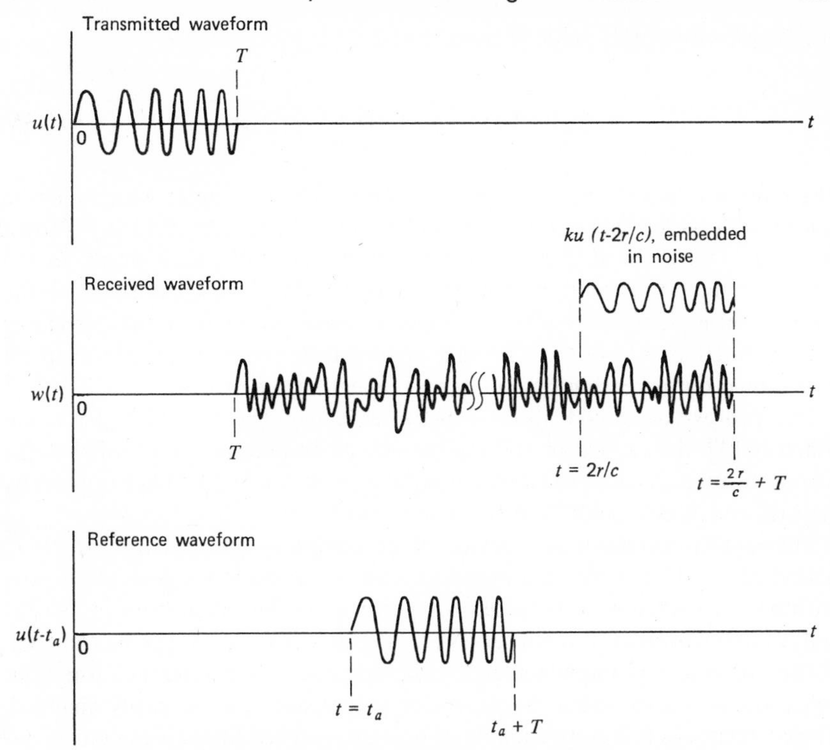

FIG. 10.13 Waveforms in the replica correlation processor.

A versatile matched filter, the replica correlator. The simple example of the previous section agrees with intuitive feelings about the behavior of filters processing signals in noise. It is possible to implement a matched filter for any signal, using correlation techniques. To show that this is so, we will consider a processor which is based upon the computation of the correlation of a reference function $u(t)$ and the signal-plus-noise waveform $w(t)$. This processor is known as a replica correlator. Beginning at time $t = 0$ a transmission $u(t)$ having duration T is made (Fig. 10.13). At $t = T$ the receiver is turned on and a waveform $w(t)$ consisting primarily of reverberation and noise, appears at the input to the receiver. At time $t = 2r/c$ a scaled version of the transmitted waveform, $ku(t - 2r/c)$, appears in the noise at the receiver input. Its position is shown above $w(t)$ in Figure 10.13, beginning at $t = 2r/c$. In reality it is embedded in the $w(t)$ waveform and its amplitude may be too small to be recognized by an observer.

The replica correlator is implemented by computing the sum of the products of $w(t)$ and $u(t - t_a)$ in the interval $t_a \leq t \leq t_a + T$. The reference function is also shown in Figure 10.13. The sum of these products is the correlator output waveform $x(t)$ value at time $t_1 + T$.

$$x(t_a + T) = \int_{t_a}^{t_a+T} w(t)u(t - t_a)\,dt \qquad (10.35)$$

The reference waveform is a replica of the transmitted pulse which may be placed with its left edge at any particular time. In Figure 10.13 it is shown at $t = t_a$. With the reference function at this position, each point on the reference function is multiplied by the corresponding point just above on the receiver input function $w(t)$. The sum of these products is the correlator output function $x(t_a + T)$. The next point on the receiver output waveform is obtained by increasing t_a slightly and repeating the process.

The process described above constitutes a matched filter for the signal which is like the replica in all respects except amplitude. A proof that the correlator used as described above is equivalent to a matched filter is given by Stewart and Westerfield[14] and by Woodward.[15]

The replica correlator is a device which computes the function $x(t_a + T)$ described by (10.35). Such a processor can be versatile if a general-purpose arithmetic package is constructed for performing the integration in (10.34) and if the function $u(t)$ which serves as its reference function can be changed. If the correlator is implemented by digital means, the reference function, hence the signal to which the processor is matched, can be easily changed. The correlator is therefore capable of being a matched filter to any particular waveform which can be used as its reference.

Implementation of the replica correlator. Several types of practical correlators have been constructed, depending upon the manner in which reference and input waveform storage is accomplished. All are time compression systems. All can be classed as one of three general types, the sum band heterodyne correlator, the difference band heterodyne correlator, and the dc correlator. All of the correlators have in common the multiplication of two band-limited waveforms, the reference and the processor input waveforms, followed by an IF (intermediate frequency) filter. This filter can be set to accept the sum band resulting from the products or the difference band. In practice, the band in which the heterodyned signal (the product of the signal and the reference) becomes a single frequency sinusoid is used. In sum band correlators this requires that the reference function be a spectrum-inverted version of the expected signal. In difference band correlators the reference function is identical to the expected signal. The dc correlator

[14] J. L. Stewart, and E. C. Westerfield, "A Theory of Active Sonar Detection," *Proc. IRD*, **47,** 872 (1959).

[15] P. M. Woodward, *Probability and Information Theory with Application to Radar*, McGraw-Hill, New York, 1953.

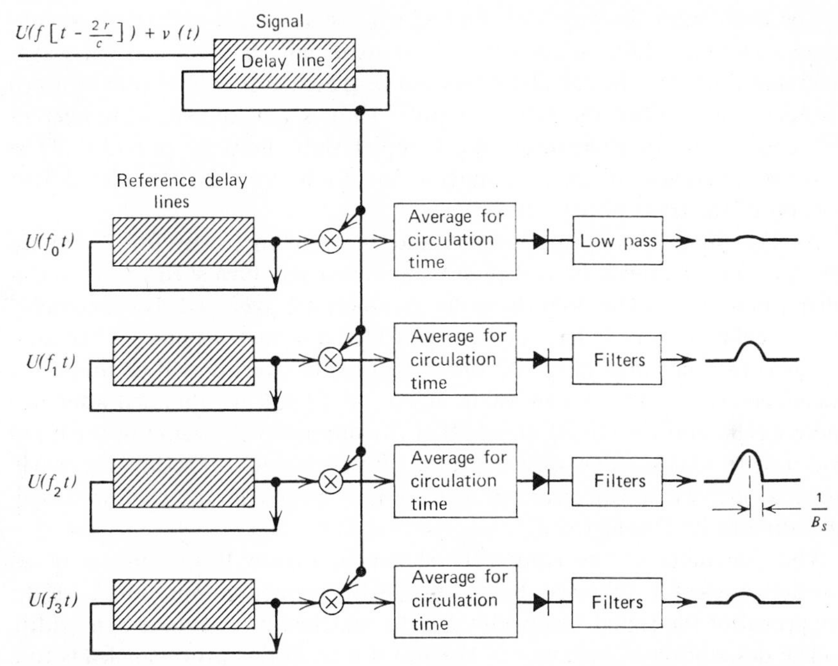

FIG. 10.14 Instrumentation: the difference band heterodyne correlator. Time resolution = 1/*B*. Frequency resolution = 1/*T*. Low pass filter cutoff = B_s. Frequency spacing = 1/*T*.

differs from the sum and difference band correlators in that the input waveforms are heterodyned to a band symmetric about zero frequency before the correlation step.

In this discussion only the difference band correlator will be considered. Figure 10.14 shows one version of the difference band correlator. The incoming waveform, shown as the sum of a reference function and a noise function $u(t)$ is read into the circulating signal delay line. The input waveform is an appropriately sampled waveform. The delay line is sufficiently long to hold all of the samples in the expected signal simultaneously. In operation all the input waveform samples in the signal delay line emerge from the output of the delay line and are replaced at the input in a time equal to the time between samples of the input waveform. Such a technique constitutes a time-compression scheme. All the samples of the signal can be used in a calculation before the next sample arrives in the real time scale. In order to update the contents of the delay line, the oldest sample in the delay line is deleted and is replaced by a new sample on each circulation. The reference delay lines are similar in behavior except that their contents recirculate without update.

The echo from the target of interest will usually have undergone an unknown Doppler shift. If only the transmitted pulse is used as a correlator reference function, the correlator will not be matched to echoes with nonzero Doppler shift. Since the actual Doppler shift is not known, a number of references, one for each resolvable Doppler shift, must be provided. The required reference for each channel is merely a properly Doppler-shifted version of the transmitted pulse.

As the input waveform samples and the reference waveform samples emerge from the delay lines they are simultaneously furnished to the multipliers, marked x. The point-by-point products are averaged over one delay line circulation time. This process provides a scaled version of the sum (10.34a) for one value of t_a. At the output of the average will be found the waveform $x(t_a + T)$. A new value of $x(t_{a'} + T)$ will be obtained after one more circulation time using an updated (by one sample) version of the input waveform. This same process occurs in all the channels. Presumably, when a signal arrives, it will produce a maximum output in the channel corresponding to its Doppler shift.

The remainder of the equipment shown in Figure 10.14 consists of an envelope detector followed by a filter with averaging time equal to the reciprocal of the signal bandwidth B_s, the matched filter output bandwidth.

The description of behavior of the waveforms in this processor leads to a simple model of its operation, one that has some intuitive appeal. The development of this description will be based upon the pseudo noise frequency modulated pulse shown in Figure 10.15.

The transmitted waveform is a constant amplitude sinusoid with duration T whose frequency is piecewise constant but changes from one value to another during a single transmission. The upper plot (*a*) in Figure 10.15 shows a schematic version of the waveform. The solid line in the center plot (*b*) shows the manner in which the transmitted frequency and the reference function frequency vary with time. The dashed plot in (*b*) shows a slightly Doppler-shifted version of the transmitted pulse which will serve as the signal. The correlator takes the product of each sample in the signal waveform and the corresponding sample in the reference waveform. Both these waveforms are sinusoids with the discontinuous frequence behavior. A processor which computes the products of two sinusoids provides a product function which is also two sinusoids, one having the frequency of the sum of the input frequencies and one having the frequency of the difference of the input frequencies. The time dependence of the sum and difference frequencies is shown in (*c*) of Figure 10.15. The sum frequency in (*c*) is the sum of the solid frequency plot and the dashed frequency plot in (*b*). Its variation is roughly double the variation of either of the plots in (*b*). The difference frequency plot in (*c*), on the other hand, is nearly a constant. It would be a

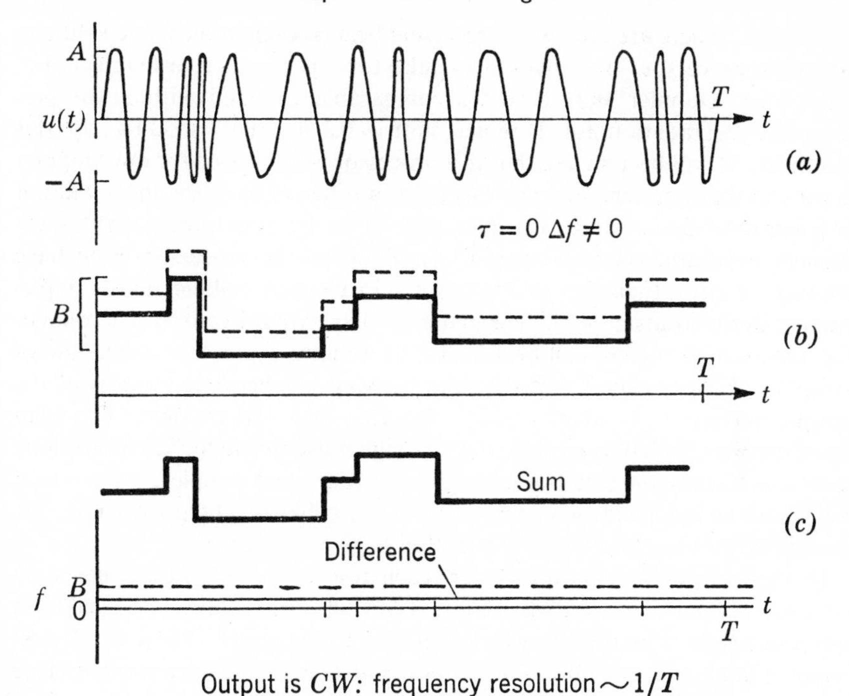

FIG. 10.15 Difference Band Correlator (Frequency Behavior).

constant except for the differential Doppler shift of the high and low frequencies in the transmitted pulse.

In the correlator processor time compression by the ratio of the sampling time τ to the signal length T is used so that all the frequencies in Figure 10.15 are raised from their normal values by a factor T/τ. The CW at the multiplier output is a "zero frequency" CW in the difference band correlator when the reference function and the signal are perfectly matched. A low pass filter can be used for the band pass filter at the output of the multiplier to remove the interference from the sum band spectral components. The bandwidth of the low pass filter is usually selected to have a time–bandwidth product $BT = 1$. The CW embedded in flat Gaussian noise at the processor output of the low pass filter provides an envelope behavior which is statistically like the probability distribution in Figure A1.5 in Appendix 1.

The principles for establishing the performance of an optimum or matched filter processor have been described. An example showing how a performance prediction is made will now be described.

An optimum processor example. In this example it will be assumed that the transmitted waveform is a linear FM sweep with duration $T = 0.25$ sec and signal bandwidth B_s-300 Hz. This transmission is made on each of 30

12° beams. There are also 30 12° receiving beams, each implemented with an optimum processor. This particular pulse form is sensitive to Doppler shift only when Doppler shift becomes comparable to bandwidth (1 dB performance degradation for Doppler shift = 0.1 B_s, that is 30 Hz, in this example). It will be assumed that this processor is to be used for low Doppler targets so that multichannel processing for a range of Doppler shifts will not be required. Since target Doppler shift is to be small compared to the signal bandwidth, the input, band pass filters to the processors need have width no greater than B_s provided own ship Doppler nullification is implemented during transmission. The matched filter output bandwidth is equal to the transmitted signal bandwidth B_s. It follows from the discussion of sampling in Appendix 2 that the time between independent samples of the output waveform is about $1/2B_s$. Because two independent, Gaussian waveforms are added to produce the Rayleigh noise envelope,[16] each envelope sample is already the result of adding two independent samples. As a result, we expect an independent sample of envelope behavior about each $1/B_s$ sec. A summary of the processor information is given in Table 10.5.

In the operation of the processor, each resolution cell is examined only once each 24 sec. The time when a target can appear in a particular resolution cell is determined by the two-way travel time to the target. There are 21,600 opportunities to make a noise mark on the display each 24 sec and therefore 21,600/24 = 900 opportunities to do so per second. The steps in predicting the performance follow.

1. Some assumptions must be made concerning the noise-marking rate or the probability that a single resolution cell will contain a mark. At this point,

TABLE 10.5

Optimum Processor Characteristics
(Matched filter followed by an envelope detector)

Beamwidth = bearing resolution = 12°
Number of beams = 30
Signal bandwidth = 300 Hz
Signal duration = 0.25 sec
Input noise bandwidth = 300 Hz
Processor output bandwidth = 300 Hz
Approximate time between output samples = 3.3 msec
Time resolution at processor output ≃ 3.3 msec
Time equivalent of search range (20 kyd) = 24 sec
Number of resolution cells = (360/12)(24/0.0033) = 21,600
Opportunities for noise mark per second = 21.600/24 = 900/sec

[16] The Rayleigh distribution described by (A1.6) of Appendix 1 results when waveform samples from sinusoids, having the same frequency but a phase difference of 90° and modulated by independent Gaussian envelopes, are added. See M. Schwarz, *Information Transmission, Modulation and noise*, McGraw-Hill, New York, 1959, p. 392.

the choice is somewhat arbitrary, but no prediction can be made unless one is chosen. For the first example it will be assumed that about 0.1% of the resolution cells on the display are marked. This choice is equivalent to the

average number of noise marks/display = 21.6
probability a particular cell is marked = 0.001
number of noise marks per second = 0.9.

2. The probability that signal will mark in the resolution cell containing signal can be determined in terms of the signal-to-noise ratio at the input to the envelope detector[17] from Figure A1.5, Appendix 1. The required signal-to-noise ratios and corresponding probabilities are shown in Table 10.6. One enters Figure A1.5 with 0.001, the noise-marking probability, and finds that the required threshold $\xi^2/2N$ is 6.9 using the noise-alone curve ($S/N = -\infty$). With this threshold, one then reads the probability that signal plus noise exceeds the same threshold as a function of signal-to-noise ratio.

3. The signal-to-noise ratio[18] required at the input to the matched filter corresponding to the $(S/N)_{in}$ at the detector input can now be obtained with (10.28). It must be remembered, however, that the signal-to-noise ratios in Table 10.6 are for average signal, not peak signal, so that the form of

TABLE 10.6

Process Performance Data
(Noise marking probability = 0.001)

$\left(\frac{N}{S}\right)_{in}$ (dB), Envelope detector	Probability signal mark	$\left(\frac{N}{S}\right)_{in}$, Matched filter
1	0.028	−17.75
2	0.042	−16.75
3	0.062	−15.75
4	0.098	−14.75
5	0.151	−13.75
6	0.230	−12.75
7	0.33	−11.75
8	0.48	−10.75
9	0.65	−9.75
10	0.79	−8.75
11	0.90	−7.75
12	0.98	−6.75

[17] The input to the envelope detector is synonymous with the output of the matched filter.
[18] The notation must not be allowed to cause confusion. In this equation $(S/N)_{in}$ is the signal-to-noise ratio at the input to the *matched filter* and $(S/N)_{out}$ is the signal-to-noise ratio at the output of the matched filter. $(S/N)_{out}$ is also the input signal-to-noise ratio to the envelope detector.

(10.28) required is (without the factor 2)

$$\left(\frac{S}{N}\right)_{\text{out}} = B_s T \left(\frac{S}{N}\right)_{\text{in}} \tag{10.36}$$

where the subscripts refer to the output and input to the matched filter. Since $(S/N)_{\text{out}}$ is also the signal-to-noise ratio at the input to the envelope detector, the required input signal-to-noise ratio for the third column in Table 10.6 is

$$\left(\frac{S}{N}\right)_{\text{in}} = \frac{1}{B_s T}\left(\frac{S}{N}\right)_{\text{out}} = \frac{1}{B_s T}\left(\frac{S}{N}\right)_{\text{in envelope detector}} \tag{10.37}$$

$B_s T = (300)(0.25) = 75$ for this processor; $10 \log BT = 18.75$. The third column, the required $(S/N)_{\text{in}}$, will be the entry in the first column of Table 10.6 minus 18.75.

Figure 10.16 is a plot showing the probability that signal plus noise will mark the display as a function of the matched filter input signal-to-noise ratio. From this plot -10.6 dB is the matched filter input signal-to-noise ratio required if the signal-plus-noise envelope at the output of the envelope detector is to mark with probability 0.5. The signal-to-noise marking probability rises to 0.90 when the input signal-to-noise rises to -8.0 dB (an increase of 2.6 dB) and falls to 0.10 when the input signal-to-noise ratio falls to -14.6 dB (a decrease of 4 dB). For this processor there is an input

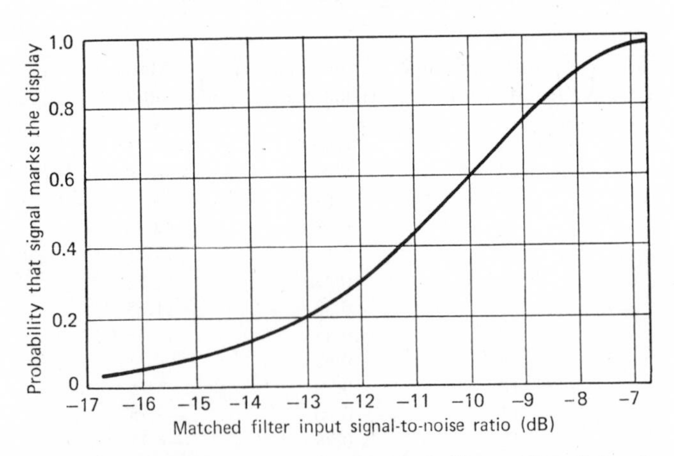

FIG. 10.16 Performance of matched filter. Signal bandwidth = 300 Hz; Noise bandwidth-300 Hz; signal duration = 0.25 sec; single cell marking probability = 0.0001; noise marks/display sweep = 22.

TABLE 10.7

Square Law Detector Characteristics

Output averaging time = 0.25 sec
Time between output-independent samples = 0.25
Number of resolution cells = (30)(24/0.25) = 2880
Number of opportunities for noise marks/sec = 120

signal-to-noise ratio interval of 6.6 dB (from −14.6 dB to −8.0 dB) in which the signal statistics changes from a situation where the signal almost never marks to a situation where it marks the display nearly every opportunity.

It is possible to determine the variation of performance with display clutter (noise marks on the display screen). If, for example, the clutter is increased by a factor 10, the total number of clutter marks on the display of the Table 10.5 processor would be 216, the number of clutter marks made per second would be 9, and the probability that any particular resolution cell would mark the display would be 0.01. Entering Figure A1.5 of Appendix 1 with the noise probability 0.01 provides 4.6 for the threshold setting and 6.1 dB for the output signal-to-noise ratio for 0.5 probability that signal plus noise will mark. Use of (10.37), as in the first example, gives the required matched filter input signal-to-noise ratio 6.1 dB − 10 log $BT = 6.1 - 18.75 = -12.75$ dB. The increase in clutter by a factor 10 lowered the required input signal-to-noise ratio by about 2.1 dB. Lowering the display clutter by a factor 10 (to about 0.0001 marking probability per resolution cell) will allow 0.50 signal plus noise marking probability to occur with −9.25 dB matched filter input signal-to-noise ratio.

Comparison of incoherent processor and the matched filter. The difference between the performance of the matched filter and the best of the incoherent processors is not very large, as can be seen from the following example. For the matched filter we will use the processor described in Table 10.5. The performance of a square law detector processing the same signals will also be determined. The special characteristics of the square law detector which are not like the Table 10.5 processor appear in Table 10.7.

In other respects the processors have the same characteristics. The performance comparison will be constructed in Table 10.8, using, at first, a common noise-marking probability. The table was begun by choosing the arbitrary set of noise marking probabilities shown in the center column. These values represent the probability that any particular resolution cell will contain a noise mark. The required output signal-to-noise ratios for 0.50 probability that signal plus noise will mark is obtained from Figure A1.4

TABLE 10.8

Signal-to-Noise Ratios for 0.50 Probability that Signal Will Mark

($B_{signal} = B_{noise} = 300$ Hz, $T = 0.25$ sec, Matched filter averaging time $1/B_s$, 30 beams)

Square law detector				Matched filter followed by envelope detector		
Clutter rate (marks/sec) $= 30\,P/T$	Required $S/N\,(\text{dB})_{in}$	Required $S/N\,(\text{dB})_{out}$	P, Noise marking probability	Required $S/N\,(\text{dB})_{out}$	Required $S/N\,(\text{dB})_{in}$	Clutter rate (marks/sec) $= 30\,B_sP$
			1×10^{-6}	11.3	−7.4	0.009
			3×10^{-6}	10.9	−7.8	0.027
1.2×10^{-3}	−3.1	12.6	1×10^{-5}	10.4	−8.7	0.09
3.6×10^{-3}	−3.3	12.1	3×10^{-5}	10.0	−8.9	0.27
1.2×10^{-2}	−3.7	11.3	1×10^{-4}	9.4	−9.3	0.90
3.6×10^{-2}	−4.0	10.7	3×10^{-4}	8.9	−9.8	2.7
1.2×10^{-1}	−4.8	9.8	1×10^{-3}	8.4	−10.3	9.0
3.6×10^{-1}	−5.2	8.9	3×10^{-3}	7.3	−11.4	27.0
1.2	−5.7	7.3	1×10^{-2}	6.2	−12.5	90.0
3.6	−6.7	5.5	3×10^{-2}	4.9	−13.8	270.0

for the square law detector and from Figure A1.5 for the matched filter followed by an envelope detector. The corresponding required input signal-to-noise ratio for the matched filter is found using (10.37), and for the square law detector by subtracting $10 \log BT = 18.75$ from the output signal-to-noise ratios and entering the processing gain plot in Figure 10.11.

At this stage of the analysis it is clear that, compared on a common marking probability per resolution cell, the matched filter system is superior to the square law detector by about 6–7 dB in required input, signal-to-noise ratio. Plots of required input signal-to-noise ratios are shown in Figure 10.17. This specific result depends upon the BT product. This comparison represents the comparative performance of the two processors when their displays exhibit the same fraction of the resolution cells in clutter. This method of comparison seems to be a straight-forward one, but there is a marked difference in the appearance of the two displays. In the display of the square law detector there are 2880 resolution cells. With $P = 0.001$, on the average only about 2.9 cells will contain a noise mark. There are then an average 2.9 marks on the display per display sweep competing with a signal mark for the attention of the sonar operator.

The situation is quite different for the matched filter processor. Its display contains 21,600 resolution cells. With $P = 0.001$, on the average about 21.6 cells will contains noise marks and there will be 21.6 marks on the display per display sweep competing with a signal mark for the attention of the operator. The increased resolution or increased output data rate associated with

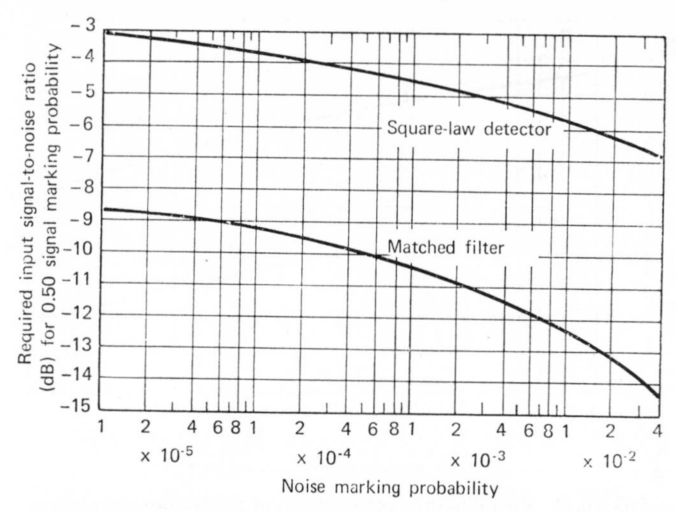

FIG. 10.17 Performance comparison of square law detector and matched filter (common single cell noise marking probability). $B_s = B_n = 300$ Hz. $T = 0.25$ sec, 30 beams.

the matched filter causes some degradation in performance at the display–operator interface. It is therefore not reasonable to compare the performances of two processors at the same single cell marking probability.

An alternate method of comparison is to compare them at the same number of marked cells per display sweep. This condition requires that both processors produce clutter or noise marks at the same rate. The noise marking rate is approximately the probability P that a particular cell will mark on noise alone times the number of opportunities to mark per second. For the processors being compared the rates are therefore 30 P/T for the square law detector and 30 PB for the matched filter processor. The factor 30 is the number of beams, T is the time between independent output samples in the square law detector, and $1/B_s$ is the reciprocal of the signal bandwidth, which is approximately the time between independent samples at the matched filter processor output. These rates, computed with the P values in Table 10.8, are listed in the columns labeled clutter rate.

Since the clutter rate values obtained in this way are not numerically equal for the two processors, a direct comparison of the processors on an equal clutter rate basis cannot be made from the tabulated values. The comparison is easily made if the required input signal-to-noise ratios for the processors are plotted as a function of clutter rate as shown in Figure 10.18. In this plot the matched filter processor is still seen to be superior to the square law detector, but its performance is only 3.5–4 dB better in terms of input signal-to-noise ratio. The requirement that the same number of clutter marks

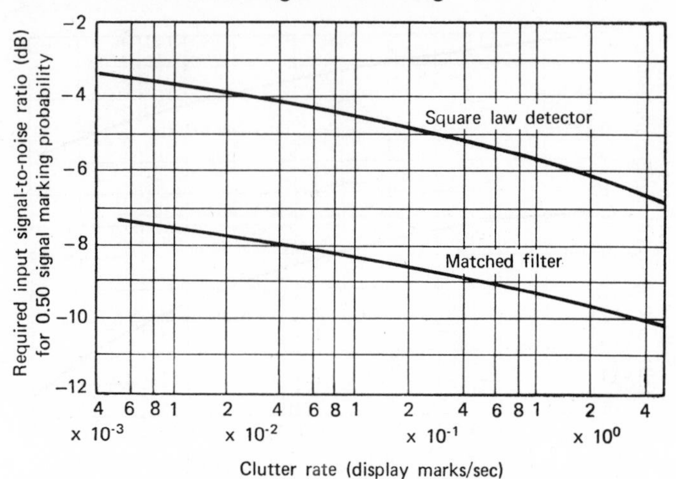

FIG. 10.18 Performance comparison of square law detector and matched filter (common noise marking rate).

appear on the display of the processors being compared has reduced the apparent advantage of the matched filter (for $BT = 75$) from the 6–7 dB range to the 3.5–4 dB range.

Bandwidth dependence. One additional comparison is useful, a comparison of the performances of the matched filter and the square law detector as a function of bandwidth B_s. This comparison is made using the same procedures used in the previous comparison except that clutter rate is fixed and

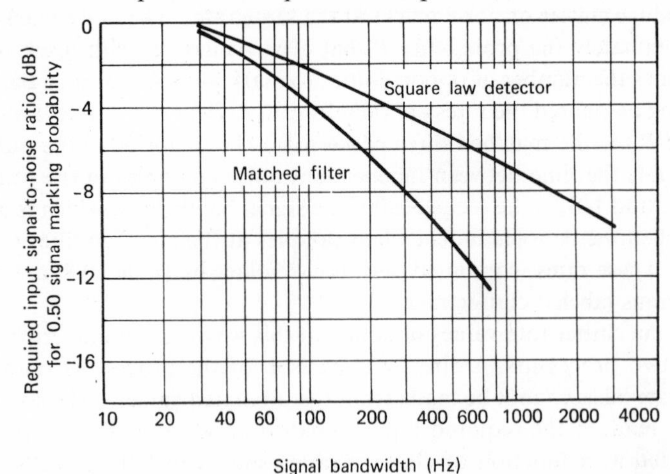

FIG. 10.19 Performance comparison of square law detector and matched filter (clutter rate 0.1/sec; $T = 0.25$ sec; 30 beams).

signal bandwidth is varied. The comparative performance is shown in Figure 10.19. It is apparent that the expected performance of the coherent processor becomes progressively better than the performance of the square law detector as signal bandwidth becomes greater.

10.2.3 Summary

The discussion of incoherent and coherent processors which has been developed was based upon idealized structure in the signals and in the interfering background. The performance predictions which have been made therefore represent limits on the best that one can do in his search for signals embedded in noise. This kind of performance prediction is an important first step in designing signal processing systems. Analyses of other processors can be prepared in a similar way using the probability distributions and principles which have been illustrated. Use of resistance–capacitance (RC) averagers instead of perfect time averages and singly or multiply tuned input filters instead of square input bands changes results slightly (perhaps 1 dB in output signal-to-noise ratio) if time constants of the RC averager are optimized with respect to the signal length and if the number of statistical degrees of freedom is used instead of the product $2\,BT$. This procedure is equivalent to defining the statistical bandwidth of a waveform.[19] These modifications, if carried out properly, make little change in the predicted performance of processors processing ideal signals from that already described.

10.3 REAL WORLD CONSIDERATIONS

Those who have experienced operations at sea readily accept the fact that the performance predicted in Section 10.2 for ideal signals and interference is optimistic compared to the performance which is really achieved in operational environments. The reasons for this difference all lie in unsuccessful efforts at making use of the knowledge of the ocean environment or ignoring that knowledge. An incomplete list of failures in making the ocean conditions like those assumed in Section 10.2 will illustrate the point (Table 10.10).

The roles of these "failures" and their possible cures make up the remainder of the discussion of signal processing. In addition a brief section on sonar design philosophy and directions for future upgrading of sonar signal processors is also included.

[19] J. S. Bendat and A. G. Piersol, *Measurement and Analysis of Random Data*, Wiley, New York, 1966, p. 265.

TABLE 10.10

Environmental Conditions Complicating the Signal-Processing Design Effort

1. Failure of AGC (automatic gain control) and TVG (time variable gain) to make interference or background level constant.
2. Failure to recognize that echoes have unknown structure related to the target structure and propagation multipath structure and that this unknown structure makes specification of a matched filter difficult, if not impossible.
3. Failure to employ displays which transfer all the waveform information to the sonar operator.

10.3.1 Normalization

A signal processor is said to be normalized when its detection performance depends only upon signal-to-noise ratio in the water and not upon the signal or noise levels in the water. Historically, normalization has been attempted by use of AGC, TVG, and waveform clipping techniques. The success of the first two (AGC and TVG) has been hampered by the fact that AGC time constants are dictated by pulse lengths and the TVG systems are not adaptable in sufficient detail to the acoustic level variation with time. Residual level variation at the gain control outputs remains several dB. Use of waveform thresholding at the display in connection with the nonstationary level degrades performance by at least the amount of the variation in level. Clipped systems are typically lower in performance than nonclipped systems. This fixed degradation, automatic normalization, and simplicity of implementation has allowed this processor to compete successfully to the present time with the square law and linear detector, which are ideally superior to the cross correlator by about 2 dB in input signal-to-noise ratio. It is not clear that this simple processor does not compete with the replica correlator (the matched filter of Section 10.2) when complexity of equipment, multipath structure, and cost are included in the analysis.

At this time lack of normalization is recognized as one of the big problems of sonar signal processors. Significant improvements in normalization require consideration of ping-to-ping averaging to define average background level for developing TVG control signals, and perhaps tighter AGC time constants based upon the output of the matched filter processor, using relatively large signal bandwidths.

10.3.2 Signal structure

Signal structure is important in predicting the performance of matched filter processors. To illustrate the seriousness of the structure factor let us suppose that two echoes, one a single arrival identical to the transmitted pulse with duration T and the other two such arrivals, each with duration T and with leading edge separation τ where $\tau \ll T$ but greater than $1/B_s$, are to be processed by a square law detector and by a replica correlator using the transmitted pulse as a reference. Let us further suppose that the average power of both echoes is the same, as might easily occur if the second target is somewhat farther away. For both these echoes, because its resolution is T seconds, the square law detector sees the same input signal-to-noise ratio and performs identically on the two echoes, at its ideal level in a stationary background. The replica correlator resolves the two arrivals in the second echo and sees each 3 dB lower in level than the one arrival in the first echo. Because of its ability to resolve, the replica correlator is not matched to the second echo and is at a disadvantage relative to an optimum processor. The 3-dB disadvantage which results from failure to match the reference to the echo removes a great deal of the expected advantage of the replica correlator over the incoherent detectors.

Increasing the signal bandwidth does not necessarily improve the performance of the replica correlation, using the transmitted pulse as a replica, as indicated in Figure 10.19, if in the process additional arrivals are resolved in the echo structure. The choice of processor is closely related to the arrival structure observed in real echoes.

10.3.3 Displays

Displays have been a weak link in the sonar detection process since the first attempts to make video displays. The insufficient dynamic range, the failure to provide an optically resolvable location for each processor resolution cell, and the failure to provide equal areas for all resolution cells have degraded sonar performance several decibels relative to what is possible. Some of these inadequacies are the result of inadequate hardware state-of-the-art, but some are the result of choice. The plan position indicator (PPI) is a good example of a display which is susceptible to all the inadequacies listed above. This fact does not mean that plan position indicators should not be employed in sonars, but they probably should not be used in the detection process.

10.3.4 Future developments

The future of sonar is undoubtedly dependent upon the implementation of real-time digital processors. It is not out of the question to consider real-time, all-digital systems in which the analog-to-digital conversion is made at the outputs of the elements of the receiving array. Such a system would carry out beam forming, temporal processing, and display control in the digital format. In such a system likelihood ratio decision processes or some form of ping-to-ping integration prior to display would be attractive in active sonars.

The choice of processor or processors is still open. On the surface the theoretical predictions are strongly in favor of the matched filter and large *BT* products, but this conclusion is based upon the assumption that multipath resolution does not decrease the per-arrival energy in the echo structure faster than bandwidth is increased. The decision to increase signal bandwidth cannot be made with assurance until the expected signal multipath structure has been determined.

APPENDIX 1

Statistical Properties of Waveforms

Bandlimited waveforms are obtained from beamformers fed by acoustic sensors in the water in a variety of operational situations. The Gaussian distribution with zero mean describes the statistics of these waveforms provided that:

1. The variance of the distribution is changed appropriately with time and beam azimuth. (The variance changes a negligible amount in times comparable to $1/B_1$, B_1 being the waveform bandwidth.)
2. Machinery sources radiating aboard ship do not dominate the acoustic fields at the hydrophones.
3. Sufficient multipath structure appears in the echo-forming process to produce Gaussian signals in active sonars. (This assumption is not as good as the other two, in general, but the assumption is far better than that signals are steady.)

The Gaussian distribution $G(x)$ for the signal-plus-noise waveform is

$$G(x) = [2\pi(\sigma_0^2 + \sigma_1^2)]^{-1/2} \int_{-\infty}^{x} \exp[-x^2/2(\sigma_0^2 + \sigma_1^2)]\, dx \qquad \text{(A1.1)}$$

in which σ_0^2 is the variance of the noise alone distribution, σ_1^2 is the variance of the signal distribution, and $\sigma_0^2 + \sigma_1^2$ is the variance of the combined waveform. $G(x)$ is the probability that a waveform measurement at a particular instant will yield a value less than or equal to x and is the area under the Gaussian density function $G'(x)$ from $-\infty$ to x. $G'(x)$ is the familiar normal curve of error:

$$G'(x) = [2\pi(\sigma_0^2 + \sigma_1^2)]^{-1/2} \exp[-x^2/2(\sigma_0^2 + \sigma_1^2)] \qquad \text{(A1.2)}$$

with variance $\sigma_0^2 + \sigma_1^2$. $G'(x)$ and $G(x)$ are illustrated in Figure A1.1.

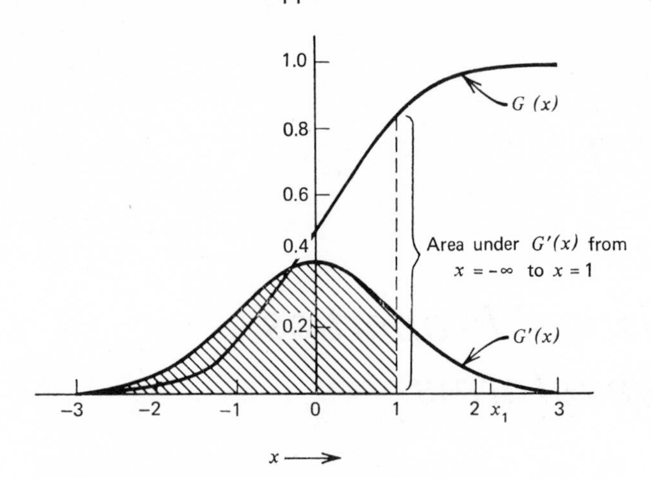

FIG. AI.I The Gaussian probability distribution $G(x)$ and density function $G'(x)$.

The expected value of x is called the mean μ of the distribution:

$$\mu = \int_0^1 x\, dG(x) = \int_{-\infty}^{\infty} xG'(x)\, dx = 0. \tag{A1.3}$$

The variance σ^2 is the expected[1] value of $(x - \langle x \rangle)^2$:

$$\sigma^2 = \langle (x - \langle x \rangle)^2 \rangle = \langle x^2 \rangle = \int_0^1 x^2\, dG(x) = \int_{-\infty}^{\infty} x^2 G'(x)\, dx^1 = \sigma_0{}^2 + \sigma_1{}^2. \tag{A1.4}$$

Since the average value of x^2 for the waveform is the power associated with the waveform (in some power unit), the sum of the average signal power S, and the average noise power N, it follows that

$$\begin{aligned} S + N &= \sigma_0{}^2 + \sigma_1{}^2, \\ S = \sigma_1{}^2, &\qquad N = \sigma_0{}^2. \end{aligned} \tag{A1.5}$$

The Gaussian distribution takes an especially simple form on a probability coordinate paper. $G(x)$ and $1 - G(x)$ are shown plotted on such a grid in Figure A1.2.

When the Gaussian signal-plus-noise waveform is narrow-band (i.e., the bandwidth-to-center frequency ratio is small compared to unity), the amplitude may be considered constant over several cycles and the waveform statistics can be described in terms of the envelope of the waveform. The envelope

[1] This integral is carried out with the help of H. B. Dwight, *Tables of Integrals*, No. 861.5, Macmillan, New York, 1947, after substituting $G'(x)$ from (A1.2).

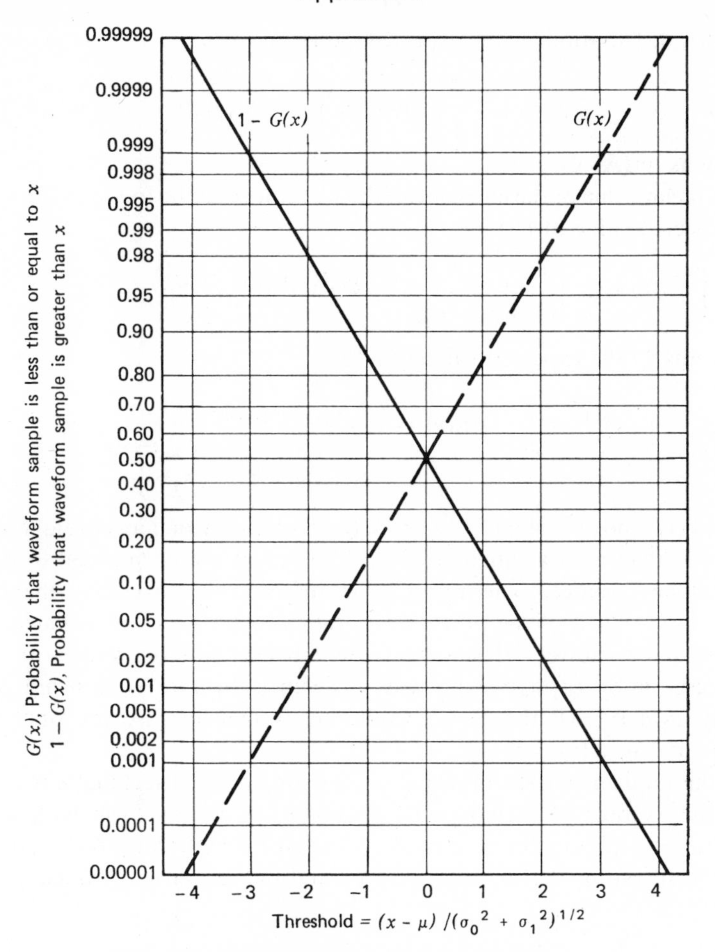

FIG. A1.2 Gaussian waveform distribution.

is described by the Rayleigh distribution,

$$R(r) = 1 - e^{-r^2/2(\sigma_0^2+\sigma_1^2)}, \tag{A1.6}$$

or its density function $R'(r)$

$$R'(r) = \frac{r}{(\sigma_0^2 + \sigma_1^2)} e^{-r^2/2(\sigma_0^2+\sigma_1^2)}. \tag{A1.7}$$

$R(r)$ is the area under the $R'(r)$ function from 0 to r. If it is recognized that the average power associated with a sinusoidal function with envelope r is $r^2/2$, then the average power, the signal-plus-noise power $(S + N)$ associated with this distribution, is the expected value[2] of $r^2/2$:

$$S + N = \langle r^2/2 \rangle = \int_0^1 \left(\frac{r^2}{2}\right) dR(r) = \int_0^\infty \frac{r^2}{2} R'(r)\, dr = \sigma_0^{\,2} + \sigma_1^{\,2} \quad \text{(A1.8)}$$

exactly as in (A1.4).

The mean value of the envelope of the waveform can be found by computing the expected value of r:

$$\langle r \rangle = \int_0^1 r\, dR(r) = \int_0^\infty r \frac{r}{\sigma_0^{\,2} + \sigma_1^{\,2}} e^{-r^2/2(\sigma_1^2+\sigma_1^2)}\, dr = \left(\frac{\pi}{2}\right)^{1/2} \sqrt{\sigma_0^{\,2} + \sigma_1^{\,2}} \quad \text{(A1.9)}$$

and similarly the expected value of r^4 is

$$\langle r^4 \rangle = \int_0^1 r^4\, dR(r) = \int_0^\infty r^4 e^{-r^2/2(\sigma_0^2+\sigma_1^2)} \frac{r\, dr}{\sigma_0^{\,2} + \sigma_1^{\,2}}$$

$$= 4 \cdot 2!\, (\sigma_0^{\,2} + \sigma_1^{\,2})^2 = 8(\sigma_0^{\,2} + \sigma_1^{\,2})^2. \quad \text{(A1.10)}$$

Figure A1.3 shows a plot of $1 - R(r)$, the probability that an envelope sample is greater than r, as a function of r^2. The curves are normalized to σ_0^2, the noise-alone variance. The lowest curve refers to the envelope behavior for noise alone. Each of the other curves refers to the envelope behavior for a specific signal power. The curves are labeled with the values of signal-to- noise ratios $(\sigma_1/\sigma_0)^2$ expressed in decibels by the relationship $S/N = 10 \log (\sigma_1/\sigma_0)^2$. All the curves would be straight lines except for the scale change at $1 - R(r) = 0.1$.

In the implementation of signal processors, a useful technique is to compute the average of N independent samples of a waveform with a given distribution. If samples of a random waveform $u(t)$, described statistically by the distribution $H_1(u)$, are measured and the sequence obtained is

$$u(t_1),\ u(t_2),\ u(t_3),\ \ldots,\ u(t_i),\ \ldots,\ u(t_n),$$

the average z_n, obtained by computing the sum,

$$z_n = \frac{1}{n} \sum_{i=1}^{n} u(t_i), \quad \text{(A1.11)}$$

is a member of a set of random numbers described statistically by a new distribution $H_n(z)$. The following statistical results[3] are very useful in predicting the behavior of averaging processors:

[2] This integral is evaluated by substituting $R'(r)$ from (A1.7) and making the substitution $r^2 = z$. The resulting integral has the form $\int_0^\infty xe^{-x/a}\, dx$, which is Dwight (loc. cit.) No. 861.2.

[3] E. Parzen, *Probability Theory and Its Application*, Wiley, New York, 1960, pp. 366, 406; A. Papoulis, *Probability, Random Variables, and Stochastic Processes*, McGraw-Hill, New York, 1966, p. 266.

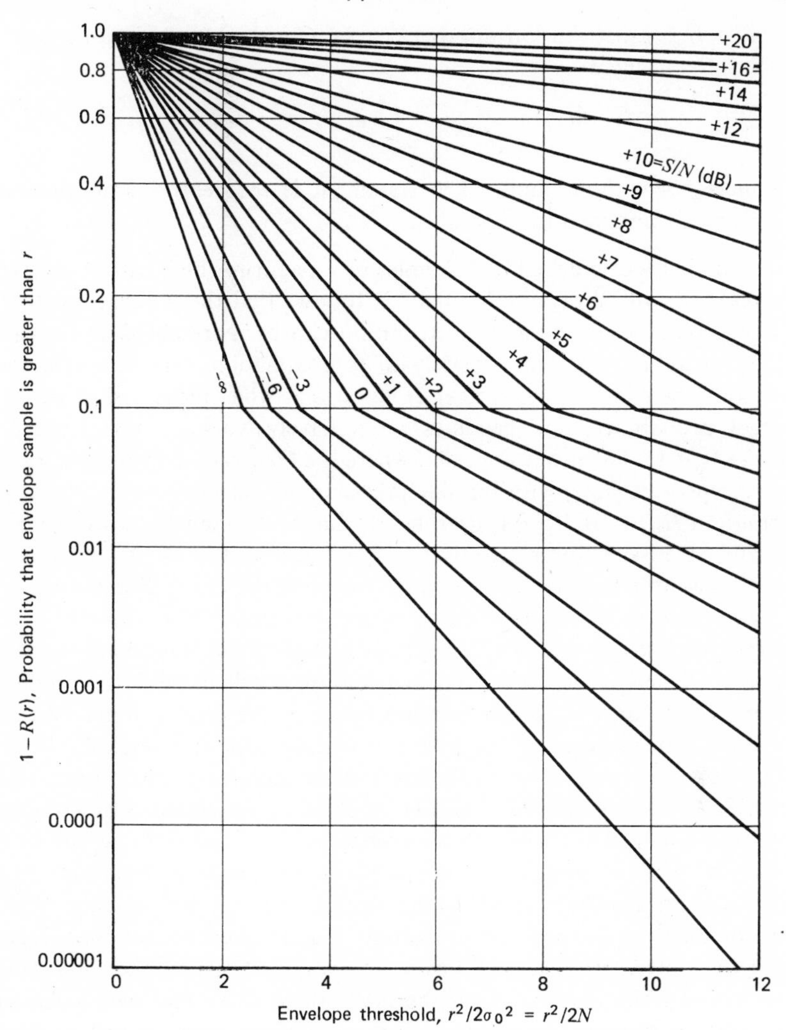

FIG. AI.3 Envelope distribution: Gaussian signal in Gaussian noise.

1. If $\mu_1 = \langle u \rangle$ is the mean of the $H_1(u)$ distribution and $\mu_n = \langle z_n \rangle$ is the mean of the $H_n(u)$ distribution, then

$$\mu_n = \mu_1. \tag{A1.12}$$

2. If $\sigma_1{}^2 = \langle (u - \mu_1)^2 \rangle$ is the variance of the $H_1(u)$ distribution and $\sigma_n{}^2 = \langle (z - \mu_n)^2 \rangle$ is the variance of the $H_n(u)$ distribution, then

$$\sigma_n{}^2 = \frac{\sigma_1{}^2}{n}. \tag{A1.13}$$

3. The $H_n(z)$ distribution approaches the Gaussian distribution, that is, $H_n(z) \rightarrow G(z)$, where

$$G(z) = \frac{1}{\sqrt{2\pi\sigma_n^2}} \int_{-\infty}^{z} e^{-(z'-\mu_n)^2/2\sigma_n^2} \, dz' \qquad \text{(A1.14)}$$

as n increases without limit. The statement is the central limit theorem. μ_n and σ_n are obtained from result 2.

The number n of independent samples of waveform which will be available for averaging will always be limited. It follows that the assumption that n is large enough so the average distribution can be represented as Gaussian by the central limit theorem is merely an approximation. The approximation is generally best in the region near $H_n(z) = 0.5$, when measured in per cent error of probability, and becomes progressively worse in this respect as $H_n(z) \rightarrow 0$ or 1. The lower limit of n where the Gaussian distribution may be used to represent $H_n(z)$ with suitable precision depends upon the form of the distribution $H_1(u)$. If $H_1(u)$ is itself nearly Gaussian, then $H_n(z)$ will be quite accurately Gaussian for relatively small n, such as 3 or 4. If $H_1(u)$ is far from Gaussian, n will have to be much greater if the $H_n(z)$ distribution is to be represented as Gaussian with the same precision. For example, if the sequence $u(t_1), u(t_2), \ldots, u(t_n)$ consists of a random sequence made up of the numbers $+1$ or -1 ($H_1(u)$ for this distribution is called the Bernoulli distribution), $H_n(z)$ for $n = 20$ can be represented by the central limit Gaussian only if an error as large as 0.09 in the probability can be tolerated.

The situation under discussion refers to conditions at the output of an averaging processor. Since, if n is large enough, the statistics at this point will be Gaussian, it is convenient to prepare a set of Gaussian curves on the probability grid normalized to the noise-alone standard deviation[4] $\sigma(N)$. [The standard deviation $\sigma(N)$ is the square root of the variance $\sigma^2(N)$.] The probability, $1 - G(x)$ (the probability that a waveform sample is greater than x), is plotted in Figure A1.4 as a function of x for a number of different mean values. The mean $\mu(N)$ of the noise-alone curve has been arbitrarily translated to zero in these plots. The noise-alone curve therefore passes through $1 - G(x) = 0.50$ at $(x - \mu(N))/\sigma(N) = 0$. The presence of a signal causes a shift in the mean from $\mu(N)$ to $\mu(S + N)$ in the averaging processors. The other curves therefore represent distributions expected for different signal sizes. Each passes through $1 - G(x) = 0.50$ at

$$\frac{x - \mu(N)}{\sigma(N)} = \frac{\mu(S + N)}{\sigma(N)}, \qquad \text{(A1.15)}$$

[4] In order to avoid confusion at a later stage μ_n and σ_n have been replaced with the symbols $\mu(N)$ and $\sigma(N)$ for noise-alone waveforms and $\mu(S + N)$ and $\sigma(S + N)$ for signal-plus-noise waveforms.

where $\mu(S+N)/\sigma(N)$ is the signal-plus-noise mean associated with curve expressed in units of noise-alone standard deviation. It is customary to use the square of the mean shift when signal arrives as the output signal power and the variance $\sigma^2(N)$ of the noise as the noise power. In decibels the output signal-to-noise ratio is

$$\frac{S}{N}(\text{dB})_{\text{out}} = 10 \log \left[\frac{\mu(S+N)-\mu(N)}{\sigma(N)}\right]^2 (\text{dB}). \tag{A1.16}$$

The signal curves in Figure A1.4 were plotted parallel to the noise-alone curve in such a way that they pass through $1 - G(x) = 0.5$ at selected values

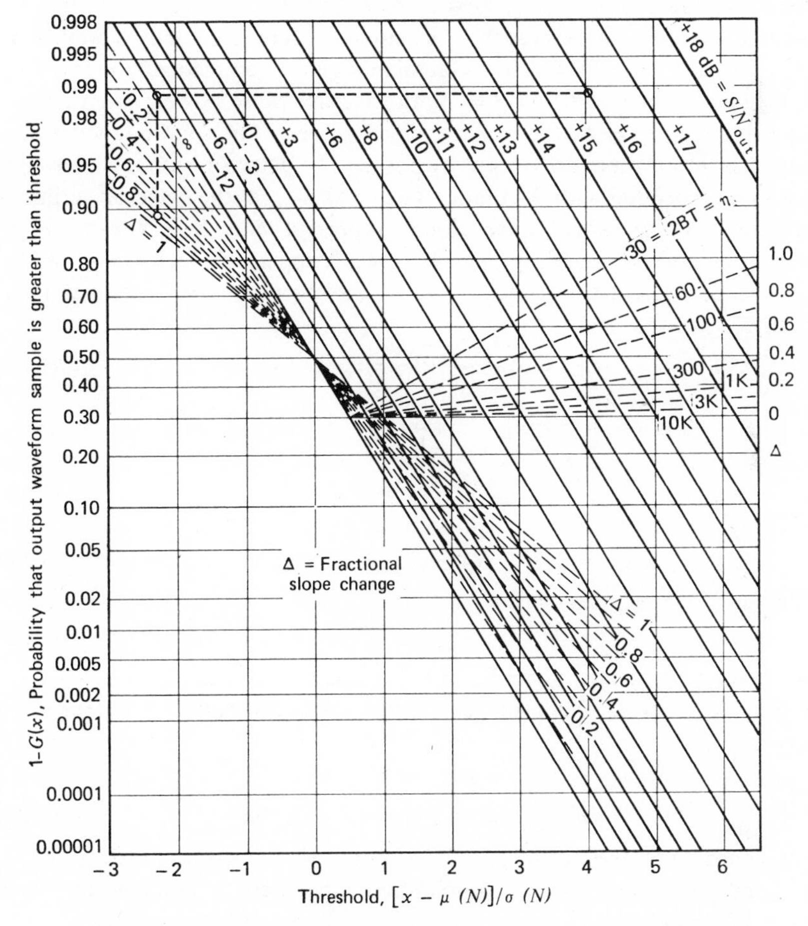

FIG. A1.4 Signal-plus-noise statistics at averager output (Square-law detector: read Δ at right; Linear detector: read half indicated Δ).

of $\mu(S + N)/\sigma(N)$. Equation A1.16 was used to compute the appropriate S/N(dB) label.

The curves are valid only for situations where the output signal-plus-noise standard deviation is the same as the output noise-alone standard deviation. This is not true, in general, but the approximation improves as the number of input samples averaged in computing an output sample is increased.

It is fairly easy to determine the conditions under which the slopes of the curves in Figure A1.4 must be modified. This is done by calculating the output signal-to-noise ratio in terms of the moments of the waveform distribution at the input to the averagers. The corrections for a square law detector-averager will be calculated as an example.

Square law detector averager: The input to the detector is a waveform with Gaussian noise in which a Gaussian signal is occasionally embedded. The variance of the noise is $\sigma_0^2 = N$, the average noise power; the variance of the signal is $\sigma_1^2 = S_1$ the average signal power. The square law detector squares the input waveform, point by point. The detector output is therefore the power associated with the input. The mean μ_{Det} at the detector output is therefore $N = \sigma_0^2$ when noise alone is present and $S + N = \sigma_0^2 + \sigma_1^2$ when signal and noise are present. This result appears in (A1.4). Similarly, the variance σ^2_{Det} of the detector output waveform can be obtained from $\langle x^4 \rangle$ since

$$\sigma^2_{\text{Det}} = \langle (x^2 - \overline{x^2})^2 \rangle = \langle x^4 \rangle - (x^2)^2. \tag{A1.17}$$

$$\langle x^4 \rangle = \frac{1}{\sqrt{2\pi}} \int_{-\infty}^{\infty} \frac{x^4}{\sqrt{\sigma_0^{\,2} + \sigma_1^{\,2}}} e^{-x^2/2(\sigma_0^2+\sigma_1^2)}\, dx$$

$$= 2(\sigma_0^{\,2} + \sigma_1^{\,2})^2 = 2(S + N)^2 \tag{A1.18}$$

and

$$\sigma^2_{\text{Det}} = 2(S + N)^2 - (S + N)^2 = (S + N)^2. \tag{A1.19}$$

For noise alone σ^2_{Det} becomes

$$\sigma^2_{\text{Det}} = N^2 \tag{A1.20}$$

With the help of Eqs. (A1.12), (A1.13), and (A1.14), these moments give, for the moments at the averager output (after averaging n independent samples)

$$\mu_{\text{output}} = \mu_{\text{Det}} = S + N \tag{A1.21}$$

$$\sigma^2_{\text{output}} = \sigma^2_{\text{Det}}/n = (S + N)^2/n \tag{A1.22}$$

and, according to the definition of the output signal-to-noise ratio (A1.16), the signal-plus-noise variance which should be used to plot the curves in Figure A1.4 is σ^2_{output},

$$\sigma^2(S + N) = \frac{N^2}{n}\left(1 + \frac{S}{N}\right)^2 \tag{A1.23}$$

instead of

$$\sigma^2(N) = \frac{N^2}{n} = \frac{\sigma^2(S + N)}{(1 + S/N)^2}, \tag{A1.24}$$

which was used. In the expressions for the moments, S and N represent the input signal and noise power. It follows that using (A1.24) to determine the slopes of the plots in Figure A1.4 gives a valid result only as long as the input signal-to-noise ratio is rather small compared to unity.

A convenient result would be an expression for $\sigma^2(S + N)$ in terms of the output signal-to-noise ratio. In the ratio form rather than the decibel form Eq. (A1.16), after substitution of $\mu(S + N)$, $\mu(N)$, and $\sigma(N)$, becomes

$$\left[\frac{S}{N}\right]_{\text{out}} = n\left[\frac{S + N - N}{N}\right]^2 = n\left(\frac{S}{N}\right)^2. \tag{A1.25}$$

This expression can be used to find a value of input signal-to-noise ratio to substitute into (A1.24):

$$\sigma^2(S + N) = \sigma^2(N)\left(1 + \frac{1}{n}\left[\frac{S}{N}\right]_{\text{out}}\right)^2. \tag{A1.26}$$

The slopes of the signal-plus-noise curves should therefore be reduced by a factor $\sigma(S + N)/\sigma(N)$, that is by a factor,

$$\frac{\sigma(S + N)}{\sigma(N)} = 1 + \frac{1}{n}\left[\frac{S}{N}\right]_{\text{out}}. \tag{A1.27}$$

The corrections are conveniently shown as the fractional change in slope

$$\Delta = \frac{1}{n}\left[\frac{S}{N}\right]_{\text{out}}. \tag{A1.28}$$

A set of plots, passing through the noise mean, with slopes differing from $\sigma(N)$ by the fraction Δ are shown dashed at the left. A nomograph based upon (A1.28) is shown as a pencil of lines through the point (0.30, 0.5) for several values of n. (Appendix 2 shows that $n = 2BT$, where B is the input bandwidth and T is the averaging time of the averager.) For a given $2BT = n$ value and output signal-to-noise ratio $(S/N)_{\text{out}}$, the fractional slope change can be read using the scale on the right at the intersection of the plot labeled n and the plot labeled S/N_{out}.

For example, if $(S/N)_{\text{out}} = 16$ dB and $2BT = 60$, $\Delta = 0.86$. At threshold $= 4.0$, $1 - G(x) = 0.989$ if no slope correction is required. To make the slope correction one follows the dotted line from 0.989 on the 16-dB curve to the left to the noise-alone plot, then vertically downward to the plot corre- ponding to $\Delta = 0.86$. The result obtained is 0.89. The slope corrections decreases $1 - G(x)$ almost 10%. It should be noticed that, for $BT >$ 5000 ($2BT = 10{,}000$), corrections in the slope have dropped below 5% for output signal-to-noise ratios less than 15 dB.

The curves in Figure A1.4 can also be used with the linear detector. They are used in exactly the same way except that the value of Δ obtained on the right scale is halved before proceeding to make the slope correction.

This nomograph is somewhat cumbersome, but it allows one to make

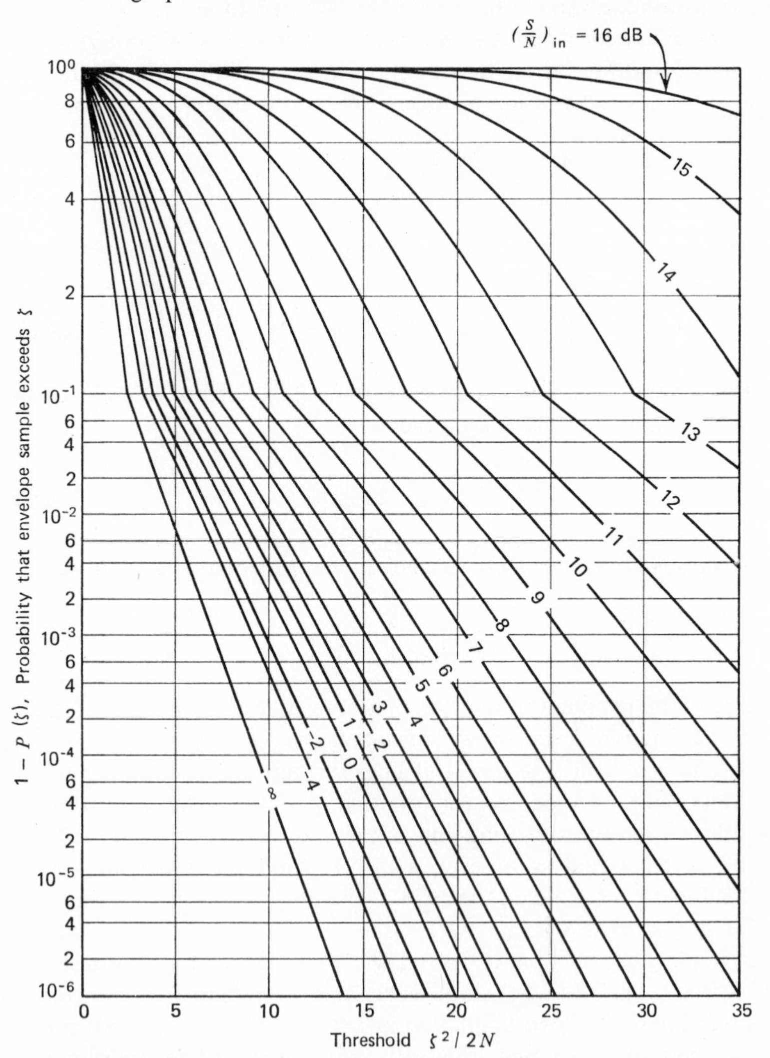

FIG. A1.5 Probability that signal-plus-noise power exceeds ζ^2 (matched filter output statistics).

rough plots of signal detection statistics without plotting out all the signal-plus-noise curves with the proper slopes. It may be used as given for BT values greater than 1500 if 2% errors in $1 - G(x)$ near $1 - G(x) = 0.80$ and 0.20 are not excessive.

One additional distribution is needed to describe the statistical behavior at the output of a matched filter. It is the distribution of the envelope of the waveform produced by a constant amplitude sinusoid embedded in a background of white, Gaussian, narrowband noise. This distribution is described by Rice.[5] The first extensive tables of the distribution were prepared and published by Marcum.[6] The distribution is plotted as a function of threshold squared in units of twice the average noise power in Figure A1.5. The parameter, $(S/N)_{\text{in}}$, is the average signal power–noise power ratio at the input to the envelope detector. The statistical distribution gives the probability that the envelope at the output of the envelope detector will exceed the threshold ζ. The $(S/N)_{\text{in}} = -\infty$ dB plot is identical to the correspondingly labeled plot in Figure A1.3 for Gaussian signals in Gaussian noise.

When these curves are used to determine the performance of a matched filter followed by a linear detector, the signal-to-noise ratio at the matched filter output (10.28) is the input signal-to-noise ratio to the envelope detector and the $(S/N)_{\text{in}}$ used in entering the curves of Figure A1.5.

[5] S. O. Rice, "Mathematical Analysis of Random Noise," in *Bell Sys. Tech. J.* **17,** 592 (1938). A concise discussion of this distribution is given in W. B. Davenport and W. L. Root, *Introduction to the Theory of Random Signals and Noise*, McGraw-Hill, New York, 1958, p. 165.

[6] J. I. Marcum, "Tables of Q Functions," Rand Research Memorandum RM 339, Rand Corporation, Santa Monica, California (DDC Document AD 116551).

APPENDIX 2

Waveform Sampling

It is necessary to determine the data rate to be associated with a given waveform in order to estimate the smoothing which can be achieved by averaging sections of a waveform. This is usually done by calculating the correlation function and determining an estimate of the separation of independent samples of the waveform. This is a method which has great generality but its development depends upon transform techniques beyond the scope of the present effort in development. For this reason a more elementary approach, an approach involving the Fourier series, will be employed to develop the sampling theorem.

The sampling theorem establishes the minimum number of quantities needed to describe an arbitrary time function. Suppose a band limited, continuous, time function[1] $u(t)$ having duration T is to be represented exactly by a Fourier series.[2] The form of the series is

$$u(t) = \frac{a_0}{2} + \sum_{h=1}^{\infty} [a_h \cos 2\pi h f_1 t + b_h \sin (2\pi h f_1 t)], \tag{A2.1}$$

in which the a_h and b_h are given by

$$a_h = \frac{2}{T} \int_0^T u(t) \cos (2\pi h f_1 t)\, dt$$

$$b_h = \frac{2}{T} \int_0^T u(t) \sin (2\pi h f_1 t)\, dt,$$

where $f_1 = 1/T$ is known as the fundamental frequency.

[1] It will become apparent in the following pages, if it is not already apparent, that a time-limited function cannot be rigorously band limited, but for each time function it is possible to specify a bandwidth B_1 in which 80% of the waveform energy is found, a bandwidth B_2 in which 95% of the energy is found, etc.

[2] This statement is true for all points of the segment of time function except the end points. At these points the series predicts a value midway between the ordinates of the end points.

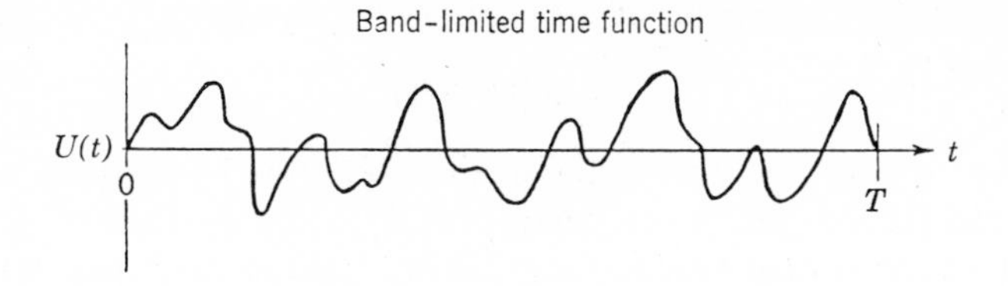

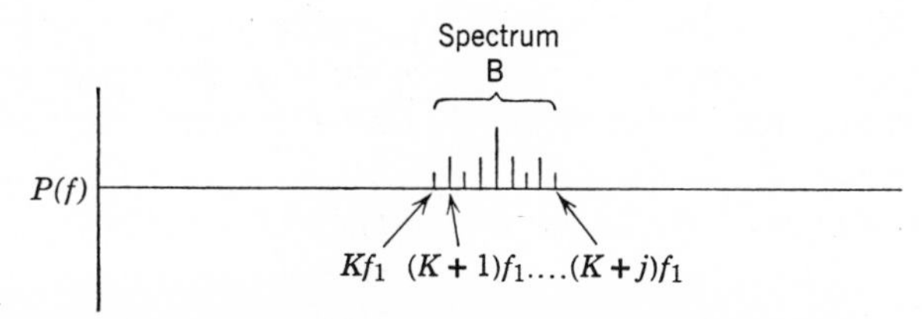

FIG. A2.1 Fourier representation of a narrow band time function. $ut = \sum_{k=h}^{k=j} (a_h \cos 2\pi hf_1 t + b_h \sin 2\pi hf_1 t)$.

Because $u(t)$ is band limited to a bandwidth B, the a_h and b_h corresponding to frequencies hf_1 outside the bandwidth B will be identically zero. This situation is pictured in Figure A2.1. The lowest frequency present in $u(t)$ is shown as kf_1 and the highest frequency present is shown as $(k+j)f_1$. In the representation of $u(t)$ by (A2.1) those terms prior to $h = k$ and following $h = k + j$ may be deleted from the sum with the result that $u(t)$ is represented by the finite series,

$$u(t) = \sum_{h=k}^{k+j} [a_h \cos (2\pi hf_1 t) + b_h \sin (2\pi hf_1 t)]. \tag{A2.2}$$

This series provides an exact representation of $u(t)$ over the interval of length T if $u(t)$ is continuous.

The question may now be asked concerning how many quantities need be specified to describe $u(t)$. There is first $f_1 = 1/T$, the reciprocal of the duration of the time function. Then there are $(j + 1)$ of the a_h and $(j + 1)$ of the b_h, a total of

$$\eta = 2(j + 1) \tag{A2.3}$$

coefficients.

This number is clearly related to the bandwidth of $u(t)$ because the bandwidth is divided into at least j intervals of width f_1, hence the equality,

$$jf_1 = B \tag{A2.4}$$

must hold. Elimination of f_1 from (A2.4) in terms of T and use of the value of j given by (A2.3) gives the result that the number of quantities which must be specified in the description of $u(t)$ satisfies the equality,

$$\eta = 2(BT + 1). \tag{A2.5}$$

The number η does not include the quantity kf_1, the lower band edge. In other words, except for lower band edge kf_1, the number of quantities needed to specify a time function $u(t)$ having bandwidth B and duration T is $(2BT + 2)$.

Equation (A2.5) is of great practical importance. It alone determines the number of digital samples which must be made when processing $u(t)$. Since $2BT + 2$ constants must be determined, $2BT + 2$ independent samples of $u(t)$ must be obtained. Suppose that $u(t_1), u(t_2), \ldots, u(t_\eta)$, equally spaced values of $u(t)$, are read from the graph of $u(t)$ and η simultaneous equations of the form,

$$\begin{aligned}
u(t_1) &= \sum_{i=k}^{k+j} [a_i \cos 2\pi i f_1 t_1 + b_i \sin 2\pi i f_1 t_1] \\
u(t_2) &= \sum_{i=k}^{k+j} [a_i \cos 2\pi i f_1 t_2 + b_i \sin 2\pi i f_1 t_2] \\
&\ \ \vdots \\
u(t_\eta) &= \sum_{i=k}^{k+j} [a_i \cos 2\pi i f_1 t_\eta + b_i \sin 2\pi i f_1 t_\eta]
\end{aligned} \tag{A2.6}$$

are written. The equally spaced samples of $u(t)$ would completely specify all of the a_i and b_i. The complete original time function could be reconstructed from (A2.2) and the a_i and b_i which were found with (A2.6). It is implied by this discussion that the equally spaced samples represent a complete set of independent samples of $u(t)$ and that additional samples would contain no new information about the sampled function.

On the strength of the preceding statements it is concluded that only a number η of independent samples of $u(t)$ can be taken in a time T and that η is $2(BT + 1)$. This statement is sometimes made in terms of the number of independent samples which may be made per unit time. If (A2.5) is divided by T, the equality becomes

$$\dot{\eta} = 2B + \frac{2}{T} \tag{A2.7}$$

in which $\dot{\eta}$ means the number of independent samples which may be made per second. This result is familiar. Just over $2B$ samples per second may be considered to be independent.

The discussion to this point has been made under the tacit assumption that a narrow-band time function was under discussion, one whose bandwidth

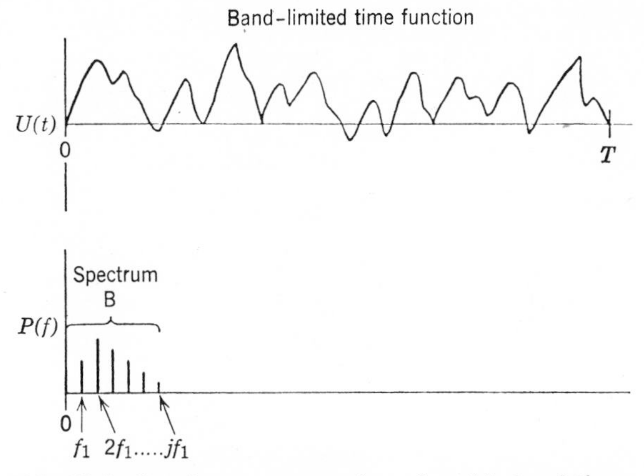

FIG. A2.2 Fourier representation of a wide band time function. $u(t) = a_0/2 + \sum_{h=1}^{j} (a_h \cos 2\pi h f_1 t + b_h \sin 2\pi h f_1 t)$.

is small compared to its center frequency. (See Fig. A2.1.) A wideband time function, one for which the bandwidth is comparable to the center frequency, is shown in Figure A2.2. For this case, as in the narrow-band case,

$$jf_1 = B. \tag{A2.8}$$

The same arguments which lead to (A2.5) can be made again.

A sampling theorem based upon this discussion may now be stated:

Theorem:

> A band-limited time function is completely specified by $2B + \epsilon$ equally spaced samples per second ($\epsilon > 0$).

This theorem may be applied to either wide-band or narrow-band signals provided that Fourier series reconstruction of the waveform takes place in subsequent signal processing of the samples.

If the bandwidth of the waveform to be sampled extends from 0 frequency to B, sampling at $2B + \epsilon$ provides just over 2 samples per cycle of the highest frequency components present in the waveform. This result is about what we would expect. It is somewhat surprising that the same sampling rate suffices for a bandwidth B (say 200 Hz) centered at 1 Mega Hz. In this example just over 400 samples/sec would be required, about one sample every 2500 cycles of the center frequency. The Fourier series constructed using these samples would reconstruct the complete waveform. This result suggests that the carrier frequency itself is unimportant. This concept is in agreement with our experience that the information in the band B may be heterodyned to a lower center frequency without loss.

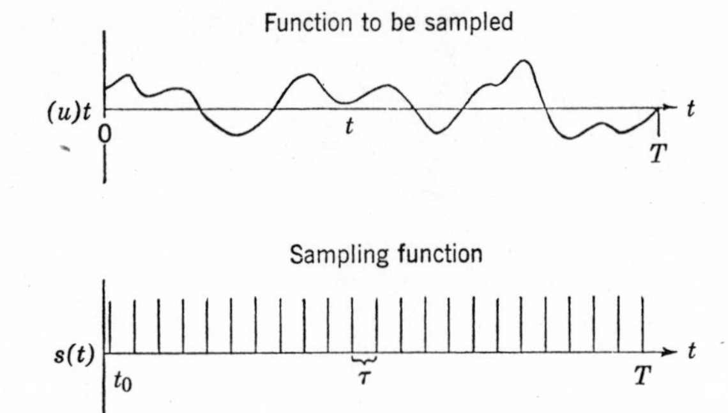

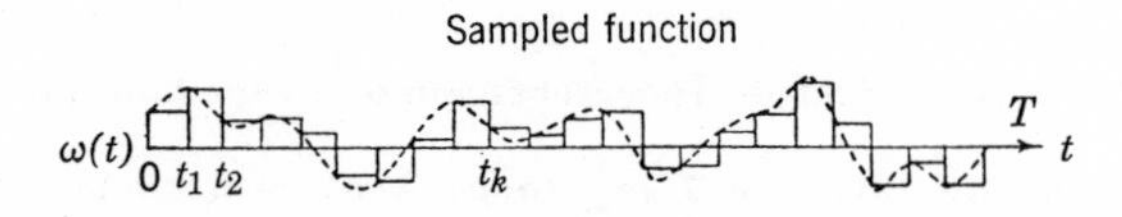

FIG. A2.3 Sampling as a product of *u*(t) and the sampling function *s*(t).

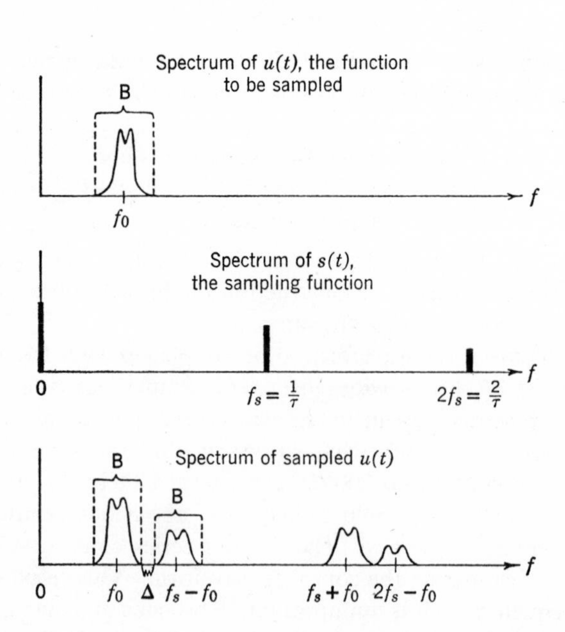

FIG. A2.4 Spectral description of sampled function.

The theorem is the optimum sampling theorem. It specifies the number of waveform samples which are required per unit time to reconstruct the sampled waveform.

The smallest number of samples required to reconstruct the function $u(t)$ and the largest number of samples of the function which are independent happen to be the same, namely $2B + \epsilon$ samples/sec. It is clear that if $2B + \epsilon$ samples/sec will allow reconstruction of the waveform, then no new information can be obtained by sampling more often. It follows that sampling more often than $2B + \epsilon$ samples/sec provides no samples independent of those already obtained. If fewer than $2B + \epsilon$ samples/sec are made, there are too few points to reconstruct the waveform by the Fourier expansion technique. It follows that sampling at a rate greater than $2B + \epsilon$ guarantees that reconstruction of $u(t)$ can be accomplished and at the same time guarantees that the samples will not all be independent in a statistical sense. On the other hand, sampling at a lower rate does not allow reconstruction of $u(t)$ but does guarantee that the samples will be independent in a statistical sense.

The actual sampling rate employed in practical processing depends upon the method which is to be used in reconstructing the time function and the functions derived from it in the computations. When sampling is done at a rate just satisfying the theorem, reconstruction of $u(t)$ from the samples must be accomplished by means of the terminated Fourier series.

If sampling according to the relaxed rule to be quoted below is carried out, reconstruction by simple filter techniques is possible. With this technique, the relaxed or "practical" sampling theorem will be made plausible by describing the spectrum of the time function to be sampled and the spectrum of the sampled function. The capability of reconstructing $u(t)$ by filter techniques will be clearly evident.

The sampling process may be considered equivalent to the product of $u(t)$, the function to be sampled, and the function $s(t)$, the sampling function. Figure A2.3 shows a time function $u(t)$ and a sampling function $s(t)$. The simplest sampling function is a series of equally spaced spikes, τ sec apart. The product of $s(t)$ and $u(t)$ will have values only at the positions of the spikes. The product function will be the sampled version of $u(t)$ and will be a series of spikes whose heights are proportional to the values of $u(t)$ at the sampling times. The sampled function then takes the form of the sequence of steps shown in the lower plot of Figure A2.3.

The spectra of $u(t)$ and $s(t)$ are shown in Figure A2.4. The spectrum of $s(t)$ can be defined if the spikes of the sampling function are constructed with fixed area under each spike. The illustration is given for a narrow-band signal. The spectrum of $u(t)$ is centered at f_0 and has a width denoted by B. The spectrum of the sampling function is a series of spikes spaced along the frequency axis at intervals $f = 1/\tau$, where τ is the sampling period.

The spectrum of the product of $u(t)$ and $s(t)$ is easily found from the identity

$$(a \cos 2\pi f_a t)(b \cos 2\pi f_b t) = \frac{ab}{2} [\cos 2\pi(f_a + f_b)t + \cos 2\pi(f_a - f_b)t]. \quad \text{(A2.9)}$$

For each frequency f_a present in $u(t)$ and for each frequency f_b present in $s(t)$, the frequencies $(f_a + f_b)$ and $(f_a - f_b)$ appear in the spectrum of the product. The lowest frequency present in the sampling function is $f = 0$; hence $f_0 - 0$ and $f_0 + 0$ both appear at f_0. The other frequencies in the bandwidth B transfer in the same way to a region around f_0 in the spectrum of the sampled function. The next lowest frequency in the sampling function is $f_s = 1/\tau$. This frequency will give rise to a band centered at $(f_s - f_0)$ and another at $(f_s + f_0)$. The next lowest frequency $2f_s$ results in a band centered at $(2f_s - f_0)$ and at $(2f_s + f_0)$ (not shown) to the right of the edge of the figure.

The two versions of $u(t)$, the original one and the sampled one, differ in that a number of higher, spurious frequencies are now present in the spectrum of the sampled version. Recovery of the original $u(t)$ from the sampled $u(t)$ can be accomplished provided none of the spurious frequencies fall in the band centered at f_0. This situation exists in the diagram; the nearest spurious band is separated from the band at f_0 by an amount Δ. Δ will be greater than zero if

$$\Delta = \left(f_s - f_0 - \frac{B}{2}\right) - \left(f_0 + \frac{B}{2}\right) > 0. \quad \text{(A2.10)}$$

This condition is satisfied if f_s, the sampling frequency, satisfies the inequality,

$$f_s > 2\left(f_0 + \frac{B}{2}\right). \quad \text{(A2.11)}$$

It will be noted that $(f_0 + B/2)$ is the highest frequency in the spectrum of $u(t)$.

In order to reproduce $u(t)$, the sampled function is played back through a narrow-band filter whose upper cut-off frequency is placed in the gap between the bands centered at f_0 and at $(f_s - f_0)$. If sufficiently steep skirted, this filter will remove a large portion of the power of the spurious spectral components and reconstruct $u(t)$ to a very good approximation. Clearly this job is easiest to do when the gap Δ is made larger. With commercially available, adjustable three-stage filters, the smallest useful value of f_s is about

$$f_s \sim 3\left(f_0 + \frac{B}{2}\right). \quad \text{(A2.12)}$$

This sampling rate provides a gap between the spectral components of $u(t)$

and the spurious components of the sampled $u(t)$ in excess of $(f_0 + B/2)$ and about 18 dB rejection of the spurious spectral components with commercial three-stage filters. In this case so-called sampling noise would be about 18 dB below the waveform which is represented by the samples.

When the spectrum of $u(t)$ extends to zero frequency, the upper frequency in $u(t)$ is B. The required f_s by (A2.10) is

$$f_s > 2B, \tag{A2.10}$$

the rate required by the ideal sampling theorem. If the sampled function is to be reconverted to analog form by means of filters, a suitable value of f_s is

$$f_s \sim 3B. \tag{A2.14}$$

The value of f_s actually used can be any size greater than three times the highest frequency in $u(t)$, but the expense of computing increases with the sampling rate (as the square of the sampling rate for correlation). It follows that extensive signal processing programs need to use the smallest feasible value of f_s. In setting up a signal processing system, one usually translates the signal band to as low a frequency as possible. In difference-band correlator applications the low edge of the band might be translated to B, the upper edge of the translated band will be $2B$. Sampling at $7B$ is adequate for this case in correlator applications.

The ideas presented concerning the "practical" sampling techniques indicate that it is possible to state a "practical sampling theorem" which will provide enough equally spaced samples to allow reconstruction of functions to a good approximation by simple filter techniques. In each application it is necessary to consider the actual calculations which are to be made and test the individual steps to make certain that the band of interest is at all times completely separated from spurious or sampling noise bands.

Practical sampling theorem:

> Sampling at a rate corresponding to 2^k times the highest frequency present in a time function will permit reconstruction of the function by simple n-stage filter techniques in which the spurious components are more than $6kn$ dB below their unfiltered values.

APPENDIX 3

Miscellaneous Tables

TABLE A3.1

Elastic Constants, Velocity of Sound, Characteristic Impedance[a]

(*a*) SOLIDS

Solid	Density kg/m³ ρ_0	Young's Modulus newtons/m² Y	Shear Modulus newtons/m² G	Bulk Modulus newtons/m² B	Poisson's Ratio σ	Velocity m/sec c		Characteristic Impedance MKS rayls $\rho_0 c$	
						Bar	Bulk	Bar	Bulk
		$\times 10^{10}$	$\times 10^{10}$	$\times 10^{10}$				$\times 10^{6}$	$\times 10^{6}$
Aluminum	2700	7.1	2.4	7.5	0.33	5150	6300	13.9	17.0
Brass	8500	10.4	3.8	13.6	0.37	3500	4700	29.8	40.0
Copper	8900	12.2	4.4	16.0	0.35	3700	5000	33.0	44.5
Iron (cast)	7700	10.5	4.4	8.6	0.28	3700	4350	28.5	33.5
Lead	11300	1.65	0.55	4.2	0.44	1200	2050	13.6	23.2
Nickel	8800	21.0	8.0	19.0	0.31	4900	5850	43.0	51.5
Silver	10500	7.8	2.8	10.5	0.37	2700	3700	28.4	39.0
Steel	7700	19.5	8.3	17.0	0.28	5050	6100	39.0	47.0
Glass (Pyrex)	2300	6.2	2.5	3.9	0.24	5200	5600	12.0	12.9
Quartz (X-cut)	2650	7.9	3.9	3.3	0.33	5450	5750	14.5	15.3
Lucite	1200	0.4	0.14	0.65	0.4	1800	2650	2.15	3.2
Concrete	2600	—	—	—	—	—	3100	—	8.0
Ice	920	—	—	—	—	—	3200	—	2.95
Cork	240	—	—	—	—	—	500	—	0.12
Oak	720	—	—	—	—	—	4000	—	2.9
Pine	450	—	—	—	—	—	3500	—	1.57
Rubber (hard)	1100	0.23	0.1	0.5	0.4	1450	2400	1.6	2.64
Rubber (soft)	950	0.0005	—	0.1	0.5	70	1050	0.065	1.0
Rubber (rho-c)	1000	—	—	0.24	—	—	1550	—	1.55

[a] From L. E. Kinsler and A. R. Frey, *Fundamentals of Acoustics*, 2nd ed., Wiley, New York, 1962, pp. 502, 503.

(*b*) LIQUIDS

Liquid	Temperature °C t	Density kg/m³ ρ_0	Bulk Modulus newtons/m² B_T	Ratio of Specific Heats γ	Velocity m/sec c	Characteristic Impedance MKS Rayls $\rho_0 c$	Coefficient of Viscosity newton sec/m² η
			$\times 10^9$			$\times 10^6$	
Water (fresh)	20	998	2.18	1.004	1481	1.48	0.001
Water (sea)	13	1026	2.28	1.01	1500	1.54	0.001
Alcohol (ethyl)	20	790	—	—	1150	0.91	0.0012
Caster oil	20	950	—	—	1540	1.45	0.96
Mercury	20	13600	25.3	1.13	1450	19.7	0.0016
Turpentine	20	870	1.07	1.27	1250	1.11	0.0015
Glycerin	20	1260	—	—	1980	2.5	1.2

(*c*) GASES
(at a pressure of 1.013×10^5 newtons/m²)

Gas	Temperature °C t	Density kg/m³ ρ_0	Ratio of Specific Heats γ	Velocity m/sec c	Characteristic Impedance MKS rayls $\rho_0 c$	Coefficient of Viscosity newton sec/m² η
Air	0	1.293	1.402	331.6	428	0.000017
Air	20	1.21	1.402	343	415	0.0000181
Oxygen	0	1.43	1.40	317.2	453	0.00002
CO_2 (low freq.)	0	1.98	1.304	258	512	0.0000145
CO_2 (high freq.)	0	1.98	1.40	268.6	532	0.0000145
Hydrogen	0	0.09	1.41	1269.5	114	0.0000088
Steam	100	0.6	1.324	404.8	242	0.000013

TABLE A3.2

Trigonometric and Hyberbolic Functions[a]

x radians	sin x	cos x	sinh x	cosh x	e^x	e^{-x}
0.0	0.0000	1.0000	0.0000	1.0000	1.0000	1.0000
0.2	0.1987	0.9801	0.2013	1.0201	1.2214	0.8187
0.4	0.3894	0.9211	0.4108	1.0811	1.4918	0.6703
0.6	0.5646	0.8253	0.6367	1.1855	1.8221	0.5488
0.8	0.7174	0.6967	0.8881	1.3374	2.2255	0.4493
1.0	0.8415	0.5403	1.1752	1.5431	2.7183	0.3679
1.2	0.9320	0.3624	1.5095	1.8107	3.3201	0.3012
1.4	0.9854	+0.1700	1.9043	2.1509	4.0552	0.2466
1.6	0.9996	−0.0292	2.3756	2.5775	4.9530	0.2019
1.8	0.9738	−0.2272	2.9422	3.1075	6.0496	0.1653
2.0	0.9093	−0.4161	3.6269	3.7622	7.3891	0.1353
2.2	0.8085	−0.5885	4.4571	4.5679	9.0250	0.1108
2.4	0.6755	−0.7374	5.4662	5.5569	11.023	0.09072
2.6	0.5155	−0.8569	6.6947	6.7690	13.464	0.07427
2.8	0.3350	−0.9422	8.1919	8.2527	16.445	0.06081
3.0	+0.1411	−0.9900	10.018	10.068	20.086	0.04979
3.2	−0.0584	−0.9983	12.246	12.287	24.533	0.04076
3.4	−0.2555	−0.9668	14.965	14.999	29.964	0.03337
3.6	−0.4425	−0.8968	18.285	18.313	36.598	0.02732
3.8	−0.6119	−0.7910	22.339	22.362	44.701	0.02237
4.0	−0.7568	−0.6536	27.290	27.308	54.598	0.01832
4.2	−0.8716	−0.4903	33.336	33.351	66.686	0.01500
4.4	−0.9516	−0.3073	40.719	40.732	81.451	0.01228
4.6	−0.9937	−0.1122	49.737	49.747	99.484	0.01005
4.8	−0.9962	+0.0875	60.751	60.759	121.51	0.00823
5.0	−0.9589	0.2837	74.203	74.210	148.41	0.00674
5.2	−0.8835	0.4685	90.633	90.639	181.27	0.00552
5.4	−0.7728	0.6347	110.70	110.71	221.41	0.00452
5.6	−0.6313	0.7756	135.21	135.22	270.43	0.00370
5.8	−0.4646	0.8855	165.15	165.15	330.30	0.00303
6.0	−0.2794	0.9602	201.71	201.72	403.43	0.00248
6.2	−0.0831	0.9965	246.37	246.38	492.75	0.00203
6.4	+0.1165	0.9932	300.92	300.92	601.85	0.00166
6.6	0.3115	0.9502	367.55	367.55	735.10	0.00136
6.8	0.4941	0.8694	448.92	448.92	897.85	0.00111
7.0	0.6570	0.7539	548.32	548.32	1096.6	0.00091
7.2	0.7937	0.6084	669.72	669.72	1339.4	0.00075
7.4	0.8987	0.4385	817.99	817.99	1636.0	0.00061
7.6	0.9679	0.2513	999.10	999.10	1998.2	0.00050
7.8	0.9985	+0.0540	1220.3	1220.3	2440.6	0.00041
8.0	0.9894	−0.1455	1490.5	1490.5	2981.0	0.00034

[a] From L. E. Kinsler and A. R. Frey, *Fundamentals of Acoustics*, 2nd ed., Wiley, New York, 1962, p. 504.

TABLE A3.3

Four-Place Table of Logarithms[a]

(*a*) SOLIDS

Solid	Density kg/m³ ρ_0	Young's Modulus newtons/m² Y	Shear Modulus newtons/m² G	Bulk Modulus newtons/m² B	Poisson's Ratio σ	Velocity m/sec c		Characteristic Impedance MKS rayls $\rho_0 c$	
						Bar	Bulk	Bar	Bulk
		$\times 10^{10}$	$\times 10^{10}$	$\times 10^{10}$				$\times 10^{6}$	$\times 10^{6}$
Aluminum	2700	7.1	2.4	7.5	0.33	5150	6300	13.9	17.0
Brass	8500	10.4	3.8	13.6	0.37	3500	4700	29.8	40.0
Copper	8900	12.2	4.4	16.0	0.35	3700	5000	33.0	44.5
Iron (cast)	7700	10.5	4.4	8.6	0.28	3700	4350	28.5	33.5
Lead	11300	1.65	0.55	4.2	0.44	1200	2050	13.6	23.2
Nickel	8800	21.0	8.0	19.0	0.31	4900	5850	43.0	51.5
Silver	10500	7.8	2.8	10.5	0.37	2700	3700	28.4	39.0
Steel	7700	19.5	8.3	17.0	0.28	5050	6100	39.0	47.0
Glass (Pyrex)	2300	6.2	2.5	3.9	0.24	5200	5600	12.0	12.9
Quartz (X-cut)	2650	7.9	3.9	3.3	0.33	5450	5750	14.5	15.3
Lucite	1200	0.4	0.14	0.65	0.4	1800	2650	2.15	3.2
Concrete	2600	—	—	—	—	—	3100	—	8.0
Ice	920	—	—	—	—	—	3200	—	2.95
Cork	240	—	—	—	—	—	500	—	0.12
Oak	720	—	—	—	—	—	4000	—	2.9
Pine	450	—	—	—	—	—	3500	—	1.57
Rubber (hard)	1100	0.23	0.1	0.5	0.4	1450	2400	1.6	2.64
Rubber (soft)	950	0.0005	—	0.1	0.5	70	1050	0.065	1.0
Rubber (rho-c)	1000	—	—	0.24	—	—	1550	—	1.55

TABLE A3.3

(continued)

(*b*) LIQUIDS

Liquid	Temperature °C t	Density kg/m³ ρ_0	Bulk Modulus newtons/m² B_T	Ratio of Specific Heats γ	Velocity m/sec c	Characteristic Impedance MKS Rayls $\rho_0 c$	Coefficient of Viscosity newton sec/m² η
			$\times 10^9$			$\times 10^6$	
Water (fresh)	20	998	2.18	1.004	1481	1.48	0.001
Water (sea)	13	1026	2.28	1.01	1500	1.54	0.001
Alcohol (ethyl)	20	790	—	—	1150	0.91	0.0012
Caster oil	20	950	—	—	1540	1.45	0.96
Mercury	20	13600	25.3	1.13	1450	19.7	0.0016
Turpentine	20	870	1.07	1.27	1250	1.11	0.0015
Glycerin	20	1260	—	—	1980	2.5	1.2

(*c*) GASES
(at a pressure of 1.013×10^5 newtons/m²)

Gas	Temperature °C t	Density kg/m³ ρ_0	Ratio of Specific Heats γ	Velocity m/sec c	Characteristic Impedance MKS rayls $\rho_0 c$	Coefficient of Viscosity newton sec/m² η
Air	0	1.293	1.402	331.6	428	0.000017
Air	20	1.21	1.402	343	415	0.0000181
Oxygen	0	1.43	1.40	317.2	453	0.00002
CO_2 (low freq.)	0	1.98	1.304	258	512	0.0000145
CO_2 (high freq.)	0	1.98	1.40	268.6	532	0.0000145
Hydrogen	0	0.09	1.41	1269.5	114	0.0000088
Steam	100	0.6	1.324	404.8	242	0.000013

[a] From C. A. Rogers, *Elements of Algebra*, Wiley, New York, 1962.

TABLE A3.4

Directivity and Impedance Functions for a Circular Piston[a]

x	Directivity Functions ($x = ka \sin \theta$)		Impedance Functions ($x = 2ka$)	
	Pressure	Intensity	Resistance	Reactance
	$\frac{2J_1(x)}{x}$	$\left[\frac{2J_1(x)}{x}\right]^2$	$R_1(x)$	$X_1(x)$
0.0	1.0000	1.0000	0.0000	0.0000
0.2	0.9950	0.9900	0.0050	0.0847
0.4	0.9802	0.9608	0.0198	0.1680
0.6	0.9557	0.9134	0.0443	0.2486
0.8	0.9221	0.8503	0.0779	0.3253
1.0	0.8801	0.7746	0.1199	0.3969
1.2	0.8305	0.6897	0.1695	0.4624
1.4	0.7743	0.5995	0.2257	0.5207
1.6	0.7124	0.5075	0.2876	0.5713
1.8	0.6461	0.4174	0.3539	0.6134
2.0	0.5767	0.3326	0.4233	0.6468
2.2	0.5054	0.2554	0.4946	0.6711
2.4	0.4335	0.1879	0.5665	0.6862
2.6	0.3622	0.1326	0.6378	0.6925
2.8	0.2927	0.0857	0.7073	0.6903
3.0	0.2260	0.0511	0.7740	0.6800
3.2	0.1633	0.0267	0.8367	0.6623
3.4	0.1054	0.0111	0.8946	0.6381
3.6	0.0530	0.0028	0.9470	0.6081
3.8	+0.0068	0.00005	0.9932	0.5733
4.0	−0.0330	0.0011	1.0330	0.5349
4.5	−0.1027	0.0104	1.1027	0.4293
5.0	−0.1310	0.0172	1.1310	0.3232
5.5	−0.1242	0.0154	1.1242	0.2299
6.0	−0.0922	0.0085	1.0922	0.1594
6.5	−0.0473	0.0022	1.0473	0.1159
7.0	−0.0013	0.00000	1.0013	0.0989
7.5	+0.0361	0.0013	0.9639	0.1036
8.0	0.0587	0.0034	0.9413	0.1219
8.5	0.0643	0.0041	0.9357	0.1457
9.0	0.0545	0.0030	0.9455	0.1663
9.5	0.0339	0.0011	0.9661	0.1782
10.0	+0.0087	0.00008	0.9913	0.1784
10.5	−0.0150	0.0002	1.0150	0.1668
11.0	−0.0321	0.0010	1.0321	0.1464
11.5	−0.0397	0.0016	1.0397	0.1216
12.0	−0.0372	0.0014	1.0372	0.0973
12.5	−0.0265	0.0007	1.0265	0.0779
13.0	−0.0108	0.0001	1.0108	0.0662
13.5	+0.0056	0.00003	0.9944	0.0631
14.0	0.0191	0.0004	0.9809	0.0676
14.5	0.0267	0.0007	0.9733	0.0770
15.0	0.0273	0.0007	0.9727	0.0880
15.5	0.0216	0.0005	0.9784	0.0973
16.0	0.0113	0.0001	0.9887	0.1021

[a] From L. E. Kinsler and A. R. Frey, *Fundamentals of Acoustics*, 2nd ed., Wiley, New York, 1962, p. 506.

TABLE A3.5

Conversion of Units[a]

A. Mechanical Units

				CGS Units		English Units	
Quantity	Symbol	Dimensional Equivalent	MKS Unit	Multiply value in MKS by:	To obtain value in CGS units:	Multiply value in MKS by:	To obtain value in English units:
Length	l	—	meter (m)	10^2	centimeter (cm)	39.37 3.28	inch (in.) feet (ft)
Mass	M	—	kilogram (kg)	10^3	gram (g)	(2.2 $\frac{2.2}{32.2} = 0.0684$	pound-mass)* slug
Time	t	—	second (sec)	1	sec	1	sec
Force	F	$[Ml/t^2]$	newton	10^5	dyne	($3.28 \times 2.2 = 7.2$ $\frac{7.2}{32.2} = 0.224$	poundal)* pound-weight
Work	$\mathcal{W}$	$[Fl]$	joule	10^7	erg	0.736	foot-pound
Power	W	$[Fl/t]$	watt	10^7	erg/sec	746	horsepower (hp)

* English absolute units, not commonly used in engineering practice.

B. Electric and Magnetic Units

				Electrostatic Units		Electromagnetic Units	
Quantity	Symbol	Dimensional Equivalent	MKS Unit	Multiply value in MKS by:	To obtain value in esu:	Multiply value in MKS by:	To obtain value in emu:
Current	I	—	ampere (amp)	3×10^9	statamp	10^{-1}	abamp
Charge	Q	$[It]$	coulomb	3×10^9	statcoulomb	10^{-1}	abcoulomb
Voltage	V	$[Fl/Q]$	volt	$\frac{1}{3} \times 10^{-2}$	erg/statcoulomb	10^8	erg/abcoulomb
Electric field strength	E	$[V/l]$	volt/m	$\frac{1}{3}$	dyne/statcoulomb	—	—
Electric displacement	D	$[Q/l^2]$	coulomb/m²	3×10^5	statcoulomb/cm²	—	—

[a] From T. F. Heuter and R. H. Bolt, *Sonics*, Wiley, New York, 1955, inside back cover.

TABLE A3.5

Conversion of Units[a]

Quantity	Symbol	Dimensional Equivalent	MKS Unit	Multiply value in MKS by:	To obtain value in CGS units:	Multiply value in MKS by:	To obtain value in English units:
Capacitance	C	$[Q/V]$	farad (F)	9×10^{11}	statcoulomb²/erg	10^{-9}	abcoulomb²/erg
Magnetic field strength	H	$[I/l]$	amp/m	—	—	79.6	oersted
Magnetic flux density	B	$[Vt/l^2]$	weber/m²	—	—	10^{-4}	gauss
Inductance	L	$[Vt/I]$	henry (h)	$\frac{1}{9} \times 10^{-11}$	erg·sec²/statcoulomb	10^{9}	erg·sec²/abcoulomb
Resistance	R	$[V/I]$	ohm (Ω)	$\frac{1}{9} \times 10^{-11}$	erg·sec/statcoulomb	10^{9}	erg·sec/abcoulomb

C. Acoustical Units

				CGS Units		English Units	
Quantity	Symbol	Dimensional Equivalent	MKS Unit	Multiply value in MKS by:	To obtain value in CGS units:	Multiply value in MKS by:	To obtain value in English units:
Area	S	$[l^2]$	m²	10^{4}	cm²	10.76	sq ft.
Density	ρ	$[M/l^3]$	kg/m³	10^{-3}	g/cm³	—	—
Particle velocity	u	$[l/t]$	m/sec	10^{2}	cm/sec	3.28	fps
Sound velocity	c	$[l/t]$	m/sec	10^{2}	cm/sec	3.28	fps
Volume velocity	Q_0	$[l^3/t]$	m³/sec	10^{6}	cm³/sec	35.29	cu ft/sec
Pressure	p	$[F/l^2]$	newton/m²	10	dyne/cm²	1.45×10^{-4}	psi (lb/sq in)
Intensity	$\mathscr{I}$	$[W/l^2]$	watt/m²	10^{3}	erg/cm²/sec	—	—
Impedance: Specific Mechanical Acoustic	 Z_{sp} Z_m Z_{ac}	 $[p/u = \rho c]$ $[F/u = \rho c S]$ $[p/Q_0 = \rho c/S]$	 MKS rayl MKS mechanical ohm MKS acoustic ohm	 10^{-1} 10^{3} 10^{-5}	 rayl mechanical ohm acoustic ohm	 — — —	 — — —

D. Physical Equivalents

Length: 1 micron (μ) = 10^{-6} m = 10^{-4} cm; 1 angstrom unit (A) = 10^{-10} m = 10^{-8} cm
Force: 1 kilogram force = 9.81 newton
Pressure: 1 atmosphere (atm) = 1.013×10^{5} newton/m²; 1 bar = 10^{5} newton/m²
1 mm Hg-column = 1.333×10^{2} newton/m²
Energy: 1 kilocalorie (kcal) = 4.185×10^{3} watt sec; 1 kilowatt hour = 860 kcal
Acceleration of Gravity: 1 g = 9.8 m/sec² = 980 cm/sec² = 32.2 fps²
Temperature: 1 degree Celsius ≃ 1 degree Centigrade (°C)
$\frac{9}{5} \times$ °C + 32 = degrees Fahrenheit (°F)

[a] From T. F. Heuter and R. H. Bolt, *Sonics*, Wiley, New York, 1955, inside back cover.

Answers to Problems

CHAPTER 1

1.2 -26.6 dB
1.5 0.304 kg
1.7 Mean diameter 0.7 m
Wall thickness 0.06 m
1.8 $Q_m = b\rho_m c_m / a\rho_0 c_0$
1.9 Velocity amplitude 0.1075 m/sec displacement amplitude 8.56×10^{-6}
1.10 $f_0 = 3370$ Hz;
mass $= 84.5$ kg;
$s = 3.8 \times 10^{10}$ N/m
$Q_m = 2.4$
$F = 2.2 \times 10^4$ N
1.11 Maximum displacement 2.54×10^{-5} m
$P_a = 82$ kW
$P_T = 106$ kW
1.12 $R_m = 113{,}000$ kg/sec
$s = 5.68 \times 10^9$ N/m

CHAPTER 2

2.1 $f_0 = 18{,}200$ Hz; $Q_m = 6$
2.2 $f_2 = 11{,}420$; $f_1 = 10{,}580$
2.3 (a) 15.9 cm
(b) 4.9 cm from mass load
2.5 7.42 cm
2.6 (a) 10.2 cm
(b) 2.55×10^{-4}
(c) 3.6
2.7 (a) 16.8 cm
(b) 1.77×10^{-4}
(c) $Q_m = 5$

CHAPTER 3

3.2 (a) 2.63×10^{-5} cm²/sec; 2.63×10^{3} cm²/sec
(b) 1.40×10^{-10}; 1.40×10^{-2}
(c) 4.80×10^{-6} cm/sec; 4.80×10^{2}
(d) 7.65×10^{-10} cm; 7.65×10^{-2} cm
3.3 0.86
3.4 1.12×10^{-3}
3.6 70 kHz

CHAPTER 4

4.1 1485 m/sec; $+0.017$ sec^{-1}; 87.4 km
4.4 7.23×10^{-14}; 8 dB
4.5 9°

CHAPTER 6

6.1 (a) $S = -26 \cdot 10^{-6}$
(b) $T = 60 \times 10^{5}$ N/m² = 870 psi
6.2 (a) 30 oersteds required
$NI = 2500$ ampere turns
(b) $\int_0^{(rD-g)} H_i\, dl + \int_0^{g} \frac{\beta_i}{u_0}\, dl = NI$
6.3 (a) 5030 Hz
(b) $L_0 = 9.48$ mH
Z core $= [RX(x + jR)]/(R^2 + x^2)$, where $R = 2\pi L_0 f_c$
At 10 kHz Z core $= 60 + j\,600\ r$
6.4 (a) $|Z_{me}|^2 = D_a mw/Q_m = (2500 \times 24 \times 31{,}400)/55$
$|Z_{me}| = 585$ ohms
$|Z_{me}| = |-(blNu^S Xh/a| = 583$ ohms
(b) $k_{eff} = 0.38$
(c) 17.3×10^{6} N/Wb
6.5 (a) $S = 2.34 \times 10^{-4}$
(b) $T = 2.57 \times 10^{7}$ N/m²
(c) $S = 1.64 \times 10^{-4}$
6.6 (a) 0.114 m
(b) 0.137 m.
6.7 $C_0 = 2.02 \times 10^{-9}\ \mu f$
$Q = 1.34 \times 10^{8}$ N/V
$M = 0.646$ kg;
$K = 3.3 \times 10^{9}$ N/m
$R_m = 231$ kg/sec
6.8 (a) 0.491 m
(b) (122,000 + 411) psi
1.94 W/cm²

CHAPTER 7

7.1 41.8°
7.2 $p = \frac{1}{3}(\cos\beta + \cos 3\beta + \cos 5\beta)$; $\beta = (\pi/2)\sin\theta$
7.3 First null at $\theta = \pm 19.5°$; 1st secondary peak at $\theta = \pm 30°$
second null at $\theta = \pm 38.2°$.
7.5 (a) 25 cm;
(b) 12.6°.
7.6 Use of slide rule gives $X_0 = 1.182$
7.8 25.5 dB
7.9 181 in.^{2}
7.10 11.5 m

CHAPTER 8

8.1 (a) 10 cycles
(b) 10 ft
8.2 For M_0 dB/(V)(μN)(m^2) = JS_0 dB/(μN)(m^2)(A)
$J = (2\lambda\gamma/\rho c) \times 10^{-12}$
8.5 (a) $Q_m = 4.7$
(b) $\eta = -1.5$ dB = 71%
(c) $M_0 = 79.4$/(V)(μbar)
8.7 (a) −97 dB/(V)(μbar)
(b) $DI = 13$ dB
(c) $M_0 = -97$ dB/(V)(μbar)
$C_0 = 1.27$ μfarads
(d) $M_0 = -71$ dB/(V)(μbar)
$C_0 = 3.17 \times 10^{-3}$ μfarads

CHAPTER 9

9.1 (b) 11.5 dB/μbar (c) 24.5 dB/μbar
9.2 (a) $S = 58.5$ dB/μbar at 1 yd
(b) $L = 0.5$ dB/μbar
(c) 7.5 dB higher
9.3 (a) −27 dB/μbar; (b) $H_m = 80.5$ dB, (c) $r_m = 3200$ yd
9.4 (a) one; (b) zero; (c) 3 dB; (d) −21 dB; (e) 13 dB
9.5 L at kHz = 29.6 dB/μbar
L at 30 kHz = 47.6 dB/μbar
9.6 $L = 54$ dB/μbar
echo level = −12 dB/μbar
9.7 (a) $S - N = 43$ dB
(b) 3800 yd
9.8 (a) 25,000 yd
(b) 54,000 yd
(c) 4055 Hz

INDEX